Probability Logics

Zoran Ognjanović · Miodrag Rašković
Zoran Marković

Probability Logics

Probability-Based Formalization of Uncertain Reasoning

Springer

Zoran Ognjanović
Mathematical Institute of the Serbian
Academy of Sciences and Arts
Belgrade
Serbia

Zoran Marković
Mathematical Institute of the Serbian
Academy of Sciences and Arts
Belgrade
Serbia

Miodrag Rašković
Mathematical Institute of the Serbian
Academy of Sciences and Arts
Belgrade
Serbia

ISBN 978-3-319-83637-9 ISBN 978-3-319-47012-2 (eBook)
DOI 10.1007/978-3-319-47012-2

Printed on acid-free paper

This Springer imprint is published by Springer Nature
The registered company is Springer International Publishing AG
The registered company address is: Gewerbestrasse 11, 6330 Cham, Switzerland

Preface

The problems of representing, and working with, uncertain knowledge are ancient problems dating, at least, from Leibnitz, and later explored by a number of distinguished scholars—Jacob Bernoulli, Abraham de Moivre, Thomas Bayes, Johann Heinrich Lambert, Pierre-Simon Laplace, Bernard Bolzano, Augustus De Morgan, George Boole, just to name a few of them. In the last decades there is a growing interest in the field connected with applications to computer science and artificial intelligence. Researchers from those areas have studied uncertain reasoning using different tools, and have used many methods for reasoning about uncertainty: Bayesian network, non-monotonic logic, Dempster–Shafer Theory, possibilistic logic, rule-based expert systems with certainty factors, argumentation systems, etc.

Some of the proposed formalisms for handling uncertain knowledge are based on probability logics. The present book grew out a sequence of papers on probability logics written by the authors since 1985. Also, some of our papers, from 2001 onwards, were coauthored by (in alphabetical order): Branko Boričić, Tatjana Davidović, Dragan Doder, Radosav Đorđević, Silvia Ghilezan, John Grant, Nebojša Ikodinović, Angelina Ilić Stepić, Jelena Ivetić, Dejan Jovanović, Ana Kaplarević-Mališić, Ioannis Kokkinis, Jozef Kratica, Petar Maksimović, Bojan Marinković, Uroš Midić, Miloš Milovanović, Miloš Milošević, Nenad Mladenović, Aleksandar Perović, Nenad Savić, Tatjana Stojanović, Thomas Studer, Siniša Tomović. Two chapters in this book, five and six, are written in collaboration with Aleksandar Perović, Dragan Doder, Angelina Ilić Stepić, and Nebojša Ikodinović.

Although the earliest of those papers were motivated by the work of H.J. Keisler on probability quantifiers, our focus in this book is on latter results about probability logics with probability operators. The aim of this book is to provide an introduction to probability logic-based formalization of uncertain reasoning. So, our primary interest is related to mathematical techniques for infinitary probability logics used to obtain results about proof-theoretical and model-theoretical issues: axiomatizations, completeness, compactness, decidability, etc., including solutions of some problems from literature. This text might serve as a base for further research projects and as a reference text for researchers wishing to use probability logic, but also as a textbook for graduate logic courses. An extensive bibliography is provided to point to related works.

The authors would like to thank to all collaborators and also to acknowledge the support obtained from the Ministry of science, Ministry of Science, Technology and Development, and Ministry of Education, Science and Technological Development of the Republic of Serbia that provided us with partial funding under grants ON0401A and ON04M02 (1996–2000), ON1379 (2002–2005), ON144013 (2006–2010), ON174026 and III044006 (2011–2016).

Belgrade, Serbia

Zoran Ognjanović
Miodrag Rašković
Zoran Marković

Contents

1 Introduction 1
1.1 What Is this Book About: Consequence Relations and Other Logical Issues 1
1.2 Finiteness Versus Infiniteness 3
1.2.1 ω-rule 4
1.2.2 Infinitary Languages 5
1.2.3 Hyperfinite Numbers and Infinitesimals 5
1.2.4 Admissible Sets 7
1.2.5 Ranges of Probability Functions 9
1.3 Modal Logics 9
1.4 Kolmogorov's Axiomatization of Probability and Probability Logics 11
1.5 An Overview of the Book 12
References 14

2 History 19
2.1 Pre-leibnitzians 20
2.2 Leibnitz 21
2.3 Jacob Bernoulli 24
2.4 Probability and Logic in the Eighteenth Century 29
2.4.1 Abraham de Moivre 29
2.4.2 Thomas Bayes 31
2.4.3 Johann Heinrich Lambert 33
2.5 Laplace and Development of Probability and Logic in the Nineteenth Century 35
2.5.1 Laplace 35
2.5.2 Bernard Bolzano, Augustus de Morgan, Antoine Cournot 37
2.5.3 George Boole 40
2.5.4 J. Venn, H. MacColl, C. Peirce, P. Poretskiy 43

2.6 Rethinking the Foundations of Logic and Probability in the Twentieth Century 47
2.6.1 Logical Interpretation of Probability 48
2.6.2 Subjective Approach to Probability 51
2.6.3 Objective Probabilities as Relative Frequencies in Infinite Sequences 54
2.6.4 Measure-Theoretic Approach to Probability 56
2.6.5 Other Ideas 60
2.7 1960s: And Finally, Logic 62
2.7.1 Probabilities in First-Order Settings 63
2.7.2 Probability Quantifiers 64
2.7.3 Probabilities in Modal Settings 66
2.7.4 Probabilistic Logical Entailment 68
References 69

3 LPP_2, a Propositional Probability Logic Without Iterations of Probability Operators 77
3.1 Syntax and Semantics 78
3.1.1 Syntax 78
3.1.2 Semantics 79
3.1.3 Atoms 81
3.2 Complete Axiomatization 82
3.3 Non-compactness 84
3.4 Soundness and Completeness 86
3.4.1 Soundness 86
3.4.2 Completeness 87
3.4.3 The Role of the Infinitary Rule 94
3.4.4 Completeness for Other Classes of Models 95
3.5 Decidability and Complexity 97
3.6 A Heuristic Approach to the $LPP_{2,\mathrm{Meas}}$-Satisfiability Problem PSAT 98
3.6.1 Other Heuristics for PSAT and Similar Problems 107
References 107

4 Probability Logics with Iterations of Probability Operators 109
4.1 Introduction 110
4.2 Syntax and Semantics of $LFOP_1$ 110
4.2.1 Syntax 110
4.2.2 Semantics 111
4.3 Axiom System Ax_{LFOP_1} 113
4.4 Soundness and Completeness 114
4.4.1 Semantical Consequences 116
4.4.2 Completeness for Other Classes of Measurable First-Order and Propositional Models 117

4.5 Modal Logics Versus Probability Logics 118
4.6 (Un)decidability 121
4.6.1 The First-Order Case 121
4.6.2 The Propositional Case 122
4.7 A Discrete Linear-Time Probabilistic Logic 125
4.7.1 Semantics 126
4.7.2 Axiomatization 127
References 130

5 Extensions of the Probability Logics $\mathbf{LPP_2}$ and $\mathbf{LFOP_1}$ 133
5.1 Generalization of the Completeness-Proof Technique 134
5.2 Logic $LPP_2^{Fr(n)}$ 135
5.3 Logic $LPP_2^{A,\omega_1,Fin}$ 137
5.4 Probability Operators of the Form Q_F 141
5.4.1 Complete Axiomatization 141
5.4.2 Decidability 142
5.4.3 The Lower and the Upper Hierarchy 143
5.4.4 Representability 144
5.4.5 The Upper Hierarchy 147
5.4.6 The Lower Hierarchy 149
5.5 Qualitative Probabilities 151
5.6 An Intuitionistic Probability Logic 152
5.6.1 Semantics 153
5.6.2 Axiomatization, Completeness, Decidability 154
5.7 Logics with Conditional Probability Operators 156
5.7.1 A Logic $LPCP_2^{[0,1]_{\mathbb{Q}(\varepsilon)},\approx}$ with Approximate Conditional Probabilities 156
5.7.2 Axiomatization 157
5.8 Polynomial Weight Formulas 158
5.9 Logics with Unordered or Partially Ordered Ranges 159
5.9.1 A Logic for Reasoning About p-adic Valued Probabilities 160
5.10 Other Extensions 162
References 162

6 Some Applications of Probability Logics 165
6.1 Nonmonotonic Reasoning and Probability Logics 165
6.1.1 System P and Rational Monotonicity 165
6.1.2 Modeling Defaults in $LPCP_2^{[0,1]_{\mathbb{Q}(\varepsilon)},\approx}$ 166
6.1.3 Approximate Defaults and $LPCP_2^{[0,1]_{\mathbb{Q}(\varepsilon)},\approx}$ 173
6.2 Logic for Reasoning About Evidence 175
6.2.1 Evidence 175
6.2.2 Axiomatizing Evidence 178

6.3 Formalization of Human Thinking Processes in $L_{\mathbb{Q}_p}$ 180
6.4 Other Applications 183
References. 184

7 Related Work 187
7.1 Papers on Completeness of Probability Logics 187
7.2 Papers on (Infinitary) Modal Logics 194
7.3 Papers on Temporal Probability Logics 194
7.4 Papers on Applications of Probability Logics 195
7.5 Books About Probability Logics. 196
References. 197

Erratum to: Extensions of the Probability Logics LPP_2 and $LFOP_1$ . . . E1

Appendix A: General Notions 201

Index 209

General Notations and Conventions

$\mathbb{N}$	The set of all natural numbers 0, 1, 2, ...
$\mathbb{Z}$	The set of all integers 0, 1, 2, –1, 2, –2, ...
$\mathbb{Q}$	The set of all rational numbers
$\mathbb{R}$	The set of all real numbers
$\mathbb{C}$	The set of all complex numbers
$^*\mathbb{N}$	The nonstandard set of natural numbers
$^*\mathbb{R}$	The set of hypereal numbers
$\mathbb{Q}(\varepsilon)$	Hardy field
$[0, 1]_{\mathbb{Q}}$	The unit interval of rational numbers
$(0, 1)_{\mathbb{Q}}$	The open unit interval of rational numbers
$[0, 1]_{\mathbb{Q}(\varepsilon)}$	The unit interval of Hardy field
$\mathbb{P}(W)$	The power set of the set W
$\lvert W \rvert$	The cardinality of the set W
$\mathrm{Subf}(\Phi)$	The set of all subformulas of the formula Φ
$T \models \Phi$	The formula Φ is a semantical consequence of the set of formulas T
$T \vdash \Phi$	The formula Φ is a syntactical consequence of the set of formulas T
iff	If and only if
wrt.	With respect to

Chapter 1
Introduction

Abstract This introduction gives an overview of some fields of mathematical logic and underlying principles (ω-rule, $L_{\omega_1\omega}$, nonstandard analysis, admissible sets, modal logics) that are used in the rest of the book. Particularly, it provides motivation for various applications of infinitary means in obtaining the presented results. A comparison between the mainstream approach to mathematical theory of probability based on Kolmogorov's axioms and probability logics is given. Finally, the organization of the book is presented.

1.1 What Is this Book About: Consequence Relations and Other Logical Issues

A significant part of mathematical logic explores consequence relations, i.e., derivations of some formulas from other formulas. In that business, it is assumed that mathematical logic should be as reliable as possible. In the first place, it means that a precise definition of a symbolic language in which formulas are formed should be given. Furthermore, semantics should be associated to the language, giving the meaning to building blocks of formulas: atomic formulas, logical connectives, and quantifiers. One can, as suitable instruments, introduce the notions of models and satisfiability relations so that a formula A is a semantical consequence of a (possibly empty) set of formulas T if A is satisfied (in a model, or, alternatively, in a world from a model) whenever all formulas from T are satisfied (in that model, or in that world). Simultaneously, inferences can be studied by means of an axiom system (consisting of axioms and inference rules) where the notion of proof should be determined yielding the notion of syntactical consequence.

A bridge which connects those semantical and syntactical approaches can be established by the soundness and completeness theorems. The usual forms of those theorems are:

- the weak (or simple) completeness: a formula is consistent iff it is satisfiable (i.e., a formula is valid iff it is provable), or

Z. Ognjanović et al., *Probability Logics*,
DOI 10.1007/978-3-319-47012-2_1

- the strong (or extended) completeness: a set of formulas is consistent iff it is satisfiable (a formula is a syntactical consequence of a set of formulas iff it is a semantical consequence of that set).

While the former statement follows trivially from the latter, the opposite direction is not straightforward. In classical propositional and first-order logics these theorems are equivalent, thanks to a significant property formulated as:

- the compactness theorem: a set of formulas is satisfiable iff every finite subset of it is satisfiable.

But, there are logics where compactness fails which complicates their analysis.

In our approach to probability logics, we extend the classical (intuitionistic, temporal, ...), propositional or first-order calculus with expressions that speak about probability, while formulas remain true or false. Thus, one is able to make statements of the form (in our notation) $P_{\geq s}\alpha$ with the intended meaning "the probability of α is at least s". Such probability operators behave like modal operators and the corresponding semantics consists of special types of Kripke models (possible worlds) with addition of probability measures defined over the worlds. We will explain in details in Sect. 3.3 that for probability logics compactness generally does not hold, and discuss some consequences of that property. For example, it is possible to construct sets of formulas that are unsatisfiable and consistent[1] with respect to finitary axiomatizations (for the notion of finitary axiom systems see Appendix 1.1.2). That can be a good reason for a logician to investigate possibilities to overcome the mentioned obstacle. On the other hand, from the point of view of applications, one can argue that, since propositional probability logics are generally decidable, all we need is an efficient implementation of a decision procedure which could solve real problems. However, as we know, propositional logic is of rather limited expressivity and in many (even real life) situations first order logic is a must. It was proved that the sets of valid formulas in probabilistic extensions of first-order logic are not recursively enumerable, so that no complete finitary axiomatization is possible at all (see Chap. 4). Hence, there are no finitary tools that allow us to adequately model reasoning in this framework. We believe that this is not only of theoretical interest, which has motivated us to investigate alternative model-theoretic and proof-theoretic methods appropriate for providing strongly complete axiomatizations for the studied systems. The main part of this book is devoted to those issues.

As one of the distinctive characteristics of our approach in exploring relationship between logic and probability,[2] we have used different aspects of infiniteness which has proved to be a powerful tool in this endeavor. At the same time, we will try to accomplish it with tools as weak as possible, i.e., to limit the use of infinitary means: we generally use countable object languages and finite formulas, while only proofs are allowed to be infinite.

[1] Contradiction cannot be deduced from the set of formulas.

[2] Actually, in Chap. 2 we will present some evidences about common roots of these two important branches of mathematics.

Other important problems which will be addressed in the book are related to decidability and complexity of probability logics. We will also describe our attempts to develop heuristically-based methods for the probability logic satisfiability problem, PSAT.

The main contribution of our work presented in this book concerns development of a new technique for proving strong completeness for non-compact probability logics which combines Henkin style procedures for classical and modal logics and which works with infinitary proofs. This method enabled us to solve some open problems, e.g., strong completeness for real-valued probabilities in the propositional and first-order framework and for polynomial weight formulas (see the Chaps. 3, 4, 5, 7). It was also applied to other non-compact logics, for example to linear and branching discrete time logics [3, 4, 31, 37, 39, 46], and logics with probability functions with partially ordered ranges, etc.

1.2 Finiteness Versus Infiniteness

Standard courses of mathematical logic, usually encompassing classical propositional and first-order logic, assume that axiom systems are finitary. Such a system is presented by a finite list of axiom schemas and inference rules (each rule with a finite number of hypothesis and one conclusion). It might create an impression that all axiom systems are finitary in the above sense. Nevertheless, infiniteness can play an important role and significantly expand expressive power of formal systems. It can be traced back to an extremely important period of development of mathematical logic, i.e., to 1930s.[3] These years brought many significant results in mathematical logic and, what we call today, theoretical computer science. One of the most prominent among them, the first Gödel's incompleteness theorem [10], says that for any consistent first order formal system, expressive enough to represent finite proofs about natural numbers, there is no recursive (finitary) complete axiomatization. It suggests that some kind of infiniteness should be involved into formal systems to study the standard model of arithmetics. Indeed, several such approaches were introduced before 1940.

The seminal work of Gerhard Gentzen [9] showed that, by associating ordinals to derivations, the consistency of the first-order arithmetic is provable in a theory with the principle of transfinite induction up to the infinite ordinal ε_0.

In his Ph.D. Thesis [60] Alan Turing considered a formal system T_0 powerful enough to represent arithmetics, and a sequence of logical theories (each theory T_{i+1} obtained from the preceding one by adding the assertion about consistency of T_i, $T_\omega = \cup T_i$, and further iterated into the transfinite). He asked whether one of the logics indexed with denumerable ordinals is complete with respect to statements true in the standard model of natural numbers. Although Turing established that $T_{\omega+1}$ proves an important subclass of true formulas (all valid Π_1 sentences, i.e., sentences

[3]It is pointed out in Chap. 2 that already Leibnitz discussed infinitary proofs.

of the form $(\forall x)A(x)$, where A is a recursive predicate), later on it was showed in [7] that this progression is not complete (already for true $\forall\exists$ sentences).

Finally (and more relevant to this text), some well-known logicians (Tarski, Hilbert, Carnap) introduced ω-rule to overcome the limitations of finitary formal systems of arithmetic [24, 59].

In Gödel's analysis of undecidability, the role of recursive (Δ_0) and recursively enumerable (Σ_1) sets (arithmetical predicates, formulas) is important. Informally speaking, a Δ_0-formula (or a bounded formula) is a formula whose all quantifiers are bounded, while a Σ_1-formula is, up to equivalence, in the form of a block of existential quantifiers applied on a Δ_0-formula, i.e., if α is a Δ_0-formula, then $\exists x_1 \ldots \exists x_n \alpha$ is a Σ_1-formula. More precisely, a Δ_0-formula is inductively defined as follows:

- Any quantifier free formula is a Δ_0-formula;
- Boolean combination of Δ_0-formulas is a Δ_0-formula;
- If α is a Δ_0-formula, then $\forall x(x \leq t \rightarrow \alpha)$ and $\exists x(x \leq t \wedge \alpha)$ are Δ_0 formulas.

In investigation presented in this book, we will be using different manifestations of infinity:

- infinitary proofs,
- infinitary formulas,
- infinitary ranges of probability functions with an infinitary property (σ-additivity),
- ranges of probability functions containing infinitely small values,[4] and
- admissible sets,

but also, where possible, their finitary counterparts will be discussed.

1.2.1 ω-rule

The basic form of this rule in the language of arithmetic $\{+, \cdot, S, 0\}$ is

- from $A(0), A(1), A(2) \ldots$, infer $(\forall x)A(x)$

where $1 = S0, 2 = SS0, \ldots$, are numerals. When one adds this rule to a usual axiom system of arithmetics (PA or Robinson arithmetic Q), a complete logic allowing proofs of infinite length is obtained [8, 58]. More recently, some versions of ω-rule (with the additional assumption that proofs of all premises $A(n)$ are recursive) suitable for effective implementation in automated deduction environments have been considered [1].

In axiom systems presented in this book several inference rules with infinite number of premisses and one conclusion, related to different aspects of probability, will be used.

[4] Infinitesimals.

1.2.2 Infinitary Languages

Probably the simplest infinitary logic is $L_{\omega_1\omega}$ which admits at most countable conjunctions and disjunctions, and finite blocks of quantifiers [25].

Note that the increased expressivity enables formal syntactical description of any countable first-order structure. For instance, the additive group $\langle \mathbb{Z}, +\rangle$ of integers can be formally coded by the following $L_{\omega_1\omega}$-sentence:

$$\phi_{\mathbb{Z}} \Leftrightarrow_{\text{def}} \forall x \left(\bigvee_{n\in\mathbb{Z}} x = c_n \right) \wedge \bigwedge_{n\neq m} c_n \neq c_m \wedge \bigwedge_{n,m\in\mathbb{Z}} c_n * c_m = c_{n+m}.$$

The underlying first-order language $L_{\mathbb{Z}}$ contains one binary function symbol $*$ and countably many constants $\{c_n \ : \ n \in \mathbb{Z}\}$. It is easy to see that an $L_{\mathbb{Z}}$-structure $\langle M, *_M\rangle$ is a model of $\phi_{\mathbb{Z}}$ iff it is isomorphic to the group $\langle \mathbb{Z}, +\rangle$.

However, the increased expressiveness comes with the price: the compactness theorem is not true for the $L_{\omega_1\omega}$. Indeed, using the same language $L_{\mathbb{Z}}$ as in the previous example, the following set of $L_{\mathbb{Z}}$-sentences

$$\left\{ \bigvee_{n\in\mathbb{Z}\setminus\{0\}} c_n = c_0 \right\} \cup \{c_n \neq c_0 \ : \ n \in \mathbb{Z} \setminus \{0\}\}$$

is finitely satisfiable, but it is not satisfiable.

As a formal theory, $L_{\omega_1\omega}$ extends classical first-order logic in the following way:

- $L_{\omega_1\omega}$ admits infinitary formulas,
- $L_{\omega_1\omega}$ has three additional axioms:
  - $\bigwedge_{i\in\mathbb{N}} \alpha_i \rightarrow \alpha_k$, for every $k \in \mathbb{N}$
  - $\neg \bigwedge_{i\in\mathbb{N}} \alpha_i \leftrightarrow \bigvee_{i\in\mathbb{N}} \neg\alpha_i$
  - $\neg \bigvee_{i\in\mathbb{N}} \alpha_i \leftrightarrow \bigwedge_{i\in\mathbb{N}} \neg\alpha_i$,
- $L_{\omega_1\omega}$ has an additional infinitary inference rule
  - From $\{\beta \rightarrow \alpha_i : i \in \mathbb{N}\}$ infer $\beta \rightarrow \bigwedge_{i\in\mathbb{N}} \alpha_i$.

For $L_{\omega_1\omega}$ strong completeness fails, and only weak completeness can be proved.

In Chap. 5 a fragment of $L_{\omega_1\omega}$ will be used in characterization of probability functions with arbitrary finite ranges.

1.2.3 Hyperfinite Numbers and Infinitesimals

The nonstandard analysis was introduced by Abraham Robinson (1918–1974) in 1961 [57]. He successfully applied the compactness theorem in order to perform the so-called rational reconstruction of the Leibnitz's differential and integral calculus. The key feature of Robinson's theory was consistent foundation of infinitesimals and hyperfinite numbers.

Suppose that S is an arbitrary set. A superstructure on S is the set

$$V(S) = V_\omega(S) = \bigcup_{n \in \omega} V_n(S),$$

where $V_0(S) = S$ and $V_{n+1}(S) = \mathbb{P}(V_n(S))$. If $S = \emptyset$, then $V(S) = V_\omega = HF$, i.e., $V(\emptyset)$ coincides with the set HF of hereditary finite[5] sets. For the nonstandard analysis the most interesting case is $S \subseteq \mathbb{R}$. Anyhow, S should be large enough to include all relevant objects within the scope of the underlying problem.

A nonstandard universe on S is a pair $\langle {}^*V(S), *\rangle$, where ${}^*V(S)$ is a proper superset of the standard universe $V(S)$ and $*$ is so-called lifting function $* : V(S) \longrightarrow {}^*V(S)$ such that

$${}^*s =_{\text{def}} *(s) = s$$

for all $s \in S$.

A set $X \in V({}^*S)$ is:

- internal, iff there is $A \in V(S)$ such that $X \in {}^*A$;
- external, iff it is not internal;
- standard, iff $X = {}^*A$ for some $A \in V(S)$.

For example, ${}^*\mathbb{N}$ is a standard set,[6] $\sin(Hx)$ is an internal set for any $H \in {}^*\mathbb{N} \setminus \mathbb{N}$, while S and $\mathbb{N}$ are external sets.

In particular, elements of the set ${}^*\mathbb{N} \setminus \mathbb{N}$ are called hyperfinite numbers. An internal set A is called hyperfinite iff there are a hyperfinite number H and an internal bijection $f : H \longrightarrow A$.

So, the notion of a hyperfinite set is a direct generalization of the notion of the finite set. Of special significance for applications of nonstandard analysis in probability theory and probability logic is the so-called hypertime interval

$$T =_{\text{def}} \left\{ \frac{n}{H} \ : \ n \leq H \text{ and } n \in {}^*\mathbb{N} \right\}.$$

Note that, in terms of the nonstandard universe, T is a hyperfinite set since it has H elements. However, T is not only an infinite set, but its cardinality is equal to continuum since there is a bijection between T and the real unit interval $[0, 1]$. This

[5]A recursive definition of HF goes as follows:

- $\emptyset \in HF$;
- $X \in HF$ iff X is finite and all its elements are also hereditary finite.

By axiom of regularity, there is no sequence of sets $\langle x_n \ : \ n \in \omega\rangle$ such that $x_{n+1} \in x_n$ for all n, so our definition is correct. In particular, $\emptyset$ is the simplest hereditary finite set.

[6]It is also a proper superset of $\mathbb{N}$, provided the usual restriction $\mathbb{N} \notin S$.

fact was used by Peter Loeb to define the Loeb measure and to establish a natural correspondence between the counting measure and the Lebesgue measure (so called Loeb construction or Loeb process) [30]. Thus, the notion of hyperfinite is a bridge between discrete and continuous.

An infinitesimal is any $\varepsilon \in {}^*\mathbb{R}$ such that

$$-\frac{1}{n} < \varepsilon < \frac{1}{n}$$

for all $n \in \mathbb{N} \setminus \{0\}$. For example, if H is a hyperfinite number, then $\frac{1}{H}$ is a positive infinitesimal.

Some of the key features of the nonstandard analysis are listed below:

- **Internal definition principle**. A set $X \in V({}^*S)$ is internal iff

  $$X = \{x \ : \ x \in A \text{ and } \alpha(x, A_1, \dots, A_n)\},$$

  where α is a Δ_0-formula and $A, A_1, \dots, A_n$ are internal sets;
- **Standard definition principle**. A set $X \in V({}^*S)$ is internal iff

  $$X = \{x \ : \ x \in A \text{ and } \alpha(x, A_1, \dots, A_n)\},$$

  where α is a Δ_0-formula and $A, A_1, \dots, A_n$ are standard sets;
- ω_1**-saturatedness**. If $\{A_n \ : \ n \in \mathbb{N}\}$ is a countable descending family of internal nonempty sets (i.e., $A_{n+1} \subseteq A_n$ for all n), then $\bigcap_{n\in\mathbb{N}} A_n \neq \emptyset$;
- **Congruence**. For any $A \in S$, $x \in A$ iff ${}^*x \in {}^*A$. Similarly, ${}^*(A \cup B) = {}^*A \cup {}^*B$, ${}^*(A \times B) = {}^*A \times {}^*B$, ${}^*(A \setminus B) = {}^*A \setminus {}^*B$ etc.;
- **Overspill**. If A is an internal set and $\mathbb{N} \cap A$ is infinite, then A contains at least one hyperfinite number;
- **Underspill**. If an internal set A contains arbitrary small hyperfinite numbers (i.e., for all hyperfinite $H \in A$ exists a hyperfinite $K \in A$ such that $K < H$), then $A \cap \mathbb{N} \neq \emptyset$.

Nonstandard notions and techniques are used in the Chaps. 5 and 6 to obtain a complete axiomatization and to prove decidability of a logic with approximate conditional probabilities.

1.2.4 Admissible Sets

The theory of admissible sets was introduced by Kenneth Jon Barwise (1942–2000) [2] in order to provide a minimal formal framework for the study of recursion theory. The notion of finiteness is generalized by so-called admissible countability or $\mathbb{A}$-finiteness for the given admissible set $\mathbb{A}$.

Admissible set theory is a fragment of Zermelo–Fraenkel set theory with the following axioms:

- **Extensionality**. $A = B$ iff they have the same elements;
- **Empty set**. $\emptyset =_{\text{def}} \{x : x \neq x\}$ is a set;
- **Pair**. If A and B are sets, then $\{A, B\} =_{\text{def}} \{x : x = A \vee x = B\}$ is also a set;
- **Union**. If A is a set, then $\bigcup A =_{\text{def}} \{x : (\exists a \in A) x \in a\}$ is also a set;
- **Δ_0-separation**. If A is a set and α is a Δ_0-formula, then $\{x : x \in A \wedge \alpha\}$ is also a set;
- **Δ_0-collection.** Suppose that $\alpha(x, y)$ is a Δ_0-formula such that for any set X there is a set Y such that $\alpha(X, Y)$ holds. Then, for any set A there is a set B such that $(\forall a \in A)(\exists b \in B)\alpha(a, b)$ is true;
- **Regularity**. The membership relation $\in$ is regular, i.e., each set has $\in$-minimal element. More precisely, for any set A exists $a \in A$ such that $a \cap A = \emptyset$;
- **Infinity**.[7] There exists set A such that $\emptyset \in A$ and $a \cup \{a\} \in A$ for all $a \in A$.

The most notable difference between the admissible set theory and *ZFC* is the absence of axioms of choice and the powerset axiom. Hence, the admissible set theory cannot be used for the study of infinitary combinatorics due to the fact that one cannot establish the hierarchy of infinite cardinals. It can be shown that certain important mathematical concepts, such as ordered pair and Cartesian product, can be coded by means of the admissible set theory.

An admissible set is any set $\mathbb{A}$ such that the pair $\langle \mathbb{A}, \in \rangle$ is a model of the admissible set theory. For the study and development of probability logic, the most important example of the admissible set is the set HC of all hereditary countable sets. Similarly to the set HF of hereditary finite sets, the set HC is inductively defined as follows:

- $HF \subseteq HC$;
- $X \in HC$ iff X is at most countable and $x \in HC$ for all $x \in X$.

As before, the axiom of regularity provides the correctness of the above definition.

The main technical aspect of the set HC of all hereditary countable sets is the fact that the admissible fragment $L_{\mathbb{A}}$ of the infinitary logic $L_{\omega_1 \omega}$ can be effectively coded in HC by means of the admissible set theory. For example, suppose that $F = \{\alpha_i : i \in I\}$ is a countable admissible set of formulas and that $f : F \longrightarrow HC$ is an admissible coding of F. If $k \in HC$ is a Gödel number (effective or recursive code) of a conjunction, then $\langle k, f \rangle \in HC$ is a Gödel number of the infinitary $L_{\mathbb{A}}$-formula $\bigwedge_{i \in I} \alpha_i$.

In other words, recursive infinitary logical constructions (formula formations, proofs, completion technique) can be represented as sets and set operations in the admissible set theory.

In particular, the elements of an admissible set $\mathbb{A}$ are called $\mathbb{A}$-finite. The most important technical tool of the admissible set theory is the Barwise compactness theorem that connects consistency with $\mathbb{A}$-finitness:

[7] This axiom is optional, i.e., some authors do not include it in the system.

Barwise compactness theorem. Suppose that $\mathbb{A}$ is a countable admissible set and that T is a Σ_1-definable[8] set of $L_{\mathbb{A}}$-sentences. Then, T is satisfiable iff each $\mathbb{A}$-finite subset of T is satisfiable.

An admissible fragment of a probabilistic counterpart of $L_{\omega_1\omega}$ is constructed in Chap. 5 to completely axiomatize probability functions with arbitrary finite ranges.

1.2.5 Ranges of Probability Functions

For our basic logics, in the Chaps. 3 and 4, we develop completion and decidability techniques wrt. the standard real-valued probability functions. However, real-valued probabilities are proved to be inadequate to model different types of uncertainty, as it is the case in default reasoning. For this purpose we consider other kinds of probability functions with various ranges:

- the finite set $\{0, \frac{1}{n}, \frac{2}{n}, \ldots, \frac{n-1}{n}, 1\}$,
- the unit interval of rational numbers $[0, 1]_{\mathbb{Q}}$, or some other recursive subsets of $[0, 1]$,
- the unit interval of Hardy field $[0, 1]_{\mathbb{Q}(\varepsilon)}$,
- some partially ordered countable commutative monoid with the least element 0, e.g., $[0, 1]_{\mathbb{Q}} \times [0, 1]_{\mathbb{Q}}$, and
- a closed ball in the field $\mathbb{Q}_p$ of p-adic numbers.

As expected, different types of ranges impose numerous challenges in axiomatizations. In this book we provide appropriate methodology to resolve those issues.

1.3 Modal Logics

Motivated by paradoxes of material implication (see Sect. 2.5.4), development of modal logics at first evolved in a pure syntactical framework. Clarence Irving Lewis (1883–1964) published a number of papers since 1912 and proposed several formal systems to axiomatize strict implication[9] understood as "it is impossible that the antecedent is true, while the consequent is false", or equivalently as "it is necessary that if the antecedent is true, then so is the consequent" [13, 29]. There are numerous modal logics, but the most studied between them are so-called normal modal logics. The simplest normal modal logic, denoted K, is axiomatized using the axiom schemata:

---

[8]There is a Σ_1-formula α such that $\langle \mathbb{A}, \in \rangle \models T = \{x : \alpha\}$.

[9]In modern notation the formal language of modal logics extends the classical propositional language with the unary necessity operator $\Box$. Then, the strict implication is written as $\Box(\alpha \to \beta)$.

1. all substitutional instances of the classical propositional tautologies, and
2. (Axiom K) $\Box(\alpha \rightarrow \beta) \rightarrow (\Box\alpha \rightarrow \Box\beta)$,

and the inference rules:

1. Modus ponens, and
2. (Necessitation) if α, then $\Box\alpha$.

Other normal modal systems extend K with additional axioms that determine properties of the modal operator $\Box$.

Today the most widely accepted semantics for modal logics was proposed in the late 1950s by Saul Kripke (1940) [28]. The semantics is based on the idea of possible worlds equipped with a relation which represents visibility or accessibility between worlds. A Kripke model for propositional modal logics is a structure $\mathbf{M} = \langle W, R, v \rangle$ such that:

- W is a nonempty set of objects called worlds,
- $R \subset W \times W$ is an accessibility relation between worlds,
- $v : W \times \phi \rightarrow \{true, false\}$ provides for each world $w \in W$ a two-valued valuation of the set ϕ of primitive propositions,

while a formula α is satisfied in a world w (denoted $w \models \alpha$):

- if $\alpha \in \phi$, $w \models \alpha$ iff $v(w)(\alpha) = true$,
- if $\alpha = \neg\beta$, $w \models \alpha$ iff $w \not\models \beta$,
- if $\alpha = \beta \wedge \gamma$, $w \models \alpha$ iff $w \models \beta$ and $w \models \gamma$, and
- if $\alpha = \Box\beta$, $w \models \alpha$ iff for every $u \in W$, if wRu, then $u \models \beta$.

Note that, since the truth value of $\Box\alpha$ in a world w depends on R, i.e., on worlds accessible from w, modal logics are not truth-functional. Modal models without particular requirements for R characterize the system K. For stronger systems, additional axioms correspond to particular properties of R, for example:

- (T) $\Box\alpha \rightarrow \alpha$ corresponds to reflexivity,
- (4) $\Box\alpha \rightarrow \Box\Box\alpha$ corresponds to transitivity,
- (B) $\alpha \rightarrow \Box\neg\Box\neg\alpha$, etc.

The operator $\Box$ can be interpreted in many ways:

- temporal: $\Box\alpha$ is read "α always holds" [50],
- epistemic: $\Box\alpha$ is read "an agent knows α" [12],
- proof-theoretical: $\Box\alpha$ is read "α is provable" [11], etc.,

which is of great importance in applications. Therefore, modal logics are today accepted as formal bases for many systems in computer science and artificial intelligence.

One of the consequences of similarities between Kripke modal models and probability models (see the Definitions 3.2 and 4.1; instead of accessibility relations those models involve probability spaces) is that probability operators are not truth-functional. Since the semantics of $\Box$ is given using universal quantification over possible worlds, probability operators can be seen as a sort of softening of the necessity operator which gives additional expressivity and inspires possible mixing of the modal and probability languages.

1.4 Kolmogorov's Axiomatization of Probability and Probability Logics

Although there are several other proposals, the axiomatization of probability based on measure-theoretic notions given by Andrei Nikolaevich Kolmogorov (1903–1987) (see Sect. 2.6.4) is generally accepted as a standard. One can legitimately ask whether it is a logic, or what is its relationship with probability logics. To clarify that questions, one should be aware of the methodology which is used in mathematical logic. As we emphasize at the beginning of this Chapter and in Appendix, mathematical logic distinguishes between:

- syntax and semantics,
- object language and meta language, and
- object level and meta-level of reasoning.

While ordinary mathematicians often do not recognize these levels and mix them into one, the primary interest of mathematical logic is to formulate and prove (at the meta-level) statements *about* syntactical and semantical notions from the object level of reasoning (e.g., object-level theorems, valid formulas and so on). So, this methodological difference forces that many questions that are in the focus of probability logics (consequence relations, completeness, compactness, decidability, complexity, etc.) are not of huge importance in probability theory.[10] For instance, we do not expect that a probabilist would be too much interested whether optimal bounds of probabilities for consequences of some uncertain premisses are effectively computable.

In that sense, we do not consider Kolmogorov's axiomatization as a logic. Kolmogorov's axiomatization is used as a basis for semantics for some of the probability logics presented in this book, but, other approaches to probability are also studied: non-real-valued probabilities, probabilities with partially ordered ranges, coherent probabilities, etc.

Finally, we would like to point out that investigations in the field of probability logics can be useful in proving new theorems about probability: e.g., Keisler in [26, 27] proves existence theorems for some stochastic differential equations which are not proved by classical methods.

[10] And vice versa—probability logics do not carefully study some of the issues in probability theory.

1.5 An Overview of the Book

We present a number of probability logics. The main differences between the logics are:

- some of the logics are infinitary, while the others are finitary,
- the corresponding languages contain different kinds of probabilistic operators, both for unconditional and conditional probability,
- some of the logics are propositional, while the others are based on first-order logic,
- for most of the logics we start from classical logic, but in some cases the basic logic can be intuitionistic or temporal,
- in some of the logics iterations of probabilistic operators are not allowed,
- for some of the logics, restrictions of the following kinds are used: only probability measures with fixed finite range are allowed in models; ranges of probability functions are rational numbers, or complex numbers, or p-adic numbers, or domains of monoids; only one probability measure on sets of possible worlds is allowed in a model; the measures are allowed to be finitely additive.

For all these logics we give the corresponding axiomatizations, prove completeness, and discuss their decidability. More precisely, we consider the following logics:

- LPP_1 (L for logic, the first P for propositional, and the second P for probability), probability logic which starts from classical propositional logic, with iterations of the probability operators and real-valued probability functions [38, 38, 42, 52],
- $LPP_1^{\mathrm{Fr}(n)}$ and LPP_1^S propositional probability logics with probability functions restricted to have ranges $\{0, 1/n, \ldots, (n-1)/n, 1\}$ and S, respectively [38, 40, 42, 52],
- $LPP_1^{A,\omega_1,\mathrm{Fin}}$, propositional probability logic with probability functions restricted to have arbitrary finite ranges [5, 52],
- LPP_1^{LTL}, probability logic similar to LPP_1, but the basic logic is discrete linear time logic LTL [37–39], and LPP_1^{BTL}, propositional discrete probabilistic branching time logic [3, 46],
- LPP_2, $LPP_2^{\mathrm{Fr}(n)}$, $LPP_2^{A,\omega_1,\mathrm{Fin}}$, and LPP_2^S, probability logics without iterations of the probability operators [38, 42, 51, 52, 52],
- $LPP_{2,P,Q,O}$, probability logic which extends LPP_2 by having a new kind of probabilistic operators of the form Q_F, with the intended meaning "the probability belongs to the set F" [18, 41],
- $LPP_{2,\preceq}$ and $LPP_{2,\preceq}^{\mathrm{Fr}(n)}$, probability logics similar to LPP_2 and $LPP_2^{\mathrm{Fr}(n)}$, but allowing reasoning about qualitative probabilities [45, 47],
- LPP_2^I, probability logics similar to LPP_2, but the basic logic is propositional intuitionistic logic [32–34],
- $LFOP_1$, $LFOP_1^{\mathrm{Fr}(n)}$, $LFOP_1^{A,\omega_1,\mathrm{Fin}}$, $LFOP_1^S$ and $LFOP_2$, the first-order counterparts of the above logics [42, 53],
- several Kolmogorov's style-conditional probability propositional and first-order logic, with or without iterations of the probability operators, with real valued

probability functions, or probability functions with the range $[0, 1]_{\mathbb{Q}(\varepsilon)}$ that can express approximate probabilities [14, 35, 36, 43, 44, 54–56],

- $LPCP_2^{\mathrm{Chr}}$, propositional conditional probability logic, which corresponds to de Finetti's view on coherent conditional probabilities [14, 15],
- $LPG_2, LCOMP_B, LCOMP_S, CPL_{\mathbb{Z}_{\mathbb{Q}_p}}, CPL_{\mathbb{Q}_p}^{fin}$, propositional probability logics with monoid-valued (complex-valued, p-adic-valued) probability functions [16, 17, 19–23],
- LWF and PWF, propositional probability logics with linear and polynomial weight formulas (introduced in [6]) with $^*\mathbb{R}$-valued probability functions [47–49].

The parts of the book can be described as follows.

Chapter 2 introduces readers to a fascinating story about interactions between mathematical logic and probability which is full of great ideas and discoveries. We will particularly try to emphasize topics that motivated our research.

The key concepts (syntax and semantics, an infinitary axiomatization, the corresponding strong completeness, decidability and complexity) of LPP_2 are presented in Chap. 3. As the semantics, we introduce a class of models that combine properties of Kripke models and probabilities defined on sets of possible worlds. We consider the class of so-called measurable models (which means that all sets of possible worlds definable by classical formulas are measurable) and some of its subclasses: in the first case all subsets of worlds are measurable, then probabilities are required to be σ-additive, while models in the last subclass satisfy that only empty set has the zero probability. The proposed axiomatization is infinitary, i.e., there is an inference rule with countably many premisses and one conclusion:

- From $A \rightarrow P_{\geq s-\frac{1}{k}}\alpha$, for every integer $k \geq \frac{1}{s}$, and $s > 0$ infer $A \rightarrow P_{\geq s}\alpha$.

The rule corresponds to the following property of real numbers: if the probability is arbitrary close to s, it is at least s. Thus, proofs with countably many formulas are allowed. We give full details of the proof of strong completeness, so that it can be, with the corresponding modifications, used as a template for the other completeness proofs presented in the book. Decidability of PSAT, the satisfiability problem for LPP_2, is proved by a reduction to linear programming problem. Since the related linear systems can be of exponential sizes, we describe some heuristic approaches to the probabilistic satisfiability problem.

Chapter 4 investigates the first-order probability logic $LFOP_1$ which allows iterations of probability operators, so that it is possible to formalize reasoning about higher order probabilities. Since validity is not even recursively enumerable in that first-order framework, the presented infinitary axiom system, obtained by adding the probability axioms and inference rules (introduced in Chap. 3) to the classical axiomatization, is a reasonable tool for formalization of the logic. We also discuss relationship between $LFOP_1$ and modal logics by analyzing some properties of first-order modal models (constant domains, rigidness of terms) from the perspective of probability logics. Then we prove (un)decidability of (some fragments of) $LFOP_1$. Finally, a logic which combines temporal and probability reasoning is introduced.

Chapter 5 covers various probability logics, i.e., variants of LPP_2 and $LFOP_1$ obtained by putting restrictions on (or by extending) ranges of probability functions and/or on the used formal languages. We consider new types of probabilistic operators:

- the conditional probability operators $CP_{\geq s}$, $CP_{\leq s}$,
- the probability operators that express imprecise probabilities $P_{\approx s}$, $CP_{\approx s}$,
- the qualitative probability operator $\preceq$,
- the probability operators of the form Q_F with the intended meaning "the probability belongs to the set F".

and alternative ranges of probability functions: finite, countable, with infinitesimals, or partially ordered. We consider inference rules that help us to syntactically define ranges of probability functions. Also, we describe the logic LPP_2^I which extends propositional intuitionistic logic.

Chapter 6 deals with applications. In the case of $LPCP_2^{[0,1]_{\mathbb{Q}(\varepsilon)},\approx}$, the range is the unit interval of a recursive non-Archimedean field which makes it possible to express statements about approximate probabilities: $CP_{\approx s}(\alpha, \beta)$ which means "the conditional probability of α given β is approximately s". Formulas of the form $CP_{\approx 1}(\alpha, \beta)$ can be used to model defaults, i.e., expressions of the form "if β, then generally α". So, we relate $LPCP_2^{[0,1]_{\mathbb{Q}(\varepsilon)},\approx}$ with the well-known system P which forms the core of default reasoning. We also discuss other applications to, for example, reasoning about evidence, modeling of the process of human thinking based on p-adic numbers, etc.

A limited number of related papers published between the 1980s and 2010s are discussed in Chap. 7.

Finally, Appendix provides an overview of some general concepts in mathematical logic (formal systems, syntax, semantics, axiom systems, proofs, completeness, etc.) and probability theory ((σ-) algebras, (σ-) additive measures, the usual and coherent concepts of probability, etc.) which could help less experienced readers to follow the rest of the text.

Each chapter ends with the list of relevant references.

The book concludes with an index.

References

1. Baker, S.: Automated Deduction in Arithmetic with the Omega Rule. AAAI Technical Report FS-93-01 (1993)
2. Barwise, K.J.: Admissible Sets and Structures. Springer, Berlin (1975)
3. Doder, D.: Primene infinitarnih logika u verovatnosno-temporalnom rezonovanju i teoriji modela Ph.D. thesis (in Serbian). University of Belgrade (2011). http://elibrary.matf.bg.ac.rs/bitstream/handle/123456789/2467/docddFR.pdf?sequence=1
4. Doder, D., Ognjanović, Z., Marković, Z.: An axiomatization of a first-order branching time temporal logic. J. Univers. Comput. Sci. **16**(11), 1439–1451 (2010)

5. Đorđević, R., Rašković, M., Ognjanović, Z.: Completeness theorem for propositional probabilistic models whose measures have only finite ranges. Arch. Math. Logic **43**, 557–563 (2004)
6. Fagin, R., Halpern, J., Megiddo, N.: A logic for reasoning about probabilities. Inf. Comput. **87**(1–2), 78–128 (1990)
7. Feferman, S.: Transfinite recursive progressions of axiomatic theories. J. Symb. Logic **27**, 259–316 (1962)
8. Franzén, T.: Transfinite progressions: a second look at completeness. Bull. Symb. Logic **10**(3), 367–389 (2004)
9. Gentzen, G.: Die Widerspruchsfreiheit der reinen Zahlentheorie. Mathematische Annalen **112**, 493–565 (1936)
10. Gödel, K.: Über formal unentscheidbare Sätze der Principia Mathematica und verwandter Systeme. I. Monatshefte für Mathematik und Physik **38**, 173–198 (1931)
11. Gödel, K.: Eine Interpretation des Intuitionistischen Aussagenkalküls. Ergebnisse eines Mathematischen Kolloquiums **4**, 39–40 (1933)
12. Hintikka, J.: Knowledge and Belief: An Introduction to the Logic of the Two Notions. Cornell University Press, New York (1962)
13. Hughes, G.E., Cresswell, M.J.: A New Introduction to Modal Logic. Routledge, Abingdon-on-Thames (1996)
14. Ikodinović, N.: Some Probability and Topological Logics. Ph.D. thesis (in Serbian). University of Kragujevac (2005). http://elibrary.matf.bg.ac.rs/bitstream/handle/123456789/194/phdNebojsaIkodinovic.pdf?sequence=1
15. Ikodinović, N., Ognjanović, Z.: A logic with coherent conditional probabilities. In: Godo, L. (ed.), Proceedings of the 8th European Conference Symbolic and Quantitative Approaches to Reasoning with Uncertainty ECSQARU 2005. Lecture Notes in Computer Science (Subseries: Lecture Notes in Artificial Intelligence), Barcelona, Spain, July 6–8, 2005, vol. 3571, pp. 726–736. Springer, Heidelberg (2005)
16. Ikodinović, N., Rašković, M., Marković, Z., Ognjanović, Z.: Measure logic. In: Proceedings of the 9th European Conference Symbolic and Quantitative Approaches to Reasoning with Uncertainty ECSQARU-2007. Lecture Notes in Computer Science (Subseries: Lecture Notes in Artificial Intelligence), Hammamet, Tunisia 2007, vol. 4724, pp. 128–138. Springer, Heidelberg (2007)
17. Ikodinović, N., Rašković, M., Marković, Z., Ognjanović, Z.: Logics with generalized measure operators. J. Multiple Valued Logic Soft Comput. **20**(5–6), 527–555 (2013)
18. Ikodinović, N., Ognjanović, Z., Rašković, M., Perović, A.: Hierarchies of probabilistic logics. Int. J. Approx. Reason. **55**(9), 1830–1842 (2014)
19. Ilić Stepić, A.: O formalizaciji p-adskih, kvalitativnih i uslovnih verovatnoća. Ph.D. thesis (in Serbian). University of Belgrade (2013). http://elibrary.matf.bg.ac.rs/bitstream/handle/123456789/2466/docangFR.pdf?sequence=1
20. Ilić Stepić, A., Ognjanović, Z.: Complex valued probability logics. Publications de l'Institut Mathematique, n.s. tome **95**(109), 73–86 (2014). http://elib.mi.sanu.ac.rs/files/journals/publ/115/n109p073.pdf
21. Ilić Stepić, A., Ognjanović, Z.: Logics for reasoning about processes of thinking with information coded by p-adic numbers. Studia Logica **103**, 145–174 (2015)
22. Ilic-Stepić, A., Ognjanović, Z., Ikodinović, N., Perović, A.: A p-adic probability logic. Math. Logic Q. **58**(4–5), 263–280 (2012)
23. Ilić-Stepić, A., Ognjanović, Z., Ikodinović, N.: Conditional p-adic probability logic. Int. J. Approx. Reason. **55**(9), 1843–1865 (2014)
24. Isaacson, D.: Some considerations on arithmetical truth and the ω-rule. In: Detlefsen, M. (ed.), Proof, Logic and Formalization, pp. 94–138. Routledge, New York (1992)
25. Jerome Keisler, H.: Model Theory for Infinitary Logic. North-Holland (1971)
26. Jerome Keisler, H.: An infinitesimal approach to stochastic analysis. In: Memoirs of the American Mathematical Society, vol. 48 (1984)

27. Jerome Keisler, H.: Stochastic differential equations with extra properties. In: Arkeryd, L., Cutland, N., Henson, W. (eds.), Proceedings of the NATO Advanced Study Institute on Nonstandard Analysis and its Applications, Nonstandard Analysis: Theory and Applications, Edinburgh, UK, June 30 July 13 (1996). (NATO ASI Ser., Ser. C, Math. Phys. Sci. 493, 259–277, 1997)
28. Kripke, S.: A completeness theorem in modal logic. J. Symbol. Logic **24**, 1–14 (1959)
29. Lewis, C.I., Langford, C.H.: Symbolic Logic. Dover (1932)
30. Loeb, P.: Conversion from nonstandard to standard measure spaces and applications in probability theory. Trans. Am. Math. Soc. **211**, 113–122 (1975)
31. Marinković, B., Ognjanović, Z., Doder, D., Perović, A.: A propositional linear time logic with time flow isomorphic to ω^2. J. Appl. Logic **12**(2), 208–229 (2014)
32. Marković, Z., Ognjanović, Z., Rašković, M.: A probabilistic extension of intuitionistic logic. Math. Logic Q. **49**, 415–424 (2003)
33. Marković, Z., Ognjanović, Z., Rašković, M.: An intuitionistic logic with probabilistic operators. Publications de L'Institute Matematique, n.s. **73**(87), 31–38 (2003). http://elib.mi.sanu.ac.rs/files/journals/publ/93/n087p031.pdf
34. Marković, Z., Ognjanović, Z., Rašković, M.: What is the proper propositional base for probabilistic logic? In: Zadeh, L.A. (ed.), Proceedings of the Information Processing and Management of Uncertainty in Knowledge-Based Systems Conference IPMU 2004, July, 4–9, pp. 443–450. Perugia, Italy (2004)
35. Milošević, M., Ognjanović, Z.: A first-order conditional probability logic. Logic J. Interest Gr. Pure and Appl. Logics **20**(1), 235–253 (2012)
36. Milošević, M., Ognjanović, Z.: A first-order conditional probability logic with iterations. Publications de L'Institute Matematique, n.s. **93**(107), 19–27 (2013). http://elib.mi.sanu.ac.rs/files/journals/publ/113/n107p019.pdf
37. Ognjanović, Z.: A logic for temporal and probabilistic reasoning. In: Workshop on Probabilistic Logic and Randomised Computation, ESSLLI 1998, Saarbruecken, Germany (1998)
38. Ognjanović, Z.: Some probability logics and their applications in computer sciences. Ph.D. thesis (in Serbian). University of Kragujevac (1999). http://elibrary.matf.bg.ac.rs/bitstream/handle/123456789/197/phdZoranOgnjanovic.pdf?sequence=1
39. Ognjanović, Z.: Discrete linear-time probabilistic logics: completeness, decidability and complexity. J. Logic Comput. **16**(2), 257–285 (2006)
40. Ognjanović, Z., Rašković, M.: A logic with higher order probabilities. Publications de L'Institute Matematique, n.s. **60**(74), 1–4 (1996). http://elib.mi.sanu.ac.rs/files/journals/publ/80/n074p001.pdf
41. Ognjanović, Z., Rašković, M.: Some probability logics with new types of probability operators. J. Logic Comput. **9**(2), 181–195 (1999)
42. Ognjanović, Z., Rašković, M.: Some first-order probability logics. Theor. Comput. Sci. **247**(1–2), 191–212 (2000)
43. Ognjanović, Z., Marković, Z., Rašković, M.: Completeness theorem for a Logic with imprecise and conditional probabilities. Publications de L'Institute Matematique Ns. **78**(92), 35–49 (2005). http://elib.mi.sanu.ac.rs/files/journals/publ/98/n092p035.pdf
44. Ognjanović, Z., Ikodinović, N.: A logic with higher order conditional probabilities. Publications de L'Institute Matematique (Beograd), ns. **82**(96) 141–154 (2007). http://elib.mi.sanu.ac.rs/files/journals/publ/102/n096p141.pdf
45. Ognjanović, Z., Perović, A., Rašković, M.: Logics with the qualitative probability operator. Logic J. IGPL **16**(2), 105–120 (2008)
46. Ognjanović, Z., Doder, D., Marković, Z.: A branching time logic with two types of probability operators. In: Benferhat, S., Grant, J. (eds.), Fifth International Conference on Scalable Uncertainty Management SUM-2011. Lectuer notes in computer science, October 10–12, 2011, Dayton, Ohio, USA, vol. 6929, pp. 219–232. Springer, Heidelberg (2011)
47. Perović, A.: Some applications of the formal method in set theory, model theory, probabilistic logics and fuzzy logics. Ph.D. thesis (in Serbian). University of Belgrade (2008). http://elibrary.matf.bg.ac.rs/bitstream/handle/123456789/100/phdAleksandarPerovic.pdf?sequence=1

48. Perović, A., Ognjanovć, Z., Rašković, M., Marković, Z.: How to restore compactness into probabilistic logics? In: Proceedings of the 11th European Conference. JELIA 2008. Lecture Notes in Computer Science, Dresden, Germany, September 28–October 1, 2008, vol. 5293, pp. 338–348. Springer, Heidelberg (2008)
49. Perović, A., Ognjanovć, Z., Rašković, M., Marković, Z.: A probabilistic logic with polynomial weight formulas. In: Hartmann, S., Kern-Isberner, G. (eds.), Proceedings of the Fifth International Symposium Foundations of Information and Knowledge Systems, FoIKS 2008. Lecture Notes in Computer Science, Pisa, Italy, February 11–15, 2008, vol. 4932, pp. 239–252. Springer, Heidelberg (2008)
50. Prior, A.N.: Time and Modality. Oxford University Press, Oxford (1957)
51. Rašković, M.: Classical logic with some probability operators. Publications de l'Institut Mathématique n.s. **53**(67), 1–3 (1993). http://elib.mi.sanu.ac.rs/files/journals/publ/73/n067p001.pdf
52. Rašković, M., Ognjanović, Z.: Some propositional probabilistic logics. In: Kovačević-Vujičić, V., Mijajlović, Ž. (eds.), Proceedings of the Kurepa's Symposium 1996, Belgrade, Scientific Review 19–20, pp. 83–90 (1996)
53. Rašković, M., Ognjanović, Z.: A first order probability logic, LP_Q. Publications de L'Institute Matematique, ns. **65**(79), 1–7 (1999). http://elib.mi.sanu.ac.rs/files/journals/publ/85/n079p001.pdf
54. Rašković, M., Ognjanović, Z., Marković, Z.: A probabilistic approach to default reasoning. In: Proceeding of the NMR '04, pp. 335–341 (2004)
55. Rašković, M., Ognjanović, Z., Marković, Z.: A logic with conditional probabilities. In Proceedings of the JELIA'04, Lecture Notes in Computer Science, vol. 3229, pp. 226–238. Springer, Heidelberg (2004)
56. Rašković, M., Marković, Z., Ognjanović, Z.: A logic with approximate conditional probabilities that can model default reasoning. International. J. Approx. Reason. **49**(1), 52–66 (2008)
57. Robinson, A.: Non-standard Analysis. North-Holland, Amsterdam (1966)
58. Shoenfield, J.R.: On a restricted ω-rule. Bull. Acad. Sc. Polon. Sci., Ser. des sc. Math., Astr. et Phys. **7**, 405–407 (1959)
59. Sundholm, G.: Proof theory: a survey of the omega-rule. D.Phil. thesis. University of Oxford (1983)
60. Turing, A.: Systems of logic based on ordinals (Ph.D. thesis). Princeton University (1938)

Chapter 2
History

Abstract We give a survey of the relationship through history between mathematical logic and probability theory. Actually, these two branches of science were strongly connected and intertwined since the appearances of the first treatises on probability in the second half of the seventeenth century, while probability theory was often considered as an extension of logic. However, in these early days and even in the first century and a half of development, the basic notions of probability theory, like probability itself, conditional probability, independence, etc., were not standardized, or their meanings were quite different from the modern ones, which allows several (sometimes opposite) interpretations of original statements. Where appropriate, we give such apparently contrary opinions of modern commentators. Thanks to numerous digital repositories, many of the old original writings are now at public disposal. In this chapter, we extensively quote these source texts to illustrate how their authors created one of the most exiting scientific fields and developed the main concepts in probability theory. In this way, the reader is offered the insight into the spirit of those great minds and their times, but it is also left to her/him to grasp the ideas in their original forms, so that she/he can assess what the authors wanted to say. The chapter covers very ambitiously the period from Aristotle and Plato until the middle of 1980s, but most of the materials were selected to emphasize the original ideas that motivated or, at least, that are somehow related to contemporary research in the field of probability logic (which also involves some results of the authors of this book). Far more extensive historical and philosophical studies can be found for example in Devlin, The unfinished game: Pascal, Fermat and the seventeenth-century letter that made the world modern, 2008, [45], Hailperin, Sentential probability logic, origins, development, current status and technical applications, 1996, [63], Shafer, Arch Hist Exact Sci, (19):309–370, 1978, [152], Shafer, Stat Sci, (21):70–98, 2006, [151], Styazhkin, History of mathematical logic from Leibnitz to Peano, 1969, [159], while the early surveys by Pierre-Simon Laplace and Isaac Todhunter in Laplace, Essai philosophique sur les probabilities, 1902, [88], and Todhunter, A History of the mathematical theory of probability, 1865, [164] might be particularly interesting.

Z. Ognjanović et al., *Probability Logics*,
DOI 10.1007/978-3-319-47012-2_2

2.1 Pre-leibnitzians

Many mathematical stories begin in ancient times, and this one is no exception. The invention of probable or, better said, plausible reasoning was attributed, in Plato's Phaedrus and Aristotle's Rhetoric, to legendary sophists Corax and Tisias in the fifth century B.C. [80]. To argue about legal, medical or political questions, as the main tool, they used the notion of eikos ($\varepsilon\iota\kappa\acute{o}\varsigma$). Plato understood eikos in terms of likeliness, i.e., weaker than and inferior to truth, while Aristotle interpreted it as either something subjectively acceptable or what usually happened. Here we can perceive early signs of subjective—objective duality in epistemic—statistical interpretations of probability. The following example from [2, Book II, 24] illustrates Aristotle's approach (translation from [80]):

> If the accused is not open to the charge, for instance if a weakling be tried for violent assault, the defence is that he was not likely to do such a thing.

Greek tradition did not associate numerical quantities to uncertain assumptions and conclusions, so we can hardly consider the eikos-arguments as something we today call formal reasoning about probabilities. In our opinion, these can be seen as early examples of default reasoning (see Sect. 6.1).

Following the template very common in Western tradition, after a number of centuries and the period of dark ages, new ideas about uncertainty appeared in the fifteenth and sixteenth centuries inspired by practical and profitable issues in gambling. For example, fra Luca Bartolomeo de Pacioli (1445–1517), Gerolamo Cardano (1501–1576), Niccolo Tartaglia (1499–1559) and others dealt with the division problem in games of chance, i.e., how to split the prize of a game between players,[1] if the first and the second player won a and b rounds, respectively, and the game did not finish. The deliberations and studies of such issues lead to well-known correspondence (1654–1660) between Pierre de Fermat (1601–1665) and Blaise Pascal (1623–1662), and Christiaan Huygens' (1629–1695) treatise De ratiociniis in ludo aleae [72, 138]. Fermat and Pascal had different approaches to solving the division problem: while Fermat analyzed the complete list of all possible outcomes of a game, Pascal introduced some kind of recursion procedure so that, relying on his arithmetical triangle, he was able to express and calculate more complex cases using simpler ones [45]. Huygens in [72] tried to produce rigorous mathematical proofs for 14 propositions, and in that way justify his solutions to gambling problems. He used (in the Latin version of the text) the word *expectatio* to denote a notion which he did not define, but which could be interpreted in the framework of a game of chance as a player's share of the stake if the game is not played [142] or as the price for a ticket in a fair lottery [60]. For example, he proved (translation from [73]):

> Proposition I. If I expect a or b, and have an equal chance of gaining either of them, my Expectation is worth $\frac{a+b}{2}$.

---

[1] We cannot resist to mention here Tartaglia's view that such a question is more legal than mathematical [150, 163].

> Proposition III. If the number of Chances I have to gain a, be p, and the number of Chances I have to gain b, be q. Supposing the Chances equal; my Expectation will then be worth $\frac{ap+bq}{p+q}$.

Before all these seventeenth century's achievements, probability was perceived, following old Greek philosophers, as something epistemic and related to arguments and opinions. Therefore, some authors [45, 60, 152] particularly emphasize Pascal's contribution to emergence of the new approach to conjecturing and uncertainty reasoning[2] based on similarities with calculus and concepts established to analyze games of chance. In the well-known Pascal's wager [3, 114], using aleatory calculus he explained that rational persons should believe in the existence of God since that choice offers infinite gain of salvation.

In 1662 appeared a book [3] known as Port-Royal Logic, or Ars Cogitandi, which had a great and long-term influence. This seems to be the first text that explicitly suggests that probability (in the epistemic sense) can be numerically quantified and calculated using methods designed for games of chance. To consider possible gains and losses, it is said, one should take into account not only the corresponding amounts, but also probabilities that they could occur (translation from [4]):

> There are certain games in which ten persons lay down a crown each, and where one only gains the whole, and all the others lose; thus, each of the players has only the chance of losing a crown, and of gaining nine by it. If we consider only the gain and loss in themselves, it might appear that all have the advantage of it; but it is necessary to consider, further, that if each may gain nine crowns, and there is only the hazard of losing one, it is also nine times more probable, in relation to each, that he will lose his crown, and not gain the nine. Thus each has for himself nine crowns to hope for, one to lose, nine degrees of probability of losing a crown, and only one of gaining the nine, which puts the matter on a perfect equality.

2.2 Leibnitz

There are somewhat different assessments [60, 63, 141] of the significance of Gottfried Wilhelm Leibnitz[3] (1646–1716) in creation and development of probability logic.[4] Some authors, e.g. [152], regard his influence as minor and justify such view by the fact that a big part of his work remained unpublished long after his death. However, as Leibnitz was one of the most famous bloggers[5] of the age, for us there is no doubt that his impact on the state of the collective scientific mind was very important. Anyway, the following are certainly true:

---

[2] Although Fermat, Pascal, Huygens, and their predecessors did not use the word probability (or an equivalent) in the contemporary sense, nor did they use that concept to quantitatively measure beliefs or uncertainty.

[3] Or: Lubeniecz, Leibniz.

[4] Or: probability theory, since for Leibnitz these two were synonymous.

[5] Or: one of the main nodes in the seventeenth century research communication network. Leibnitz exchanged more than 15000 letters with more than 1000 persons [101].

- In his dissertation [90] from 1665 Leibnitz used numbers from the interval [0, 1] to represent legal conditional rights, so that 0 and 1 denote non-existence of rights and absolute rights, respectively, and fractions stand for different degrees of certainty.[6]
- Leibnitz used the word probability and advocated the concept of numerical quantification of probable. Having probability conclusions as conditional, i.e., as subjective and relative to the existing knowledge, he tried to measure knowledge [97].
- Leibnitz gave a definition of probability, relaying on equally possible cases, as the ratio of favorable cases to the total number of cases [93].
- Leibnitz understood moral certainty as something that is "infinitely probable" [98] (see the interpretation of $CP_{\approx 1}$ in Sect. 5.7).
- Leibnitz established a program for development of probability logic and had a huge impact on its realization.
- Probability, or better said probability logic, had the central role in his attempts to create a powerful universal calculus.

In his great opus Leibnitz particularly focused on analogies between the processes of thinking and computation. Already in [92], even before he started deep study in mathematics, Leibnitz emphasized the combinatorial nature of cognitive processes (translation from [82]):

> Thomas Hobbes, everywhere a profound examiner of principles, rightly stated that everything done by our mind is a computation.

Starting from his thesis [90] Leibnitz publicized ideas to develop a doctrine—or a new kind of logic—of *de gradibus probabilitatis* (degrees of probability)[7] [98, 140, 141]. Leibnitz spent several years (1672–1676) in Paris and taught by Huygens became familiar with his and Pascal's works. Leibnitz recognized that their techniques were suitable for realization of his plan. His work in this field was motivated and always concerned with jurisprudence, so he used the new ideas in a debate about inheritance of throne of Poland in 1669 [60] and to justify Huygens' method by principles of jurisprudence [152]. Leibnitz's position was clearly described in his New Essays on Human Understanding finished in 1704 (translations from [63, 100, 161]):

> (66) As for the inevitability of the result, and degrees of probability, we do not yet possess the branch of logic that would let them be estimated.
>
> (372) Perhaps opinion based on likelihood also deserves the name of knowledge; otherwise, nearly all historical knowledge will collapse, and a good deal more ...probability or likelihood is broader: it must be drawn from the nature of things; and the opinion of weighty authorities is one of the things which can contribute to the likelihood of an opinion, but it does not produce the entire likelihood by itself.
>
> (464) The entire form of judicial procedures is, in fact, nothing but a kind of logic, applied to legal questions.

---

[6] According to Keynes [78], Leibnitz in [91] used 0 for impossibility and 1 for certainty, while variable intermediate degrees of probability were denoted by $\frac{1}{2}$.

[7] Friedrich Nitzsche (1645–1702) in a letter from 1670 suggested Leibnitz to realize these ideas. Schneider emphasized in [140] that Leibnitz was not able to provide numerical methods to calculate probabilities, and that, following Skeptics who had a continuum of possible modalities, considered qualitative gradation of the probable.

> (466) I have said more than once that we need a new kind of logic, concerned with degrees of probability, since Aristotle in his Topics couldn't have been further from it. He is satisfied with arranging a few familiar rules according to common patterns; these could serve on the occasion when one is concerned with amplifying a discourse so as to give it some likelihood. No effort is made to provide a balance necessary to weight the likelihoods in order to obtain a firm judgement. Anyone wanting to deal with this question would do well to pursue the investigation of games of chance. In general, I wish that some skilful mathematician were interested in producing a detailed study of all kinds of games, carefully reasoned and with full particulars. This would be of great value in improving discovery techniques, since the human mind appears to better advantage in games than in the most serious pursuits.

Once realized, all these ideas would lead to a formal system—universal language and a powerful logical calculus—which could be the basis for all sciences and replace arguments by formal computation [95] (translation from [101]):

> If this is done, whenever controversies arise, there will be no more need for arguing among two philosophers than among two mathematicians. For it will suffice to take pens into the hand and to sit down by the abacus, saying to each other (and if they wish also to a friend called for help): Let us calculate!

Since calculations can be considered as proofs in a logical framework so important in Leibnitz's work, it is interesting to consider the following passage from [99] (translation from [166]):

> And here is discovered the inner distinction between necessary and contingent truths, which no one will easily understand unless he has some tincture of Mathematics namely, that in necessary propositions one arrives, by an analysis continued to some point, at an identical equation (and this very thing is to demonstrate a truth in geometrical rigor); but in contingent propositions the analysis proceeds to infinity by reasons of reasons, so that indeed one never has a full demonstration, although there is always, underneath, a reason for the truth, even if, it is perfectly understood only by God, who along goes through an infinite series in one act of the mind.

There are different opinions on success [58, 59] or failure [166] of Leibnitz's concept of infinitary proofs, but nevertheless an interesting analogy can be driven between this idea and some of the notions (infinitary rules and inferences) presented in this book.

Leibnitz distinguished between forward and backward calculations of probabilities, namely calculations of probabilities of consequences given probabilities of causes on the one hand, and calculations of probabilities of causes given probabilities of consequences, on the other [159]. He was concerned with the latter, which he explained in letters exchanged with Jacob Bernoulli (see Sect. 2.3).

Finally, we should mentioned that in analyzing the nature of continuum and developing the differential and integral calculi [94, 96] Leibnitz extensively used another concept we employ in this book—infinitesimal numbers. For Leibnitz, infinitesimals were ideal entities, positive but smaller than every $\frac{1}{n}$. His Law of Continuity ensured that infinitesimals were governed by the same arithmetic laws as the real numbers, except that they obviously violate Archimedes' principle.[8]

---

[8]Every number can be reached by adding (finitely many times) 1's.

2.3 Jacob Bernoulli

The significance of the book Ars Conjectandi [8] is such that it earned its author Jacob Bernoulli (1654–1705) the title of the founder of probability theory [141]. Some of his achievements are the following:

- He placed the notion of probability, understood as degree of certainty, in the very focus of scientific research.
- He introduced two notions of probability: a priori, in the absence of prior knowledge of an event, and a posteriori probabilities, with such knowledge taken into account.
- He also distinguished between another two notions of probability: subjective and objective probability.
- He improved and generalized existing techniques for calculating probabilities of compound events from probabilities of their constituents.
- He stated and proved his Golden theorem (also known as the (Weak) Law of large numbers, or just Bernoulli's Theorem).
- He realized an important part of Leibnitz's program for developing a new, probability logic, etc.

According to Bernoulli's scientific diary, Meditationes, he studied probability in the 1680s under the influence of Pascal, Port-Royal Logic[9] and Huygens, and developed most of his important results at that time. A very illustrative example Bernoulli discussed in Meditationes was related to a problem concerning a marriage contract between Titius and Caja. Bernoulli analyzed how, under the assumption that the married couple had children and the wife died before the husband, a division of the estate could be realized between Titius and the children depending on whether married couple's fathers died or not before Caja. Bernoulli provided a discussion on chances that one of those three lived longer than the others. To do so, he listed all possible orders of death and, taking into account that Caja was younger and considering experiences from real-life examples, evaluated her certainty to die first, second and third as $\frac{1}{5}$, i.e., (translation from [139]):

> One probability, five of which make the entire certainty,

$\frac{4}{15}$ and $\frac{8}{15}$, respectively. Then he explained the concept of a posteriori probabilities (translation from [141]):

> The reason that in card and dice games which are governed solely by chance the expectation can be precisely and scientifically determined is that we can perceive actually and clearly the number of cases in which gain or loss must follow infallibly and that these cases behave indifferently and can each occur with equal facility or when one is more probable than another we can at least determine scientifically by how much it is more probable. But what mortal, I pray you, counts the number of cases, diseases or other circumstances to which

[9] Even the name Ars Conjectandi was reminiscence of Ars cogitandi. In Chap. IV of Part IV, Bernoulli wrote (translation from [10]) "the celebrated author of the Ars cogitandi, a man of great insight and intelligence". Another translations can be found in [9, 11].

now the old men, now the young men are made subject, and knows whether or not these will be overtaken by death, and determines how much more probable is that one will be taken unawares than another, since all these depend on causes that are completely hidden and beyond our knowledge.

Generally in civic and moral affaires things are to be understood, in which we of course know that the one thing is more probable, better or more advisable than another; but by what degree of probability or goodness they exceed others we determine only according to probability, not exactly. The surest way of estimating probabilities in these cases is not a priori, that is by cause, but a posteriori, that is, from the frequently observed event in similar examples.

Bernoulli wrote a good part of Ars Conjectandi in the 1680s. He was never fully satisfied with the text and the book, which he was constantly changing and improving but failed to finish in his lifetime, was printed by his nephew Nicholas only eight years after Jacob's death. In the book, Bernoulli transformed several particular approaches useful for calculating chances in gambling games to a mathematical calculus applicable to uncertain reasoning in many real-life fields and to a large extent realized what Leibnitz dreamed of. Ars Conjectandi had four parts:

- In the first part, Bernoulli revised Huygens' text [72] offering new annotations, numerous comments and generalizations, and provided solutions to problems from Huygens' text.
- The second part presented results about theory of permutations and combinations like, for example, the first proof of the binomial theorem for positive integral powers.[10]
- In the third part Bernoulli solved 24 problems related to games of chances.
- Finally, the forth part, The use and applications of the previous study to civil, moral, and economic problems, was the most innovative and important part of the book.

While in the first three parts Bernoulli followed the approach of Huygens, in the last part he changed his language and wrote about probabilities. In Chap. I of Part IV, Bernoulli introduced the main notions and, following Leibnitz, defined probabilities using equally possible cases and explicitly mentioned the subjective and objective conceptions of probabilities (translation from [11]):

Certainty of some thing is considered either objectively and in itself and means none other than its real existence at present or in the future; or subjectively, depending on us, and consists in the measure of our knowledge of this existence …As to probability, this is the degree of certainty, and it differs from the latter as a part from the whole. Namely, if the integral and absolute certainty, which we designate by letter α or by unity 1, will be thought to consist, for example, of five probabilities, as though of five parts, three of which favor the existence or realization of some event, with the other ones, however, being against it, we will say that this event has $3/5\alpha$, or $3/5$, of certainty. …Possible is that which has at least a low degree of certainty whereas the impossible has either no, or an infinitely small certainty …Morally certain is that whose probability is almost equal to complete certainty so that the difference

[10] Particular cases of the Binomial Theorem, $(a+x)^n = \sum_{k=0}^{n} \binom{n}{k} a^{n-k} x^k$, were known from ancient times, while in a more general form, with fractional powers, Theorem was stated without proof in 1676 by Isaac Newton in letters to Leibnitz.

> is insensible ...Necessary is that, which cannot fail to exist at present, in the future or past ...[Contingent] is that which can either exist or not exist at present, in the past or future ...what seems to be contingent to one person at a certain moment, will be thought necessary to someone else (or even to the same person) at another time after the causes become known. And so, contingency mainly depends on our knowledge ...

In the next two chapters, Bernoulli introduced some notions that one would expect today in formal logical systems. This great novelty, as Hailperin noted in [63], has not been properly recognized by later commentators. Bernoulli first emphasized that making conjectures meant art of measuring, as exactly as possible, probabilities of things using numbers and weights[11] of arguments (dis)proving those things.

Since the corresponding notions and techniques were not fully developed at the time, it is not quite clear how Bernoulli understood arguments. Hailperin gave two possible interpretations for an argument [63]:

- it is a statement which, in the cases when it is true, justifies the conclusion, and
- it is a pair of two statements, so that there is a deduction from the former to the latter.

In the sequel Bernoulli listed general postulates[12] about applying arguments. For example (translation from [11]):

> We ought to consider not only the arguments which prove a thing but also all those which can lead to a contrary conclusion, so that, after duly discussing the former and the latter, it will become clear which of them have more weight. It is asked, with respect to a friend very long absent from his fatherland, may we declare him dead? The following arguments favor an answer in the affirmative: During the entire twenty years, in spite of all efforts, we have been unable to find out anything about him; the lives of travelers are exposed to very many dangers from which those remaining at home are exempted ...[we should] oppose them by the following supporting the contrary ...Perhaps Barbarians held him captive so that he was unable to write ...many people are known to have returned unharmed after having been absent even longer ...
>
> However, since complete certitude can only seldom be attained, necessity and custom desire that that, which is only morally certain, be considered as absolutely certain. Therefore, it would be helpful if the authorities determine certain boundaries for moral certainty ...so that a judge, unable to show preference to either side, will always have firm indications to conform with when pronouncing a sentence.

He concluded this discussion by saying (formulated in modern terms) that he was aware that the list of axioms was not complete.

To calculate probabilities generated by arguments (he called them also "degrees of certainty"), Bernoulli used ideas from the former parts of the book, i.e., the methods proposed by Huygens. He supposed that all cases are equally possible, while if they are not, more frequent cases should be reduced to simpler, equally possible, cases. It seems that Bernoulli somehow assumed that those equipossible cases were mutually exclusive, but he did not mention that explicitly. Bernoulli distinguished arguments that:

---

[11]The weight was understood as "the force of the proof".

[12]He called them rules or axioms.

- exist necessarily and provide evidences contingently,
- exist contingently and provide evidences necessarily, and
- exist and provide evidences contingently,

and illustrated them in the following way. The situation that his brother did not write him for a long time could be caused by brother's laziness, busyness, or even death. The first argument exists necessarily (since he knows the brother), but provides the evidence contingently (laziness does not prevent writing); the next one exists and provides evidence contingently (the brother could have or haven't a job), while the last one obviously exists contingently and provides evidence necessarily. It should be noted that later authors, for example, Lambert, Bolzano, and De Morgan, usually considered only necessary inferences. Furthermore, an argument can be:

- pure, which means that in some cases it proves a thing, while in the other case it does not prove anything, or
- mixed, if in some cases it proves a thing, while in the other case it proves the contrary.

In the simplest case, if there are $a = b+c$ cases, and an argument exists contingently (i.e., in b cases) and necessarily provides evidence for a thing (he denoted that by 1), while it does not exist in c cases when it fails to provide anything (denoted by 0):

> Which is, by Coroll. I, Prop. III of the first part, worth $\frac{b \cdot 1 + c \cdot 0}{a} = \frac{b}{a}$, so that such argument proves $\frac{b}{a}$ of the thing, or of the certainty of the thing.

After that he proceeded to more complicated situations, including combinations of two or more arguments that can be contingent and mixed [63, 152]. For example, let there be two arguments, and:

- b and e be the numbers of cases in which the first and the second argument prove a thing,
- c and f be the numbers of cases in which the first and the second argument prove nothing, or prove the opposite of the thing, and
- $a = b + c$, and $d = e + f$,

then the weight of the combination of two arguments is calculated as:

$$\frac{(d-f) \cdot 1 + f \cdot ((a-c) : a)}{d} = \frac{ad - cf}{ad} = 1 - \frac{cf}{ad}.$$

In such a complex setting, Shafer pointed in [152], there are situations in which probabilities might be non-additive (i.e., the sum of probabilities of a thing and its opposite is not 1). Indeed, it was written in [8] (translation from [10]):

> But if besides the proofs which serve to prove the thing other pure proofs offer themselves, proofs by which the opposite of the thing is advised, the proofs of both kinds must be weighed separately according to the preceding rules in order that there then may exist a ratio between the probability of the thing and the probability of the opposite of the thing. Whence it must be noted that if the proofs adduced for each side are strong enough, it can happen that the absolute probability of each side notably exceeds half the certainty; i.e., that each of the

> alternatives is rendered probable, although relatively speaking one is less probable than the other. And so it can happen that a thing has 2/3 certainty and its opposite possesses 3/4 certainty …

As a picturesque example, Bernoulli considered whether a document was antedated. He stated a denying argument that the document was signed by a notary and that between 50 notaries only one might be engaged in fraud. On the other hand, the affirmative arguments were that the public reputation of the notary who signed the document was bad, and that he could expect benefit from the possible cheat. Bernoulli derived that the probability that the document was valid could be estimated as 49/50 of certainty, while 999/1000 of certainty valued the opposite. He concluded that, since he knew that the notary was dishonest, the former possibility could be dismissed. In Hailperin's interpretation [63] the probabilities of the conclusion "the document is antedated" and its negation are not absolute but conditional. Since the corresponding conditions are different[13] he does not see it as an issue.

Chapter 4, similarly as in the quotation from Meditationes, distinguished between things for which the number of (un)favorable cases could be determined "by the production of nature or the free will of people" and those depending on hidden causes for which the number of (un)favorable cases could not be discovered a priori. Bernoulli emphasized that in the latter circumstances the corresponding probability could be estimated a posteriori from repeated observations as the ratio between the number of favorable cases and the number of all cases. Those estimations would be contained in intervals with boundaries that could be brought closer and closer by increasing the number of experiments and in this process moral certainty could be reached. In Chap. 5 Bernoulli proved the result,[14] he was developing for 20 years and which he highly appreciated (translation from [11]):

> Both its novelty and its very great usefulness, coupled with its just as great difficulty, can exceed in weight and value all the remaining chapters of this thesis …
>
> Main Theorem: Finally, the theorem follows upon which all that has been said is based, but whose proof now is given solely by the application of the lemmas stated above. In order that I may avoid being tedious, I will call those cases in which a certain event can happen successful or fertile cases; and those cases sterile in which the same event cannot happen. Also, I will call those trials successful or fertile in which any of the fertile cases is perceived; and those trials unsuccessful or sterile in which any of the sterile cases is observed. Therefore, let the

[13]More formally, the probabilities are of the forms $P(C|H_1)$ and $P(\neg C|H_2)$, where C, H_1, and H_2 are the conclusion and hypothesis, respectively.

[14]Later called the (Weak) Law of large numbers by Poisson (1781–1840). In modern notation, the statement can be formulated as: Let p be the probability of an event, ε a small positive number, and c a large positive number, then n can be calculated such that

$$P\left(\left|\frac{S_n}{n} - p\right| > \varepsilon\right) < \frac{1}{c}.$$

where S_n is the number of success in n binomial trials.

As an additional argument that Bernoulli did not consider non-additive probabilities, Hailperin argues that using this theorem a posteriori probabilities are obtained from frequencies and that the sum of the relative frequencies of an event and its complement is 1.

> number of fertile cases to the number of sterile cases be exactly or approximately in the ratio r to s, and hence the ratio of fertile cases to all the cases will be $\frac{r}{r+s}$ or $\frac{r}{t}$; which is within the limits $\frac{r+1}{t}$ and $\frac{r-1}{t}$. It must be shown that so many trials can be run such that it will be more probable than any given times (e.g., c times) that the number of fertile observations will fall within these limits rather than outside these limits—i.e., it will be c times more likely than not that the number of fertile observations to the number of all the observations will be in a ratio neither greater than $\frac{r+1}{t}$ nor less than $\frac{r-1}{t}$.

Bernoulli believed that from this celebrated theorem a solution for the inverse problem[15] would follow. In several letters exchanged with Bernoulli, Leibnitz expressed his concerns whether contingent events determined by infinitely many conditions could be properly characterized by finite numbers of experiments, while Bernoulli wrote (translation from [60]):

> I can already determine how many observations must be made in order that it is 100 times, 1000 times, 10000 times, etc., more likely than not—and this is moral certainty—that the ratio between the number of cases which I estimate is legitimate and genuine.

However, as Hacking pointed out in [60], this is not the case, namely the theorem only enables to compute a conditional probability[16] that the probability p of an event would be within some experimentally determined limits, given the probability p.

2.4 Probability and Logic in the Eighteenth Century

Many famous mathematicians in one way or another continued Bernoulli's work on probability theory: the members of his family—Nicolaus I (1687–1759), Nicolaus II (1695–1726) and Daniel Bernoulli (1700–1782), Pierre Rémond de Montmort (1678–1719),[17] Leonhard Euler (1707–1783), Joseph-Louis Lagrange (1736–1813), Nicolas de Condorcet (1743–1794), Johann Carl Friedrich Gauss (1777–1855), etc.

2.4.1 Abraham de Moivre

A pearl amongst outstanding eighteenth century achievements in the field was Abraham de Moivre's (1667–1754) book The Doctrine of Chances: a method of calculating the probabilities of events in play [38]. In the 1738 edition de Moivre included the chapter "A method of approximating the sum of the terms of the binomial $\overline{a+b}^n$ expanded into a series, from whence are deduced some practical rules to estimate the degree of assent which is to be given to experiments" that introduced the concept of the normal distribution as an approximation of the binomial distribution

[15] To determine probabilities a posteriori from samples.

[16] More formally: $P(p \in [S_n \pm \varepsilon] | p)$.

[17] Montmort published in 1708 a book, Essay d'analyse sur les jeux de hasard [39], which accelerated the appearance of Jacob Bernoulli's Ars Conjectandi.

and was the first example of what would later be called the central limit theorem.[18] De Moivre, inspired by Jacob Bernoulli's approach to determine probabilities from observations, realized that as a result of great importance (all quotations are from the 1756 edition):

> ...I'll take the liberty to say, that this is the hardest Problem that can be proposed on the subject of chance, for which reason I have reserved it for the last, but I hope to be forgiven if my solution is not fitted to the capacity of all readers; however I shall derive from it some conclusions that may be of use to every body ...Altho' the solution of problems of chance often requires that several terms of the binomial $(a+b)^n$ be added together, nevertheless in very high powers the thing appears so laborious, and of so great difficulty, that few people have undertaken that task; for besides James and Nicolas Bernoulli, two great Mathematicians, I know of no body that has attempted it; in which, tho' they have shewn very great skill, and have the praise which is due to their industry, yet some things were farther required; for what they have done is not so much an approximation as the determining very wide limits, within which they demonstrated that the sum of the terms was contained.

and, similarly as Jacob Bernoulli, believed that it could also be used to solve the inverse problem:

> As, upon the supposition of a certain determinate law according to which any event is to happen, we demonstrate that the ratio of happenings will continually approach to that law, as the experiments or observations are multiplied: so, conversely, if from numberless observations we find the Ratio of the events to converge to a determinate quantity, as to the ratio of P to Q; then we conclude that this ratio expresses the determinate law according to which the event is to happen ...Again, as it is thus demonstrable that there are, in the constitution of things, certain laws according to which events happen, it is no less evident from observation, that those laws serve to wise, useful and beneficent purposes; to preserve the stedfast order of the universe, to propagate the several species of beings, and furnish to the sentient kind such degrees of happiness as are suited to their state ...And hence, if we blind not ourselves with metaphysical dust, we shall be led, by a short and obvious way, to the acknowledgment of the great maker and governour of all.

Under the impression of importance of de Moivre's result, which gave mathematical formulation to an empirical phenomenon of statistical regularity, it has been usually forgotten that his book offered other original insights that established directions for further researchers. The book began with basic definitions and rules that illustrated de Moivre's understanding of reasoning about chances. Following Bernoulli, de Moivre defined probability as a ratio:

> Wherefore, if we constitute a fraction whereof the numerator be the number of chances whereby an event may happen, and the denominator the number of all the chances whereby it may either happen or fail, that fraction will be a proper designation of the probability of happening.

and considered it the primary notion of his theory. Undoubtedly, for de Moivre probabilities were additive:

---

[18]It seems that this name was introduced by György Pölya in 1920 [74]. It means that the limit theorem is of central importance in probability.

> The fractions which represent the Probabilities of happening and failing, being added together, their sum will always be equal to unity, since the sum of their numerators will be equal to their common denominator: now it being a certainty that an event will either happen or fail, it follows that certainty, which may be conceived under the notion of an infinitely great degree of probability, is fitly represented by unity. These things will easily be apprehended, if it be considered, that the word probability includes a double idea; first, of the number of chances whereby an event may happen; secondly, of the number of chances whereby it may either happen or fail.

Expectation was not the primary notion, but a defined one:

> In all cases, the expectation of obtaining any sum is estimated by multiplying the value of the sum expected by the fraction which represents the probability of obtaining it. Thus, if I have 3 chances in 5 to obtain 100^L I say that the present value of my expectation is the product of 100^L by the fraction $\frac{3}{5}$, and consequently that my expectation is worth 60^L.

De Moivre was the first who considered (in)dependent events:

> Two events are independent, when they have no connexion one with the other, and that the happening of one neither forwards nor obstructs the happening of the other. Two events are dependent, when they are so connected together as that the probability of either's happening is altered by the happening of the other.

and gave the multiplication rule to calculate probabilities of compound events:

> Suppose there is a heap of 13 cards of one colour, and another heap of 13 cards of another colour, what is the probability that taking a card at a venture out of each heap, I shall take the two aces? The Probability of taking the ace out of the first heap is $\frac{1}{13}$: now it being very plain that the taking or not taking the ace out of the first heap has no influence in the taking or not taking the ace out of the second; it follows, that supposing that ace taken out, the probability of taking the ace out of the second will also be $\frac{1}{13}$; and therefore, those two Events being independent, the probability of their both happening will be $\frac{1}{13} \times \frac{1}{13} = \frac{1}{169}$ ...From whence it may be inferred, that the probability of the happening of two events dependent, is the product of the probability of the happening of one of them, by the probability which the other will have of happening, when the first is considered as having happened[19]; and the same rule will extend to the happening of as many events as may be assigned.

In the above quotation, as Hailperin pointed in [63], de Moivre was the first to consider the notion of conditional probability, although he did not name it.

2.4.2 *Thomas Bayes*

Still without a name, conditional probabilities[20] influenced to a great extent another prominent milestone in development of probability theory—Bayes' rule. The paper

[19] In modern notation: $P(A \wedge B) = P(A) \cdot P(B|A)$. Hailperin noted in [63] that this rule can be also applied to independent events since in that case $P(B|A) = P(B)$, and $P(A \wedge B) = P(A) \cdot P(B|A) = P(A) \cdot P(B)$.

[20] It seems that George Boole in [14] was the first who used this name. According to Shafer [149], Harold Jeffreys introduced the notation $P(A|B)$ only in 1931.

entitled "An essay towards solving a problem in the doctrine of chances" [7] written by Thomas Bayes (1701–1761) was posthumously published, and may have been even repaired[21] by Richard Price (1723–1791). Bayes' aim was to discuss:

> PROBLEM: Given the number of times in which an unknown event has happened and failed: Required the chance that the probability of its happening in a single trial lies somewhere between any two degrees of probability that can be named.

that can be seen as a version of the inverse problem. Assuming uniform a priori probabilities, he proposed how to modify initial beliefs by obtaining new information from experiments. In the introduction to the paper Price emphasized great importance of that result and differences between it and what Bernoulli and de Moivre discovered:

> These observations prove that the problem enquired after in this essay is no less important than it is curious. It may be safely added, I fancy, that it is also a problem that has never before been solved. Mr De Moivre, indeed, the great improver of this part of mathematics, has in his Laws of Chance, after Bernoulli, and to a greater degree of exactness, given rules to find the probability there is, that if a very great number of trials be made concerning any event, the proportion of the number of times it will happen, to the number of times it will fail in those trials, should differ less than by small assigned limits from the proportion of the probability of its happening to the probability of its failing in one single trial. But I know of no person who has shewn how to deduce the solution of the converse problem to this; namely, 'the number of times an unknown event has happened and failed being given, to find the chance that the probability of its happening should lie somewhere between any two named degrees of probability'.

Bayes did not follow Leibnitz, Bernoulli, and de Moivre, but defined:

> The probability of any event is the ratio between the value at which an expectation depending on the happening of the event ought to be computed, and the value of the thing expected upon it's happening.

As proper terminology and techniques were not fully developed at the time, Bayes had a lot of trouble formulating and proving statements that are today treated as elementary.[22] For example:

> Prop. 3: The probability that two subsequent events will both happen is a ratio compounded of the probability of the 1st, and the probability of the 2nd on supposition the 1st happens ...COROLLARY. Hence if of two subsequent events the probability of the 1st be a/N, and the probability of both together be P/N, then the probability of the 2nd on supposition the 1st happens is P/a.
>
> Prop. 4: If there be two subsequent events to be determined every day, and each day the probability of the 2nd is b/N and the probability of both P/N, and I am to receive N if both the events happen the first day on which the 2nd does; I say, according to these conditions, the probability of my obtaining N is P/b. ...COR. Suppose after the expectation given me in the foregoing proposition, and before it is at all known whether the 1st event has happened or not, I should find that the 2nd event has happened; from hence I can only infer that the event is determined on which my expectation depended, and have no reason to esteem the value of my expectation either greater or less than it was before.

---

[21] At least Price withdrew the original introduction and wrote his own, added some remarks and Appendix.

[22] If properly understood.

> Prop. 5 If there be two subsequent events, the probability of the 2nd b/N and the probability of both together P/N, and it being first discovered that the 2nd event has happened, from hence I guess that the 1st event has also happened, the probability I am in the right is P/b.[23]

It is interesting to mention here that Shafer writes in [149] that these statements show that Bayes mixed the concepts of time and probability in the notion of subsequent events, which is absent from the contemporary timeless understanding of conditional probability.

The last proposition in the paper provided a solution to the presented problem and estimated, using geometrical approach in terms of the ratio of areas of some rectangles, the chance that the unknown prior probability x of an event which happened p times in $p+q$ experiments belonged to the interval $[x_1, x_2]$. In modern notation, it can be written as in [35]:

$$P(x_1 \le x \le x_2 | p, p+q) = \frac{\int_{x_1}^{x_2} \binom{p+q}{p} x^p (1-x)^q dx}{\int_0^1 \binom{p+q}{p} x^p (1-x)^q dx}.$$

Ironically, one cannot find Bayes' rule[24] in Bayes' paper. His work was further developed by Laplace.

2.4.3 *Johann Heinrich Lambert*

Jacob Bernoulli's work on combinations of arguments was further enhanced in the chapter Von dem Wahrscheinlichen from Neues organon [83] written by Johann Heinrich Lambert (1728–1777). Lambert was particularly interested in probabilities of propositions and probabilities inferred from them, which is quite close to the modern approach formalized in probability logics. His work is deeply analyzed in [63, 152].

---

[23] Richard Price's remark: What is proved by Mr Bayes in this and the preceding proposition is the same with the answer to the following question. What is the probability that a certain event, when it happens, will be accompanied with another to be determined at the same time? In this case, as one of the events is given, nothing can be due for the expectation of it; and, consequently, the value of an expectation depending on the happening of both events must be the same with the value of an expectation depending on the happening of one of them. In other words; the probability that, when one of two events happens, the other will, is the same with the probability of this other. Call x then the probability of this other, and if b/N be the probability of the given event, and p/N the probability of both, because $p/N = (b/N) \times x$, $x = p/b =$ the probability mentioned in these propositions.

[24] A common formulation is

$$P(A_i \mid B) = \frac{P(B \mid A_i)\, P(A_i)}{\sum_j P(B \mid A_j)\, P(A_j)}$$

where $\{A_j\}$ is a partition of the sample space.

Lambert considered propositions of the form

$$A \text{ is } B$$

where A and B are predicates. B can be characterized by an argument C,[25] where C can be a predicate or a combination (conjunction) of predicates. Then, if A is C is necessary, it follows necessarily that A is B. However, if C does not characterize B, i.e., something different from B can be also C, by enumeration or calculation one can find a ratio of the number of favorable cases to the number of all cases, which is the corresponding degree of probability. To determine probabilities, Lambert used an analogy between his descriptions of problems and aleatory calculus.

For example, suppose that there are several not characterizing arguments of B, i.e., the premises of the form B is C_i are contingent, while A is C_i is necessary. An argument C_i is modeled by a heap of tickets, where each marked ticket denotes a particularly favorable case for C_i is B, and similarly unmarked tickets symbolize unfavorable cases. For every argument for B there is a separate heap and one takes a ticket from each heap. The probability that only unmarked tickets are chosen corresponds to the case that all arguments do not imply B. Assuming independence of arguments, which Lambert did explicitly, the probability that arguments do not prove B is calculated as the product p of the individual probabilities of unmarked tickets, while $1 - p$ is the degree of probability that the arguments prove[26] A is B.

Lambert discussed probabilistic propositions of the forms $\frac{3}{4}A$ are B, $C\,\frac{2}{3}$is D and E is $\frac{1}{2}F$ that mean $\frac{3}{4}$ of objects that are A are also B, the probability that C is D is $\frac{2}{3}$ and E has $\frac{1}{2}$ attributes of F. Then, an uncertain inference is

$$\begin{aligned} \frac{3}{4}A \quad &\text{are} \quad B \\ C \quad &\text{is} \quad A \\ \text{hence, } C \ \tfrac{3}{4} \ &\text{is} \quad B, \end{aligned}$$

where C appears to be a typical member of A, i.e., that A has a uniform distribution.

In a more complex example, Lambert considered inferences of the form

$$\begin{aligned} &\left(\frac{2}{3}a + \frac{1}{4}e + \frac{1}{12}u\right) A \qquad \text{are } B \\ &\qquad\qquad\qquad\qquad C \ \text{ is} \left(\frac{3}{5}a + \frac{2}{5}u\right) A \\ \text{hence, } C &\left(\frac{2}{5}a + \frac{3}{20}e + \frac{9}{20}u\right) \quad \text{is } B, \end{aligned}$$

[25] In modern notation: $B \leftrightarrow C$.

[26] Hailperin objects that the calculated probability of the form $P(\bigvee_i (C_i \text{ is } B))$ is only the lower bound for the probability that A is B.

where a, e, and u stand for affirmative, negative and undetermined cases, respectively. So, $(\frac{2}{3}a + \frac{1}{4}e + \frac{1}{12}u)$ in the first row means that $\frac{2}{3}$ of objects that are A are certainly B, that $\frac{1}{4}$ of objects that are A are certainly not B, while the status of $\frac{1}{12}$ of objects that are A is not determined. Hailperin's interpretation of this inference is

$$\begin{aligned} \frac{2}{3} &\leq P(x \in B | x \in A) \leq \frac{9}{12} \\ \frac{3}{5} &\leq P(C \in A) \\ \text{hence, } \frac{2}{5} &\leq P(C \in B) \leq \frac{17}{20}, \end{aligned}$$

where the undetermined cases are added to both the affirmative and negative cases which produces inequalities.

Shafer in [152] emphasized that Lambert, similarly to Bernoulli, used non-additive probabilities as can be illustrated by:

$$\begin{array}{rcl} \frac{3}{4}A & \text{are} & B \\ C & \text{is} & \frac{2}{3}A \\ \text{hence, } C\ \frac{1}{2} & \text{is} & B, \end{array} \qquad\qquad \begin{array}{rcl} \frac{1}{4}A & \text{are not} & B \\ C & \text{is} & \frac{2}{3}A \\ \text{hence, } C\ \frac{1}{6} & \text{is not} & B, \end{array}$$

and (translation from [152]):

> §193. …Thus the probability that the syllogism's conclusion is negative is $\frac{1}{6}$, whereas the probability that it is affirmative is $\frac{1}{2}$. Both probabilities together yield $\frac{1}{6} + \frac{1}{2} = \frac{2}{3}$ …

while, similarly as above, Hailperin does not agree with Shafer's explanation and offers another interpretation where the conclusions are

$$P(C \in B \wedge C \in A) = \frac{1}{2} \qquad\qquad P(C \notin B \wedge C \in A) = \frac{1}{6}$$

respectively, which is the same as

$$P(C \in A) = \frac{2}{3} = P(C \in B \wedge C \in A) + P(C \notin B \wedge C \in A).$$

2.5 Laplace and Development of Probability and Logic in the Nineteenth Century

2.5.1 *Laplace*

For almost half a century Pierre–Simon Laplace (1749–1827) was the leading developer of probability theory, from 1774 and his first paper Mémoire sur la probabilité des causes par les événements [84, 111, 158] to 1824, when he added the last supplement to his monumental book[27] Théorie Analytique des Probabilités [87]. The book comprised 40 years of his numerous contributions to probability theory, particularly to:

- the inverse problem, where he gave the "Bayes' formula" for a nonuniform distribution of causes, and
- the central limit theorem, where he generalized de Moivre's result on the asymptotic normality of the binomial distributions to the case of sums of random variables with the same distributions.

Laplace's epistemic approach to probability was clearly formulated already in [85] (translation from [110]):

> We owe to the frailty of the human mind one of the most delicate and ingenious of mathematical theories, namely the science of chance or probabilities.

His point of view was elaborated in a series of principles in Essai philosophique sur les probabilités [86, 88], a popular introduction to the second edition of Théorie Analytique des Probabilités. Laplace, following his great ancestors, first states that probability is the ratio of the number of favorable cases to the number of all possible cases, assuming that all cases are equipossible. If they are not, one has to carefully determine possible cases, so that probabilities of complex cases will be the appropriate sums of probabilities of mutually exclusive simpler cases. Then he emphasizes importance of the notion of (in)dependent events (the next four translations from [88]):

> One of the most important points of the theory of probabilities and that which lends the most to illusions is the manner in which these probabilities increase or diminish by their mutual combination. If the events are independent of one another, the probability of their combined existence is the product of their respective probabilities. Thus the probability of throwing one ace with a single die is $\frac{1}{6}$; that of throwing two aces in throwing two dice at the same time is $\frac{1}{36}$ ...We cannot better compare this diminution of the probability than with the extinction of the light of objects by the interposition of several pieces of glass. A relatively small number of pieces suffices to take away the view of an object that a single piece allows us to perceive in a distinct manner. The historians do not appear to have paid sufficient attention to this degradation of the probability of events when seen across a great number of successive generations; many historical events reputed as certain would be at least doubtful if they were submitted to this test. In the purely mathematical sciences the most distant consequences participate in the certainty of the principle from which they are derived. In the applications of analysis to physics the results have all the certainty of facts or experiences. But in the moral sciences, where each inference is deduced from that which precedes it only in a probable manner, however probable these deductions may be, the chance of error increases with their number and ultimately surpasses the chance of truth in the consequences very remote from the principle.

Without the proper notation, Laplace expresses in words the standard definition for conditional probabilities, i.e., $P(B|A) = \frac{P(A \cap B)}{P(A)}$, that for two mutually depended

[27] "...the most influential book on probability and statistics ever written", claims Anders Hald in [67].

events the probability of the compound event is $P(A \cap B) = P(A)P(B|A)$, and finally Bayes' rule in the contemporary form:

> Each of the causes to which an observed event may be attributed is indicated with just as much likelihood as there is probability that the event will take place, supposing the event to be constant. The probability of the existence of any one of these causes is then a fraction whose numerator is the probability of the event resulting from this cause and whose denominator is the sum of the similar probabilities relative to all the causes; if these various causes, considered á priori, are unequally probable, it is necessary, in place of the probability of the event resulting from each cause, to employ the product of this probability by the possibility of the cause itself. This is the fundamental principle of this branch of the analysis of chances which consists in passing from events to causes.
>
> This principle gives the reason why we attribute regular events to a particular cause. Some philosophers have thought that these events are less possible than others and that at the play of heads and tails, for example, the combination in which heads occurs twenty successive times is less easy in its nature than those where heads and tails are mixed in an irregular manner. But this opinion supposes that past events have an influence on the possibility of future events, which is not at all admissible. The regular combinations occur more rarely only because they are less numerous. If we seek a cause wherever we perceive symmetry, it is not that we regard a symmetrical event as less possible than the others, but, since this event ought to be the effect of a regular cause or that of chance, the first of these suppositions is more probable than the second.

An important part of Laplace's work was devoted to applications of uncertain reasoning. He considered problems in demography (for example to o determine the size of the French population), astronomical observations, jurisprudence. Concerning mistakes of witnesses and juries Laplace estimated chances that accused persons were wrongly judged and concluded:

> The impossibility of amending these errors is the strongest argument of the philosophers who have wished to proscribe the penalty of death.

2.5.2 *Bernard Bolzano, Augustus de Morgan, Antoine Cournot*

Bernard Bolzano (1781–1848) described probability as a part of logic in Wissenschaftslehre [13]. He discussed a notion[28] called validity (in [63]), and satisfiability (in [136]), of propositions. In fact, those propositions were predicates with one free variable. The degree of validity of a proposition $A(x)$ can be seen as its absolute probability, i.e., as the usual ratio of favorable cases to the total number of cases $\frac{|x \in U : A(x)|}{|U|}$, for the corresponding set U of objects. Probability was used to measure a relation (relative validity in [63], and comparative satisfiability in [136]) between propositions. The relative validity of a proposition M with respect to a set of propositions $A, B, C, \ldots$ was actually the conditional probability $P(M|A \wedge B \wedge C \wedge \ldots)$. To calculate the degree of probability of M wrt. $A, B, C, \ldots$, Bolzano divided U into

[28]Güiltigkeit.

equipossible mutually exclusive K_1, K_2, ..., K_k, and counted the number m of K's which validated $M \wedge A \wedge B \wedge C \wedge \ldots$, and the number k of K's which validated $A \wedge B \wedge C \wedge \ldots$. Then, the sought degree of probability was $\frac{m}{k}$. He considered an inference rule of the form (translation from [63]):

> For if we accept propositions $A, B, C, D, \ldots$ then this proposition tells us whether we should also accept M. If M becomes true in more than half the cases in which $A, B, C, D, \ldots$ are true, the truth of $A, B, C, D, \ldots$ entitles us to accept M as well, otherwise not.

which Hailperin interpreted as

> If $P(M|A \wedge B \wedge C \wedge \ldots) \approx 1$
> and $A \wedge B \wedge C \wedge \ldots$,
> therefore accept M,

understanding "more than half" as "practically all". Bolzano formulated some theorems about conditional probability (i.e., about relative validity), given below in modern notation:

- $P(M|A \wedge B \wedge C \wedge \ldots) \leq 1$,
- $P(\neg M|A \wedge B \wedge C \wedge \ldots) = 1 - P(M|A \wedge B \wedge C \wedge \ldots)$, etc.

Hailperin found several mistakes in those statements, like, for example, in

- $P(M \wedge N|A \wedge B \wedge \cdots \wedge D \wedge E \wedge \ldots) = P(M|A \wedge B \wedge \ldots) \cdot P(N|D \wedge E \wedge \ldots)$

which is valid only under the restricted assumption about independence of propositions. On the other hand, he emphasized that Bolzano recognized another common error in the contemporary methods in uncertain reasoning that the probability of a conclusion is equal to the product of the probabilities of the corresponding premises, namely that:

- If M follows from A, B, ..., C, then $P(M) = P(A) \cdot P(B) \cdots P(C)$

which is due to two causes. First, the probability of M is only bounded from below by the probability of the conjunction of the premises. That issue will be addressed by a generalization of the modal axiom K in Sect. 3.4 (Lemma 3.1(1)). And second, without independence of the premises, the product of the probabilities of the premises would not be equal to the probability of the conjunction of the premises.

Augustus De Morgan (1806–1871) devoted a significant part of his work to analyzing relationships between logic and probability. He had important influence on Isaac Todhunter (1820–1884), who published extensive History of the Mathematical Theory of Probability from the Time of Pascal to that of Laplace [164], and on George Boole. De Morgan's Theory of Probabilities [40] was the first review on probability theory and its applications in English. He gave in [41, 42, 175] an exposition of Laplacian theory of probability concluding that:

> If we can make a few reflecting individuals understand, that, be the theory of probabilities true or false, valuable or useless, its merits must be settled by reference to something more than the consideration of a few games at cards, we shall have done all which we ventured to propose to ourselves.

De Morgan had a subjective, epistemic approach to probability (quotations from his Formal logic, or, the calculus of inference, necessary and probable [43]):

> We have lower grades of knowledge, which we usually call degrees of beliefs but they are really degrees of knowledge ...It may seem a strange thing to treat knowledge as a magnitude, in the same manner as length, or weight, or surface. This is what all writers do who treat of probability, and what all their readers have done, long before they ever saw a book on the subject. But it is not customary to make the statement so openly as I now do: and I consider that some justification of it is necessary. By degree of probability we really mean, or ought to mean, degree of belief ...I will take it then that all the grades of knowledge, from knowledge of impossibility to knowledge of necessity, are capable of being quantitatively conceived,

and shared with Bolzano the view that numerical probability. i.e., probabilistic inference, is a part of formal logic:

> The old doctrine of modals is made to give place to the numerical theory of probability. Many will object to this theory as extralogical ...I cannot understand why the study of the effect which partial belief of the premises produces with respect to the conclusion, should be separated from that of the consequences of supposing the former to be absolutely true,

which was elaborated in chapters On the numerically definite syllogism, On probability and On probable inference of [43]. Since De Morgan conceived probability as a measure of knowledge, he had to justify the basic principles of probability. For example:

> Now, whether we shall proceed, or stop short at this point, depends upon our acceptance or non-acceptance of the following POSTULATE:-
> When any number of events are disjunctively possible, so that one of them may happen, but not more than one, the measure of our belief that one out of any some of them will happen, ought to be the amount of the measures of our separate beliefs in each one of those some. I mean that any one would say, A, B, C, being things of which not more than one can happen, my belief that one of the three will happen is the sum of my separate beliefs in A, and in B, and in C. This is the postulate on which the balance depends; and there is a similar postulate before we can use the physical balance. The only difference (and that but apparent) is that we are to speak of weights collectively, and of events disjunctively. The weight of the (conjunctive) mass is the sum of the weights of its parts: the credibility of the (disjunctive) event is the sum of the credibilities of its components ...The laws of matter and mind being both what they are, the connexion between physical collection and mental summation is, I grant, necessary: the simplest of manual, and the simplest of mental, operations, are and, with us, must be, concomitants ...

He discussed uncertain reasoning with necessary valid inferences (he called them arguments) and probable premises (testimonies) by analyzing some examples. Hailperin in [63] noticed some systematical failures in the work of De Morgan: disregarding of (in)dependence of components of compound events, proclaiming the probability of premises which necessarily imply a conclusion to be the probability of the conclusion, etc.

> Problem 3. Arguments being supposed logically good, and the probabilities of their proving their conclusions (that is, of all their premises being true) being called their validities, let there be a conclusion for which a number of arguments are presented, of validities a, b, c, &c. Required the probability that the conclusion is proved ...Testimonies are all true together or all false together: but one of the arguments may be perfectly sound, though all the rest

> be preposterous. The question then is, what is the chance that one or more of the arguments proves its conclusion. That all shall fail, the probability is[29] $a'b'c' \dots$ that all shall not fail, the probability is $1 - a'b'c' \dots$

By denoting the conclusion, premises, their probabilities and arguments as C, A_i, $P(A_i) = a_i$, and $A_i \rightarrow C$, respectively, Hailperin pointed out that De Morgan's result is correct if $P(\wedge_i \neg A_i) = \Pi_i P(\neg A_i)$, i.e. if the premises are independent.

> Problem 6. Given an assertion, A, which has the probability a; what does that probability become, when it is made known that there is the probability m that B is a necessary consequence of A, B having the probability b? And what does the probability of B then become?

Hailperin translated this as: if $P(A) = a$, $P(B) = b$, what are the conditional probabilities $P(A|P(A \rightarrow B) = m)$ and $P(B|P(A \rightarrow B) = m)$? Again, under the assumption of independence which was not mentioned, De Morgan represented events (with probabilities that should be found) as conjunctions of other possibly negated events with known probabilities and calculated odds $P(A) : P(\neg A) = (a(1 - mb'))/a'$, and $P(B) : P(\neg B) = b/(b'(1 - ma))$.

Antoine Cournot (1801–1877) in [31] laid the foundation for frequentism in probability by stating (translation from [151]):

> The physically impossible event is therefore the one that has infinitely small probability, and only this remark gives substance—objective and phenomenal value—to the theory of mathematical probability.

Later, in the twentieth century, Lévy wrote that the principle combined with Bernoulli's theorem gives objective probabilities of events [102], while Kolmogorov and Borel understood it as the principle which connects the mathematical theory of probability with the real world [151].

2.5.3 George Boole

> The design of the following treatise is to investigate the fundamental laws of those operations of the mind by which reasoning is performed; to give expression to them in the symbolical language of a Calculus, and upon this foundation to establish the science of Logic and construct its method; to make that method itself the basis of a general method for the application of the mathematical doctrine of Probabilities; and, finally, to collect from the various elements of truth brought to view in the course of these inquiries some probable intimations concerning the nature and constitution of the human mind ...The general doctrine and method of Logic above explained form also the basis of a theory and corresponding method of Probabilities ...Hence the subject of Probabilities belongs equally to the science of Number and to that of Logic.

wrote George Boole (1815–1864) in An investigation of the laws of thought [14] where almost half of the text was devoted to the relationship between logic and

---

[29] a' means $1 - a$.

probability. Boole's understanding of probability was that it is founded upon partial knowledge about the relative frequency of occurrences of events. Instead of considering the numerical probability of the occurrence of an event, Boole often expressed it as the probability of the truth of the proposition declaring that the event will occur. Hailperin argued in [62] that, although the other authors—as, for example, Lambert and Bolzano—shared the same approach, Boole's development of logical calculus was the enabling technology providing advantages of that substitution. Boole listed seven principles as the basis for the above mentioned doctrine. The principles, mostly taken from Laplace, corresponded to additivity of probabilities, probabilities of (in)dependent events, Bayes' theorem for a priori equally probable causes, etc. For example, Principle 4 defined conditional probability, but provided neither the name nor the proper symbolism:

> 4th. The probability that if an event, E, take place, an event, F, will also take place, is equal to the probability of the concurrence of the events E and F, divided by the probability of the occurrence of E.

Boole asserted the relative sufficiency of the principles, namely that by combining these principles one can calculate the probability of a compound event that depends on a set of independent events with known probabilities. He worked with conditional propositions of the form: if the proposition X is true, the proposition Y is true, and calculated their probabilities, which he called (but in only one place in the book) conditional probabilities.

Already in the introduction Boole stated one of his main goals:

> …it is not clear that any advance has been made toward the solution of what may be regarded as the general problem of the science, viz.: Given the probabilities of any events, simple or compound, conditioned or unconditioned: required the probability of any other event equally arbitrary in expression and conception. In the statement of this question it is not even postulated that the events whose probabilities are given, and the one whose probability is sought, should involve some common elements, because it is the office of a method to determine whether the data of a problem are sufficient for the end in view, and to indicate, when they are not so, wherein the deficiency consists. This problem, in the most unrestricted form of its statement, is resolvable by the method of the present treatise …

Boole's method constitutes essentially the decision procedure for probability logic.[30] The method basically consisted of two phases:

- first, using his algebraic laws written in the symbolical language of the logical calculus, Boole expressed the event with the unknown probability as a combination of other events, and
- second, he constructed a function of known probabilities to calculate the desired probability.

The following example[31] was used to illustrate the method:

---

[30] That would, more than one century later, be rediscovered independently several times [69].

[31] In the example, and elsewhere in the book, Boole used the notations of the form:

- 1 to indicate that all the elements of the class must be taken,

Ex. 1. The probability that it thunders upon a given day is p, the probability that it both thunders and hails is q, but of the connexion of the two phænomena of thunder and hail, nothing further is supposed to be known. Required the probability that it hails on the proposed day. Let x represent the event—It thunders. Let y represent the event —It hails. Then xy will represent the event—It thunders and hails; and the data of the problem are

$$\text{Prob., } x = p, \text{ Prob., } xy = q.$$

There being here but one compound event xy involved, assume, according to the rule,

$$\text{Prob., } xy = u. \tag{2.1}$$

Our data then become

$$\text{Prob., } x = p, \text{ Prob., } xy = q. \tag{2.2}$$

and it is required to find Prob.,y. Now (1) gives

$$y = \frac{u}{xy} + \frac{1}{0}u(1-x) + 0(1-u)x + \frac{0}{0}(1-u)(1-x).$$

Hence (XVII. 17) we find

$$V = ux + (1-u)x + (1-u)(1-x),$$

$$Vx = ux + (1-u)x = x, \quad Vu = ux;$$

and the equations of the General Rule, viz.,

$$\frac{V_x}{p} = \frac{V_u}{q} = V.$$

$$\text{Prob., } y = \frac{A + cC}{V}$$

become, on substitution, and observing that $A = ux, C = (1-u)(1-x)$, and that V reduces to $x + (1-u)(1-x)$,

$$\frac{x}{p} = \frac{ux}{q} = x + (1-u)(1-x),$$

$$\text{Prob., } y = \frac{ux + c(1-u)(1-x)}{x + (1-u)(1-x)},$$

from which we readily deduce, by elimination of x and u,

$$\text{Prob., } y = q + c(l-p). \tag{2.3}$$

(Footnote 31 continued)

- 0, on the other hand, means that elements of the class must not be taken,
- $\frac{0}{0}$ to represent an indefinite class, i.e., that all, some, or none of the elements of the class must be taken, and
- $\frac{1}{0}$, as the algebraic symbol of infinity, implies that the corresponding class is 0.

Thus, according to Hailperin, the form

$$w = A + 0B + \frac{0}{0}C + \frac{1}{0}D$$

can be interpreted as $w = A + vC$, and $D = 0$, where v is indefinite.

> In this result c represents the unknown probability that if the event $(1-xy)(1-x)$ happen, the event y will happen. Now $(l-u)(l-x) = (l-xy)(1-x) = 1-x$, on actual multiplication. Hence c is the unknown probability that if it do not thunder, it will hail.
>
> The general solution (2.3) may therefore be interpreted as follows: The probability that it hails is equal to the probability that it thunders and hails, q, together with the probability that it does not thunder, $1 - p$, multiplied by the probability c, that if it does not thunder it will hail. And common reasoning verifies this result.
>
> If c cannot be numerically determined, we find, on assigning to it the limiting values 0 and 1, the following limits of Prob.,y, viz.:
>
> $$\text{Inferior limit} = q. \text{ Superior limit} = q + 1 - p.$$

On a more fine-grained level, the method was not completely correct. It was critically analyzed already by Alexander Macfarlane [62, 107] in 1879, and Platon Sereyevich Poretskiy in 1887 [159]. Much later, a systematic consideration was given in [61, 62], where Hailperin discussed the appropriate interpretations and corrections using the linear programming approach, and presented the best possible probability boundaries for logical functions of events. Hailperin criticized Boole for:

- restricting disjunction to be exclusive, so that he could apply additivity to calculate probabilities of disjunctively compound events, and
- assuming that all events can be expressed as combinations of simple stochastically independent events, he obtained probabilities of conjunctively compound events as products of probabilities of their constituents.

Furthermore, Hailperin noted that Boole did not fully distinguish conditional probabilities from probabilities of conditionals (in modern notation $P(B|A)$ and $P(B \to A)$, respectively). Still, Hailperin and many others recognized the full worth of Boole's ideas and confirmed that The laws of thought was the most successful follow-up of the idea about probability logic for almost 200 years since Leibnitz announced the challenging program.

2.5.4 *J. Venn, H. MacColl, C. Peirce, P. Poretskiy*

John Venn (1834–1923) in [167] examined laws according to which beliefs (acquired by inferences based on induction or analogy, that depend on testimonies or memories) of different strengths could be obtained. While rejecting the subjective view of probability and promoting the frequentism (citations from [167]):

> The probability is nothing but that proportion, and is unquestionably in this case derived from no other source but the statistics themselves …

Venn proposed logical treatment of probability:

> With what may be called the Material view of Logic as opposed to the Formal or Conceptualist, with that which regards it as taking cognisance of laws of things and not of the laws of our own minds in thinking about things, I am in entire accordance. Of the province of

> Logic, regarded from this point of view, and under its widest aspect, Probability may, in my opinion, be considered to be a portion. The principal objects of this Essay are to ascertain how great a portion it comprises, where we are to draw the boundary between it and the contiguous branches of the general science of evidence, what are the ultimate foundations upon which its rules rest, what the nature of the evidence they are capable of affording, and to what class of subjects they may most fitly be applied.

One of the main problems in Venn's approach [167, 168] was to properly introduce the concept of infinite series of things or events with continually increasing uniformity while having only arbitrarily long, but still finite, series. That issue was later addressed by Hans Reichenbach [159].

Charles S. Peirce (1839–1887) first, like Venn, had an objectivist approach to probability, but in his latter days he proposed some kind of propensity interpretation of probability. He assigned probability to an argument, with premises and a conclusion, rather than to a proposition or an event, and in that sense all probabilities are conditional probabilities [20]. Peirce introduced a symbol for conditional probability,[32] considered its properties, for example[33]:

- $ab_a = ba_b$
- $(1 - b)_a = 1 - b_a$,

clearly distinguished conditional probability from probability of conditional [63], while in calculations did not generally assume independence of events, as Boole did. Among his numerous contributions to logic, Peirce analyzed Boole's work on logic and probability, and noticed several mistakes and limitations. For example, in [115], he wrote:

> It being known what would be the probability of Y, if X were to happen, and what would be the probability of Z, if Y were to happen; what would be the probability of Z, if X were to happen? But even this problem has been wrongly solved by him. For, according to his solution, where $p = Y|X|$, $q = Z|Y|$, $r = Z|X|$, r must be at least as large as the product of p and q. But if X be the event that a certain man is a negro, Y the event that he is born in Massachusetts, and Z the event that he is a white man, then neither p nor q is zero, and yet r vanishes.

Peirce distinguished:

- deduction,
- induction, and
- retroduction (abduction)

as fundamentally different kinds of reasoning [26]:

> Deduction is that mode of reasoning which examines the state of things asserted in the premisses, forms a diagram of that state of things, perceives in the parts of that diagram relations not explicitly mentioned in the premisses, satisfies itself by mental experiments upon the diagram that these relations would always subsist, or at least would do so in a certain proportion of cases, and concludes their necessary, or probable, truth.

---

[32] b_a denotes the frequency of b's among the a's.

[33] In modern notation: $P(A)P(B|A) = P(B)P(A|B)$ and $P(\neg B|A) = 1 - P(B|A)$.

> Induction is that mode of reasoning which adopts a conclusion as approximate, because it results from a method of inference which must generally lead to the truth in the long run. For example, a ship enters port laden with coffee. I go aboard and sample the coffee. Perhaps I do not examine over a hundred beans, but they have been taken from the middle, top, and bottom of bags in every part of the hold. I conclude by induction that the whole cargo has approximately the same value per bean as the hundred beans of my sample. All that induction can do is to ascertain the value of a ratio.
>
> Retroduction is the provisional adoption of a hypothesis, because every possible consequence of it is capable of experimental verification, so that the persevering application of the same method may be expected to reveal its disagreement with facts, if it does so disagree. For example, all the operations of chemistry fail to decompose hydrogen, lithium, glucinum, boron, carbon, nitrogen, oxygen, fluorine, sodium, ...gold, mercury, thallium, lead, bismuth, thorium, and uranium. We provisionally suppose these bodies to be simple; for if not, similar experimentation will detect their compound nature, if it can be detected at all.

which could be illustrated by:

- by induction, from the hypothesis that beans are from this bag, and the result that these beans are white, the rule that all the beans from this bag are white is obtained, while
- by retroduction, from the rule that all the beans from this bag are white, and the result that these beans are white, the hypothesis that these beans are from this bag is formulated.

He discriminated between the different kinds of imperfection—probability, verisimilitude or likelihood, and plausibility and associated them with deduction, induction and abduction, respectively [116]:

> By Plausible, I mean that a theory that has not yet been subjected to any test, although more or less surprising phenomena have occurred which it would explain if it were true, is in itself of such a character as to recommend it for further examination or, if it be highly plausible, justify us in seriously inclining toward belief in it, as long as the phenomena be inexplicable otherwise.
>
> I call that theory likely which is not yet proved but is supported by such evidence that if the rest of the conceivably possible evidence should turn out upon examination to be of a similar character, the theory would be conclusively proved.

Hugh MacColl (1837–1909), independently of Peirce, provided a symbol for conditional probability and investigated the corresponding properties. He attached probabilities to propositions and not to arguments [63]:

> The symbol x_a denotes the chance that the statement x is true on the assumption that the statement a is true.

Besides the standard truth values he suggested, with probabilistic motivation, the modalities certain, impossible and variable, and classified statements in pure logic as[34] (the next two citations from [104]):

[34] He mixed here syntax and semantics.

In pure or abstract logic statements are represented by single letters, and we classify them according to attributes as true, false, certain, impossible, variable, respectively denoted by the five Greek letters $\tau, \iota, \epsilon, \eta, \theta$ ...The symbol A^{τ} only asserts that A is true in a particular case or instance. The symbol A^{ϵ} asserts more than this: it asserts that A is certain, that A is always true (or true in every case) within the limits of our data and definitions, that its probability is 1 ...The symbol A^{η} asserts more than this; it asserts that A contradicts some datum or definition, that its probability is 0 ...The symbol A^{θ} (A is a variable) is equivalent to $A^{-\eta}A^{-\epsilon}$; it asserts that A is neither impossible nor certain, that is, that A is possible but uncertain. In other words, A^{θ} asserts that the probability of A is neither 0 nor 1, but some proper fraction between the two ...The symbol $A^B : C^D$ is called an implication, and means $(A^B C^{-D})^{\eta}$, or its synonym $(A^{-B} + C^D)^{\epsilon}$ ...Let the symbol π temporarily denote the word possible, let p denote probable, let q denote improbable, and let u denote uncertain ...We shall then, by definition, have $(A^{\pi} = A^{-\eta})$ and $(A^{u} = A^{-\epsilon})$ while A^p and A^q will respectively assert that the chance of A is greater than $\frac{1}{2}$, that it is less than $\frac{1}{2}$.

MacColl contributed both to classical and to non-classical modal logic but for long time his achievements were often forgotten and attributed to others. His interpretation of syllogisms as inferences defined in terms of implication, for example:

Barbara asserts that "If every X is Y, and every Y is Z, then every X is Z", which is equivalent to $(S^X : S^Y)(S^Y : S^Z) : (S^X : S^Z)$.

allowed him to develop, contemporaneously with Frege, propositional logic as a branch of logic independent of the class calculus or term logic of the traditional syllogisms [159]. MacColl was the first who discussed the paradoxes of material implication [105]:

For nearly thirty years I have been vainly trying to convince them that this assumed invariable equivalence between a conditional (or implication) and a disjunctive is an error ...Take the two statements "He is a doctor" and "He is red-haired", each of which, is a variable, because it may be true or false. Is it really the fact that one of these statements implies the other? Speaking of any Englishman taken at random out of those now living, can we truly say of him "If he is a doctor he is red-haired", or "if he is red-haired he is a doctor?"

This dissatisfaction with the material implication led to the notion of the strict implication [106, 121]:

...with Mr Russell the proposition A implies B means $(AB')^{\iota}$, whereas with me it means[35] $(AB')^{\eta}$.

While Rescher listed MacColl as one of the founders of many-valued logics in [134], Simons argued against that, and understood it as a modal probability logic [154].

Platon Sergeyevich Poretskiy (1846–1907) in [120] tried

...to give scientific form to Boole's deep, but vague and unproved, idea of applying mathematical logic to the theory of probability.

He developed a calculus of logical equations and applied it to problems in the probability theory [159]. In Russia, in the second half of the nineteenth century and the first decades of the twentieth century, the research related to the probability theory

[35] In modern terms: A and $\neg B$ is false, A and $\neg B$ is impossible, respectively.

was focused on other issues [22, 147]: Pafnuty Lvovich Chebyshev (1821–1894) and his students and followers[36] generalized Laplace's and Poisson's law of large numbers [12, 27–29, 109, 165], gave the central limit theorem [28], introduced and analyzed sequences of random variables in which the future variable is determined by the present variable only, i.e., Markov chains [108], but their great achievements are out of the scope of this text.

2.6 Rethinking the Foundations of Logic and Probability in the Twentieth Century

The second half of nineteenth and the beginning of twentieth century were marked by significant efforts to establish foundations of mathematics. These efforts were caused by three crises in mathematics in nineteenth century, which for the first time opened a question of the nature of mathematical truth and reliability of mathematical knowledge. The first crisis was in Analysis, resulting from careless use of infinitesimals, for which there was no firm foundation and this resulted in suspicions or incorrect proofs. The second crisis was the discovery of non-Euclidean geometries (and proofs of their equiconsistency with Euclidean geometry), which brought an end to the general belief that Euclidean geometry is "The Truth" about space. The third developed toward the end of nineteenth century. Just as the set theory started looking not only as a powerful tool but also as a good candidate for the foundations of mathematics, numerous paradoxes started appearing. While the first crisis seemed to be resolved by replacing infinitesimals by epsilon-delta technique, and by defining real numbers from rationals using Cauchy sequences or Dedekind cuts, the second and especially the third crisis demanded more radical solutions.

Gottlob Frege (1848–1925) criticized mathematicians, who were mostly satisfied with deriving real numbers from natural numbers, and embarked on an ambitious project of deriving the whole of mathematics from some elementary logical principles, inventing in the process the predicate (quantifier) logic [52]. Frege introduced binary, ternary, etc., relations, which enabled him to introduce the second, third, etc., quantifiers, bringing thus logic out of the dead end into which Aristotle's syllogisms stranded it. Namely, already scholastics realized that the main task in the development of logic should be to somehow introduce the "second generality", i.e., the second quantifier, into syllogisms. However, the influence of Aristotle's analysis of language and his assertion that the basic element to which all language can be reduced is "subject-predicate", meaning that all language is based on unary predicates, was so strong that nobody, even such great minds as Leibnitz, did dare think out of that box. It seems incredible that such a simple and obvious step was sufficient to bring logic out of more than 2000 years of virtual standstill, but Frege was the first

---

[36]Andrei Andreevich Markov (1856–1922), Alexander Alexandrovich Chuprov (1874–1926), Sergei Natanovich Bernstein (1880–1968), Yakov Viktorovich Uspenski (1883–1947), just to mentioned some of them.

who dared make it. His second step was even bolder: he derived the set theory (and from it the arithmetic of natural numbers) from just two obvious logical axioms. One was that two sets are equal just in case they have the same members and the other was that for every property, there is the set of all objects having that property. As Frege's book was being printed, a Ph.D. student Bertrand Russell derived a paradox from the second axiom—so-called Russell's paradox, destroying thus the Frege's dream. On the other hand Russell popularized Frege's predicate calculus (which had been mostly ignored, due to his awkward two-dimensional notation) and published, with Whitehead, immensely influential Principia Mathematica [172] which tried to resurrect the Frege's program and which played very significant role in the rise of mathematical logic in the first decades of twentieth century, culminating with Kurt Gödel's (1904–1977) proof of completeness of first order logic [57].

It is interesting that from the viewpoint of this book, which is devoted to connections between logic and probability, Frege's influence might be considered as negative. Namely, Frege's interest being in founding mathematics, and mathematical truths being necessary (not contingent), there was no room for probability in his approach. He considered propositions to be names for truth or falsity and those truth values had a special status that had nothing to do with probabilities. Ironically, a century after Frege, towards the end of twentieth century, "proofs with probability" appeared in mathematics. Some statements in number theory were shown to be true—with very high probability, e.g., Robert Solovay and Volker Strassen developed a probabilistic test to check if a number is composite, or (with high probability) prime [156, 157]. The rapid development of logic in the first half of the twentieth century was the development of logic of necessary mathematical truth, and its elegance and effectiveness completely eclipsed the probability logic despite the efforts of Keynes, Reichenbach, Carnap and others that continued Boole's approach connecting probability and logic.

At the same time, beside the classical definition where the probability of an event is the quotient of the numbers of the favorable and all possible outcomes, some alternative views were proposed: frequency, logical, and subjective interpretations, etc. Particularly, influential was the measure-theoretical approach which resulted in Andrei Nikolaevich Kolmogorov's axiomatic system for probability. Since those works the mainstreams of mathematical logic and probability theory were almost completely separated until the middle of the 1960s.

2.6.1 Logical Interpretation of Probability

In this context probability logic can be seen as a generalization of deductive logic which formalizes a broader notion of inference based on the notion of confirmation[37] which one set of propositions, i.e., evidences or premises, brings to another set of propositions, conclusions.

[37] Viewed as a generalization of the notion of classical implication.

John Maynard Keynes (1883–1946), following Leibnitz,[38] conceived of probability as a part of logic, i.e., as a relation between propositions which represents an objective estimation of non-personal knowledge (citations from [78]):

> The Theory of Probability is concerned with that part [of our knowledge] which we obtain by argument, and it treats of the different degrees in which the results so obtained are conclusive or inconclusive …The Theory of Probability is logical, therefore, because it is concerned with the degree of belief which it is rational to entertain in given conditions, and not merely with the actual beliefs of particular individuals, which may or may not be rational. Given the body of direct knowledge which constitutes our ultimate premisses, this theory tells us what further rational beliefs, certain or probable, can be derived by valid argument from our direct knowledge. This involves purely logical relations between the propositions which embody our direct knowledge and the propositions about which we seek indirect knowledge …Let our premisses consist of any set of propositions h, and our conclusion consist of any set of propositions a, then, if a knowledge of h justifies a rational belief in a of degree α, we say that there is a probability-relation of degree α between a and h[39] …Between two sets of propositions, therefore, there exists a relation, in virtue of which, if we know the first, we can attach to the latter some degree of rational belief. This relation is the subject-matter of the logic of probability.

For Keynes, there exists a unique rational probability relation between sets of premises and conclusions, so that from valid premises, one could in a rational way obtain a belief in the conclusion. It means that the conclusion sometimes, or even very often, only partially follows from its premises. Thus probability extends classical logic. In the same time he did not limit himself to give only numerical meaning to probabilities and that arbitrary probabilities must be comparable.[40] He stated that, although each probability is on a path between impossibility and certainty, different probabilities can lie on different paths:

> I believe, therefore, that the practice of underwriters weakens rather than supports the contention that all probabilities can be measured and estimated numerically …It is usually held that each additional instance increases the generalisation's probability. A conclusion, which is based on three experiments in which the unessential conditions are varied, is more trustworthy than if it were based on two. But what reason or principle can be adduced for attributing a numerical measure to the increase? …We can say that one thing is more like a second object than it is like a third; but there will very seldom be any meaning in saying that it is twice as like. Probability is, so far as measurement is concerned, closely analogous to similarity. …Some sets of probabilities we can place in an ordered series, in which we can say of any pair that one is nearer than the other to certainty, that the argument in one case is nearer proof than in the other, and that there is more reason for one conclusion than for the other. But we can only build up these ordered series in special cases. If we are given two distinct arguments, there is no general presumption that their two probabilities and certainty can be placed in an order.

---

[38] At the very beginning of the book, Keynes quoted Leibnitz's demand for new logic which involves probability reasoning.

[39] This will be written $a/h = \alpha$.

[40] Namely, he believed that probabilities of events or propositions in its widest sense cannot be always associated with numbers from the unit interval of reals.

Incomparable probabilities can arise from uncertainty that can be estimated by overlapping intervals that are not mutually comparable, or when uncertainty is evaluated in terms of vectors:

> Is our expectation of rain, when we start out for a walk, always more likely than not, or less likely than not, or as likely as not ...If the barometer is high, but the clouds are black, it is not always necessary that one should prevail over another in our minds ...

Keynes saw probability between sets of propositions as an undefined primitive concept and tried to formalize it. He presented a system of axioms, e.g.,

- Provided that a and h are propositions or conjunctions of propositions or disjunctions of propositions, and that h is not an inconsistent conjunction, there exists one and only one relation of probability P between a as conclusion and h as premiss. Thus any conclusion a bears to any consistent premiss h one and only one relation of probability.
- Axiom of equivalence: If $(a \equiv b)/h = 1$, and x is a proposition, $x/ah = x/bh$.
- $(\bar{\bar{a}}a \equiv a)/h = 1$.
- $ab/h + a\bar{b}/h = a/h$.
- $ab/h = a/bh \times b/h = b/ah \times a/h$, etc.,

and then proved a number of theorems about probabilities, for example:

- $a/h + \bar{a}/h = 1$.
- If $a/h = 1$, then $a/bh = 1$ if bh is not inconsistent.

Hailperin objects to this formal framework [63]. He explains that Keynes' formalization does not fulfill modern requirements, i.e., that there is no well defined syntax, that there are no inference rules, that some definitions are not eliminable, etc. Furthermore, the axioms do not allow iterations of probabilities, namely it is not possible to write something like $(a/b = c/d)/e$. Finally, while some authors interpret Keynes' concept as degree of confirmation, his axioms do not seem strong enough to characterize notions wider than conditional probabilities.

Rudolf Carnap's (1891–1970) work on logical foundations of probability was also an attempt to develop a pure logical concept of probability [25]. He was among the first researchers who clearly acknowledged that there are two distinct concepts of probability (citation from [24]):

> Among the various meanings in which the word 'probability' is used in everyday language, in the discussion of scientists, and in the theories of probability, there are especially two which must be clearly distinguished. We shall use for them the terms 'probability$_1$' and 'probability$_2$'. Probability$_1$ is a logical concept, a certain logical relation between two sentences (or, alternatively, between two propositions); it is the same as the concept of degree of confirmation. I shall write briefly "c" for "degree of confirmation", and "$c(h, e)$" for "the degree of confirmation of the hypothesis h on the evidence e", the evidence is usually a report on the results of our observations. On the other hand, probability$_2$ is an empirical concept; it is the relative frequency in the long run of one property with respect to another. The controversy between the so-called logical conception of probability, as represented e.g. by Keynes, and Jeffreys, and others, and the frequency conception, maintained e.g. by v. Mises and Reichenbach, seems to me futile. These two theories deal with two different probability concepts which are both of great importance for science. Therefore, the theories are not

> incompatible, but rather supplement each other. In a certain sense we might regard deductive logic as the theory of L-implication (logical implication, entailment). And inductive logic may be construed as the theory of degree of confirmation, which is, so to speak, partial L-implication. "e L-implies h" says that h is implicitly given with e, in other words, that the whole logical content of h is contained in e. On the other hand, "$c(h, e) = 3/4$" says that h is not entirely given with e but that the assumption of h is supported to the degree $3/4$ by the observational evidence expressed in e ...Inductive logic is constructed out of deductive logic by the introduction of the concept of degree of confirmation.

In the framework of probability$_1$, Carnap connected the concepts of inductive reasoning, probability, and confirmation and considered that c-functions should obey the generally accepted properties of confirmation [63], so that if some c-values are given, some others can be derived. Carnap fixed a finitary unary first order language L_N with constants $a_1, a_2, \ldots a_N$ to express h and e. He considered an arbitrary non-negative measure m on conjunctions of possible negated ground atomic formulas, with the only constraint that their sum is 1, and then, using additivity, extended it to all sentences. Then, if $m(e) \neq 0$, $c(h, e)$ is defined as $m(e.h)/m(e)$, while m- and c-values for infinitary system are determined as limits of the values for finite systems. Carnap studied properties of c, for example how degrees of confirmation decrease in chains of inferences, or:

- If $c(h, e) = 1$ and $c(i, e) > 0$, then $c(h, e.i) = 1$.

In Appendix of the first edition of [25], he announced:

> In Volume II a quantitative system of inductive logic will be constructed, based upon an explicit definition of a particular c-function c^* and containing theorems concerning the various kinds of inductive inference and especially of statistical inference in terms of c^*.

The idea was to, out of an infinite number of c-functions, choose one particular function adequate as a concept of degree of confirmation which would enable us to compute the c^*-value for every given sentence. However, later he abandoned that idea [63].

Even though Carnap's work was not completely successful, it stimulated a line of research on probabilistic first-order logics with more expressive languages than Carnap's [53, 54, 144, 160].

2.6.2 *Subjective Approach to Probability*

In [18], Émil Borel (1871–1956) criticized Keynes' approach, and and argued for the existence of different meanings of probability depending on the context. Borel, as a subjectivist, allowed that different persons with the same knowledge could evaluate probabilities differently, and proposed betting as a means to measure someone's subjective degree of belief (translation from [55]):

> ...exactly the same characteristics as the evaluation of prices by the method of exchange. If one desires to know the price of a ton of coal, it suffices to offer successively greater and

> greater sums to the person who possesses the coal; at a certain sum he will decide to sell it. Inversely if the possessor of the coal offers his coal, he will find it sold if he lowers his demands sufficiently.

On the other hand, following Henri Poincaré's (1854–1912) ideas [171], he accepted also objective probabilities in science, where probabilities could be identified with statistically stable frequencies (translation from [19]):

> There is no difference in nature between objective and subjective probability, only a difference of degree. A result in the calculus of probabilities deserves to be called objective, when the probability is sufficiently large to be practically equivalent to certainty. It matters little whether one is predicting future events or reviewing past events; one may equally aver that a probabilistic law will be, or has been, confirmed.

Frank Plumpton Ramsey (1903–1930) in [122] regarded probability theory as a part of logic of partial belief and inconclusive argument. He did not reduce probability to logic and admitted that the meaning of probability in other fields could be different. Ramsey was a student of Keynes, but did not accept Keynes' objective approach to probability and doubted existence of his probability relations. Ramsey insisted that he does not perceive probability relations. For example, he argued that there is no relation of that kind between propositions "This is red" and "This is blue". The focus of his examination was on probabilities comprehended as partial subjective beliefs, and on the logic of partial belief. One of the main issues in this approach was how to regard beliefs quantitatively so that they could be appropriately related to probability. To develop a theory of quantities of beliefs, Ramsey assumed that a person acts in the way she/he thinks most likely to realize her/his desires, and used betting to measure beliefs (citations from [122]):

> The old-established way of measuring a person's belief is to propose a bet, and see what are the lowest odds which he will accept …We thus define degree of belief in a way which presupposes the use of the mathematical expectation …By proposing a bet on p we give the subject a possible course of action from which so much extra good will result to him if p is true and so much extra bad if p is false. Supposing, the bet to be in goods and bads instead of in money, he will take a bet at any better odds than those corresponding to his state of belief; in fact his state of belief is measured by the odds he will just take …

which might be seen as reminiscence of Huygens' approach and Bayes' definition of probability. Ramsey showed that consistent degrees of belief must follow the laws of probability [122], e.g.:

> (1) Degree of belief in p + degree of belief in $\overline{p} = 1$
> …
> (4) Degree of belief in $(p \wedge q)$ + degree of belief in $(p \wedge \overline{q})$ = degree of belief in p.
> …We find, therefore, that a precise account of the nature of partial belief reveals that the laws of probability are laws of consistency, an extension to partial beliefs of formal logic, the logic of consistency.

Defining probability along the lines of Ramsey's approach, Bruno de Finetti (1906–1985) emphasized his subjective interpretation of probability [37]:

> According to whether an individual evaluates $P(E'|E'')$ as greater than, smaller than, or equal to $P(E')$, we will say that he judges the two events to be in a positive or negative correlation, or as independent: it follows that the notion of independence or dependence of two events has itself only a subjective meaning, relative to the particular function P which represents the opinion of a given individual ...
>
> In what precedes I have only summarized ...what ought to be understood, from the subjectivistic point of view, by "logical laws of probability" and the way in which they can be proved. These laws are the conditions which characterize coherent opinions (that is, opinions admissible in their own right) and which distinguish them from others that are intrinsically contradictory. The choice of one of these admissible opinions from among all the others is not objective at all and does not enter into the logic of the probable ...

that all probabilities are conditional, existing only as someone's description of an uncertain world.[41] Then, the role of probability theory is to coherently manage opinions [30], which is analogous to the satisfiability checking problem in probability logic (see Sect. 3.5). In de Finetti's view, this offered more freedom than the objectivistic approach, since for him it was possible to evaluate the probability over any set of events, while objectivists needed an unnecessarily complex mathematical structure in the background (citations from [36]):

> Concerning a known evaluation of probability, over any set of events whatsoever, and interpretable as the opinion of an individual, real or hypothetical, we can only judge whether, or not, it is coherent ...Such a condition of coherence should, therefore, be the weakest one if we want it to be the strongest in terms of absolute validity. In fact, it must only exclude the absolutely inadmissible evaluations; i.e. those that one cannot help but judge contradictory.

Another de Finetti's distinguishing characteristic was his strong support of finite additivity of probability. He posed the question whether it is possible to give zero probability to all the events in an infinite partition. The first answer was negative, i.e., probability is σ-additive, so in every infinite partition there must be an at most countable number of events with positive probabilities and the sum of those probabilities is 1. Here the zero probability may or may not mean impossibility. A more general view allows uncountable partitions, in which case the sum of an uncountable many zeroes can be positive. However, de Finetti had the following opinion:

> A: Yes. Probability is finitely additive. The union of an infinite number of incompatible events of zero probability can always have positive probability, and can even be the certain event ...Let me say at once that the thesis we support here is that of A, finite additivity; explicitly, the probability of a union of incompatible events is greater than or equal to the supremum of the sums of a finite number of them.

[41] This has an interesting consequence: a conditioning event can have the zero probability, which is not possible in the standard approach, where conditional probabilities are defined as quotients of absolute probabilities.

2.6.3 *Objective Probabilities as Relative Frequencies in Infinite Sequences*

Based on Cournot's principle about events with infinitely small probabilities, and Bernoulli's theorem, the frequency interpretation seems to be an objective way to determine the meaning of probability values as limits of relative frequencies in infinite sequences of events [102, 151]. Still it opens many questions, and not the least of them concerns estimation of limits by finite sequences. Richard von Mises (1883–1953) restricted the types of sequences that would be appropriate to characterize probabilities to random sequences in which it would not be possible to acquire gains by betting on the next item given the previous outcomes [169, 170]. However, his notion of collectives, infinite sequences of outcomes with warranted limiting values of the relevant frequencies and invariance of limits in subsequences, just shifted the existing issue to another one: the existence of collectives.

Hans Reichenbach (1891–1953) considered a broader class of the so-called normal sequences [132] but his definition was far from being precise [56], and we will not give it here, since for this text what is relevant is that Reichenbach used probabilities to replace the standard truth values in a logic where inferences should be represented axiomatically (citations from [132]):

> It will be shown that the analysis of probability statements referring to physical reality leads into an extension of logic, into a *probability logic*, that can be constructed by a transcription of the calculus of probability; and that statements that are merely probable cannot be regarded as assertions in the sense of classical logic, but occupy an essentially different logical position. Within this wider frame, transcending that of traditional logic, the problem of probability finds its ultimate solution ...This ...has a mathematical advantage in that it presents the content of the mathematical discipline of probability in a logically ordered form.

which is similar to Keynes' logical approach, but here probabilities assigned to propositions are limiting frequencies of events (that correspond to propositions) in sequences. For example, let A and B denote the classes (i.e., sequences) of events "the die is thrown" and "1 is obtained", respectively. Then:

> Probability statements therefore have the character of an implication; they contain a first term, and a second term, and the relation of probability is asserted to hold between these terms. This relation may be called *probability implication* ...the probability implication expresses statements of the kind "if a is true, then b is probable to the degree p".

Then for the events $x_i \in A$, and $y_i \in B$, the probability statement is written[42] as

$$(i)(x_i \in A \underset{p}{\ni} y_i \in B) \quad \text{or} \quad A \underset{p}{\ni} B \quad \text{or} \quad P(A, B) = p \quad \text{or} \quad P(fx_i, gy_i) = p,$$

while $P(A)$ denotes the probability of A.

Reichenbach proposed a formal system with four groups of axioms[43]:

[42]Meaning: for all x_i and all y_i, if $x_i \in A$, then $y_i \in B$ with probability p.

[43]We give the axioms in the original form, i.e., $A \supset B$, $\overline{A}$, $A.B$, $A \equiv B$ denote $A \rightarrow B$, $\neg A$, $A \wedge B$, and $A \leftrightarrow B$, respectively.

I Univocality

$$(p \neq q) \supset [(A \underset{p}{\ni} B).(A \underset{q}{\ni} B) \equiv (\overline{A})]$$

II Normalization

1. $(A \supset B) \supset (\exists p)[(A \underset{p}{\ni} B).(p = 1)]$
2. $\overline{(\overline{A})}.(A \underset{p}{\ni} B) \supset (p \geqq 0)$

III Theorem of addition

$$(A \underset{p}{\ni} B).(A \underset{q}{\ni} C).(A.B \supset \overline{C})] \supset (\exists r)[(A \underset{r}{\ni} (B \vee C)).(r = p + q)]$$

IV Theorem of multiplication

$$(A \underset{p}{\ni} B).(A.B \underset{u}{\ni} C)] \supset (\exists w)(A \underset{w}{\ni} B.C).(w = p \cdot u).$$

These axioms say that the probability values are unique[44] (I), from the real unit interval (II), that probabilities are finitely additive (III), while the axiom IV correspond to the rule $P(CB|A) = P(C|BA)P(B|A)$ [56].

The truth-table for Reichenbach's logic:

$P(A)$	$P(B)$	$P(A, B)$	$P(\overline{A})$	$P(A \vee B)$	$P(A.B)$	$P(A \supset B)$	$P(A \equiv B)$	$P(B, A)$
p	q	u	$1 - p$	$p + q - pu$	pu	$1 - p + pu$	$1 - p - q + 2pu$	pu/q

with the constraints:

- $P(A, A) = 1$ and
- $\frac{p+q-1}{p} \leqq u \leqq \frac{q}{p}$

means that the logic is not truth-functional. For example, $P(A \vee B)$ depends on the values $P(A)$ and $P(B)$, but also on the third value u that is not determined uniquely by the probabilities of A and B.

Reichenbach's position was that probability is reducible to frequency. To bridge the gap between his deductive system and the frequency-based approach to probability, Reichenbach added to the axiomatic system:

> RULE OF INDUCTION. If an initial section of n elements of a sequence x_i is given, resulting in the frequency f^n, and if, furthermore, nothing is known about the probability of the second level for the occurrence of a certain limit p, we posit that the frequency f^i $(i > n)$ will approach a limit p within $f^n \pm \delta$ when the sequence is continued.

The rule might seem in spirit of the law of large numbers, but the crucial difference between the two is that Rule of induction relies on finite sequences. Reichenbach

[44] A is an empty class (denoted by $\overline{A}$), if B follows from A with two different probabilities.

tried to justify it using a strong assumption about existence of the so-called higher order probabilities that guarantee that the limiting relative frequency of an event is in a small interval and represents the true probability. Anyhow, Rule of induction was seen as a part of the meta-level which could not be mixed with the object language:

> When we use the logical concept of probability, the rule of induction must be regarded as a rule of derivation, belonging in the metalanguage. The rule enables us to go from given statements about frequencies in observed initial sections to statements about the limit of the frequencies for the whole sequence. It is comparable to the rule of inference of deductive logic, but differs from it in that the conclusion is not tautologically implied by the premisses. The inductive inference, therefore, leads to something new; it is not empty like the deductive inference, but supplies an addition to the content of knowledge.

Besides numerous philosophical objections about [48, 56, 65]: accessibility of the truth in the real world, possibility to reduce different forms of uncertainty to only one kind of probability, problematic assumption of existence of limits of relative frequencies, dependability of limits upon the arrangement of events ordered in sequences, etc., there is logically founded criticism, too. Eberhardt and Glymour in [56] and Hailperin in [63] point out that:

- Reichenbach's axiomatization is neither strong enough to characterize σ-additive probabilities nor sets of limiting relative frequencies satisfy countable additivity [56],
- the syntax is not precisely specified, for example it is not clear whether iterations of the probability implication is allowed, i.e., what is the meaning of $(A \underset{p}{\ni} B) \underset{q}{\ni} C$,,
- no definition of the consequence relation is given, etc.

2.6.4 Measure-Theoretic Approach to Probability

Theory of measure and integration was initiated by Émil Borel [15] and Henri Lebesgue (1875–1941) [89] and further developed by Constantine Carathéodory (1873–1950), Johann Radon (1887–1956), Maurice Fréchet (1878–1973), Otto Nikodym (1887–1974), Percy Daniell (1889–1946), etc. Their results, following the analogy of events and their probabilities with sets of real numbers and their measures, provided important tools for probability theory. For example, it became possible to analyze limiting behaviors of relative frequencies.

In [15, 16] a countably additive extension of length of intervals,[45] today named after Borel, was introduced.[46] The measurable sets were defined to be closed intervals, their complements and at most countable unions, while the measure of a countable union of pairwise disjoint closed intervals was the sum of the lengths of the intervals.

---

[45]Countably additive probabilities were discussed for the first time by Anders Wiman (1865–1959) [47] in [174].

[46]It was not without criticism. Arthur Schoenflies (1853–1928) in [143] objected that Borel's approach to measurability was ad hoc, while σ-additivity was introduced by a definition, and motivated only to achieve a specific goal (translation from [47]):

In [16] Borel used Lebesgue's integral to prove that the event of randomly choosing a rational number in [0, 1] has the zero probability. However, at the time, he was not fully satisfied with applications of measure theory to probability, so in [17] he did not have a measure-theoretical approach, and used limits to obtain probabilities of events in infinite sequences of trials [151]. That seminal paper, although considered as the transition point between classical and modern probability theory, caused a disagreement in interpretations of Borel's way of reasoning and proving statements [5, 64, 146]. Borel introduced denumerable probabilities as a reasonable generalization of finite probabilities which avoids the continuous case (translation from [5]):

> The cardinality of denumerable sets alone being what we may know in a positive manner, the latter alone intervenes effectively in our reasonings ...I believe that ...the continuum will prove to have been a transitory instrument, whose present-day utility is not negligible (we shall supply examples at once), but it will come to be regarded only as a means of studying denumerable sets, which constitute the sole reality that we are capable of attaining.

He also considered probabilities that are only finitely additive[47] and was in principle not against them, but concluded that (translation from [171]):

> ...such a hypothesis does not seem logically absurd to me, but I have not encountered circumstances where its introduction would have been advantageous.

Borel analyzed countably long sequences of Bernoulli trials,[48] and formulated three questions [64]:

- What is the probability that the case of success never occurs?
- What is the probability of exactly k successes?
- What is the probability that success occurs an infinite number of times?

In the first problem he extended, to the countable version, the classical rule of compound probability for independent events

$$P\left(\bigwedge_{i=1}^{\infty} A_i\right) = \Pi_{i=1}^{\infty} P(A_i)$$

and justified it by analyzing convergence of the infinite sum $\sum_{i=1}^{\infty} P(A_i)$:

- if the sum is finite, the sought probability is well defined and belongs to (0, 1), and

(Footnote 46 continued)

> Above all, it only has the nature of a postulate because we cannot decide if a property which can be verified by a finite sum, can be extended over an infinite number of terms by an axiom, but by deep examination alone.

[47] For example: for every natural number the probability to be picked out from the set of all natural numbers is 0, and the probability of the whole set is 1.

[48] Independent trials, each trial A_i with exactly two possible outcomes, with the respective probabilities $P(A_l) = p_l$ and $P(\neg A_l) = 1 - p_l$ of success and of failure in the lth trial, respectively.

- if the sum is divergent, the infinite product tends towards 0, which is the value given to the sought probability, but Borel understood that in that case the corresponding event is not impossible.

The second problem required generalization of the principle of total probability to infinite case of countable additivity

$$P\left(\bigvee_{i=1}^{\infty} A_i\right) = \sum_{i=1}^{\infty} P(A_i).$$

Borel laconically justified this by an analogy to the previous case. Regarding the third problem, Borel concluded[49] that the probability of infinite number of successes is 1 if $\sum_{i=1}^{\infty} p_i < \infty$, while if the sum is divergent, the probability is 0. This result was applied to prove that almost all real numbers from the unit interval are normal, i.e., that in the corresponding binary expansions the proportion of 1s converges to 1/2. Generally, Borel used the notion of an almost sure probabilistic occurrence as a proof of existence.

Although Borel himself was not interested in formalization of probability, his paper motivated other authors to try to apply measure theory in establishing an axiomatic foundation of probability theory, e.g., Ugo Broggi (1880–1965), Sergei Bernstein (1880–1968), Evgeny Slutsky (1880–1948), Hugo Steinhaus (1887–1972), Stanisław Ulam (1909–1984), etc. [151]. It culminated with the famous Foundations of the theory of probability by Andrei Nikolaevich Kolmogorov (citations from [79]):

> The purpose of this monograph is to give an axiomatic foundation for the theory of probability. The author set himself the task of putting in their natural place, among the general notions of modern mathematics, the basic concepts of probability theory–concepts which until recently were considered to be quite peculiar. This task would have been a rather hopeless one before the introduction of Lebesgue's theories of measure and integration. However, after Lebesgue's publication of his investigations, the analogies between measure of a set and probability of an event, and between integral of a function and mathematical expectation of a random variable, became apparent.

Kolmogorov started with elementary theory of probability which deals with a finite number of events only. He defined an abstract structure such that the corresponding relations on its elements are determined by a set of axioms:

> Let E be a collection of elements $\xi, \eta, \zeta, \ldots$, which we shall call elementary events, and $\mathfrak{F}$ a set of subsets of E; the elements of the set $\mathfrak{F}$ will be called random events.

[49] This is known as Borels zero-one law, or Borel–Cantelli lemma in recognition of the independent proof (for general independently and identically distributed random variables) by Francesco Cantelli (1875–1966) [23]. This lemma is about almost sure convergence (except for a set of sequence of the probability zero) and can be seen as the initial version of the strong law of large numbers.

I. $\mathfrak{F}$ is a field[50] of sets.

II. $\mathfrak{F}$ contains the set E.

III. To each set $A \in \mathfrak{F}$ is assigned a non-negative real number $P(A)$. This number $P(A)$ is called the probability of the event A.

IV. $P(E)$ equals 1.

V. If A and B have no element in common, then

$$P(A+B) = P(A) + P(B)$$

A system of sets, $\mathfrak{F}$, together with a definite assignment of numbers $P(A)$, satisfying Axioms I-V, is called a field of probability.

and proved its consistency by considering a singleton E, $\mathfrak{F} = \{E, \emptyset\}$, and $P(E) = 1$, $P(\emptyset) = 0$, that satisfy the axioms. He also defined the conditional probability of the event B under the condition A that has a positive probability:

$$P_A(B) = \frac{P(AB)}{P(A)}.$$

and then proved a number of theorems, for example Multiplication theorem,[51] Theorem on total probability,[52] Bayes' theorem,[53] etc. Then Kolmogorov considered the infinitary case:

The field $\mathfrak{F}$ is called a Borel field, if all countable sums of the sets A_n from $\mathfrak{F}$ belong to $\mathfrak{F}$ ...A field of probability is a Borel field of probability if the corresponding field $\mathfrak{F}$ is a Borel field ...Given a field of probability $(\mathfrak{F}, P)$. As is known, there exists a smallest Borel field $B\mathfrak{F}$ containing $\mathfrak{F}$.

and introduced the additional Axiom of Continuity:

VI. For a decreasing sequence of events $A_1 \supset A_2 \supset \ldots \supset A_n \supset \ldots$ of $\mathfrak{F}$ for which $A_1 A_2 \cdots = 0$, the following holds:

$$\lim P(A_n) = 0, \qquad n \to \infty.$$

which implies generalized addition theorem, i.e., that the probability $P(A)$ is a completely additive set function on $\mathfrak{F}$:

If $A_1, A_2, \ldots, A_n, \ldots$and A belong to $\mathfrak{F}$, then from $A = A_1 + A_2 + \ldots + A_n + \ldots$ follows the equation $P(A) = \sum_n P(A_n)$

---

[50] Cf. Hausdorff, Mengenlehre, 1927, p. 78. A system of sets is called a field if the sum, product, and difference of two sets of the system also belong to the same system. Every nonempty field contains the null set 0. Using Hausdorff's notation, we designate the product of A and B by AB; the sum by $A+B$ in the case where $AB = 0$; and in the general case by $A \dotplus B$; the difference of A and B by $A - B$. The set $E - A$, which is the complement of A, will be denoted by $\overline{A}$ ....

[51] $P(A_1 A_2 \ldots A_n) = P(A_1) P_{A_1}(A_2) P_{A_1 A_2}(A_3) \ldots P_{A_1 A_2 \ldots A_{n-1}}(A_n)$.

[52] For mutually exclusive $A_1, A_2, \ldots, A_n$ such that $A_1 + A_2 + \ldots + A_n = E$, and an arbitrary X: $P(X) = P(A_1)P_{A_1}(X) + P(A_2)P_{A_2}(X) + \cdots + P(A_n)P_{A_n}(X)$.

[53] For $A_1 + A_2 + \cdots + A_n = E$ and an arbitrary X: $P_X(A_i) = \frac{P(A_i)P_{A_i}(X)}{P(A_1)P_{A_1}(X) + P(A_2)P_{A_2}(X) + \cdots + P(A_n)P_{A_n}(X)}$.

Kolmogorov explained that he adopted this extension as an ideal, but useful generalization:

> Since the new axiom is essential for infinite fields of probability only, it is almost impossible to elucidate its empirical meaning …For, in describing any observable random process we can obtain only finite fields of probability. Infinite fields of probability occur only as idealized models of real random processes. We limit ourselves, arbitrarily, to only those models which satisfy Axiom VI …
>
> Only in the case of Borel fields of probability do we obtain full freedom of action, without danger of the occurrence of events having no probability …Even if the sets (events) A of $\mathfrak{F}$ can be interpreted as actual and (perhaps only approximately) observable events, it does not, of course, follow from this that the sets of the extended field $B\mathfrak{F}$ reasonably admit of such an interpretation. Thus there is the possibility that while a field of probability $(\mathfrak{F}, P)$ may be regarded as the image (idealized, however) of actual random events, the extended field of probability $(B\mathfrak{F}, P)$ will still remain merely a mathematical structure. Thus sets of $(B\mathfrak{F}, P)$ are generally merely ideal events to which nothing corresponds in the outside world. However, if reasoning which utilizes the probabilities of such ideal events leads us to a determination of the probability of an actual event of $\mathfrak{F}$, then, from an empirical point of view also, this determination will automatically fail to be contradictory.

Kolmogorov's understanding of applicability of the theory of probability to the world of actual events was motivated by von Mises' frequentist approach. Kolmogorov explained the relationship between probability and the real world using a (finite) system capable of unlimited repetition of trials, for example, of flipping a coin. Then the ratio between a large number n of repetitions of the experiment and m, the number of occurrences of the sought event A, could be used for empirical deduction of the axioms I–V. Thus it is possible to use it as a practical estimation of $P(A)$, the probability of A. Finally, he noticed that:

> …it does not follow that in a very large number of series of n tests each, in each the ratio m/n will differ only slightly from $P(A)$.
>
> …To an impossible event (an empty set) corresponds, in accordance with our axioms, the probability $P(\emptyset) = 0$, but the converse is not true: $P(A) = 0$ does not imply the impossibility of A …all we can assert is that …event A is practically impossible. It does not at all assert, however, that in a sufficiently long series of tests the event A will not occur.

In the decades that followed the appearance of [79], Kolmogorov's axiomatic approach has become standard and established modern probability theory, where the probability is determined by the above axioms.

2.6.5 *Other Ideas*

There are plenty of authors whose approaches to probability are more or less compatible with the above mentioned ones. In [75], Harold Jeffreys (1891–1989) developed Bayesian statistics based on the idea that uncertainty can be described by statements about probabilities, with the central role of Bayes's rule in knowledge updating.

Richard Cox (1898–1991) in [32, 34] tried to formulate rules that guarantee that probabilities are measurable, and that some probable inferences are more convincing than the others. Starting from such intuitively plausible principles, for example (citation from [34]):

> The probability of an inference on given evidence determines the probability of its contradictory on the same evidence.

he provided the foundation for logical interpretation of probability by proving that there is a measure, i.e., a function of propositions, which satisfies the usual probability axioms [153].

György Pólya (1887–1985) discussed qualitative conditions for plausible reasoning (citation from [117]):

> We have here a pattern of plausible inference:

$$\frac{\begin{array}{c} A \text{ implies } B \\ B \text{ true} \end{array}}{A \text{ more credible}}$$

and investigated the relationship between plausible reasoning and the calculus of probability.

Edwin Jaynes (1922–1998), influenced by the results of Cox and Pólya, strongly defended the principle of (citation from [74]):

> …assigning probabilities by logical analysis of incomplete information.

Leonard Savage (1917–1971) constructed probabilities from subjective preferences and then developed applications to decision theory [66, 137].

Karl Popper (1902–1994) promoted a kind of objective approach to probability called propensity interpretation[54] [66, 118, 119]. For him probability represents an experiment's tendency to produce given types of outcomes, e.g., the probability of an outcome is a propensity of the corresponding experiment to generate a sequence of outcomes of a particular type with a limiting relative frequency. For example, the properties of a coin (it is homogeneous, with a head and a tail) have the effect that the limit of the relative frequencies of heads and tails in long sequences are one half.

Patrick Suppes (1922–2014) in [160] considered two instances of the so-called statistical syllogism:

- if $P(A|B) = r$ and B is true, then $P(A) = r$, and
- if $P(A|C) = s$ and C is true, then $P(A) = s$,

which makes a paradoxical conclusion, if $r \neq s$. While the classical approach to resolve the problem relied on the concept of total evidence, i.e., that total available evidence has to be used in making the conclusion, Suppes proposed a more general rule in probabilistic inference (citations from [160]):

---

[54]Reference [60] indicates that Peirce, in the last period of his work, adopted a similar opinion. Also, similarities between Popers's propensity approach and ideas formulated by Cardano and Galileo Galilei's (1564–1642) are discussed.

$$\frac{\begin{array}{c}P(A|B) \geqq r\\ P(B) \geqq \rho\end{array}}{P(A) \geqq r\rho.}$$

which, in combination with the theorem on total probability implies that from the premises $P(A|B) = r$, $P(B) = \rho$, $P(A|C) = s$, $P(C) = \sigma$, one concludes $P(A) \geq \max\{r\rho, s\sigma\}$. Then he proved validity of other similar rules, e.g.:

$$\frac{\begin{array}{c}P(B \to A) \geqq 1-\varepsilon\\ P(B) \geqq 1-\varepsilon\end{array}}{P(A) \geqq 1-2\varepsilon} \qquad \frac{\begin{array}{c}P(A|B) \geqq 1-\varepsilon\\ P(B) \geqq 1-\varepsilon\end{array}}{P(A) \geqq (1-\varepsilon)^2}$$

which he generalized to:

Theorem 1. If $P(A) \geqq 1 - \varepsilon$ and A logically implies B then $P(B) \geqq 1 - \varepsilon$.

Theorem 2. If each of the premises $A_1, \ldots, A_n$ has probability of at least $1 - \varepsilon$ and these premises logically imply B then $P(B) \geqq 1 - n\varepsilon$. Moreover, in general the lower bound $1 - n\varepsilon$ cannot be improved on, i.e., equality holds in some cases whenever $1 n\varepsilon \geqq 0$.

As all that is not enough, in [65] another approaches are listed:

- Terrence Fine (1939) proposed comparative probabilities [50],
- David Cox (1924) used complex valued probabilities [33],
- Alfréd Rényi (1921–1970) allowed ∞ to be between possible values of probabilities [133],
- Paul Dirac (1902–1984) and Eugene Wigner (1902–1995) considered negative probabilities [46, 173],
- David Lewis (1941–2001) and Brian Skyrms (1938) pointed out that infinitesimal probabilities are essential [103, 155],
- Arthur Dempster (1929) and Glenn Shafer (1946) promoted non-additive probabilities [44, 148],
- Ernest Adams (1926–2009) considered entailments with probabilities almost 1 [1], etc.

2.7 1960s: And Finally, Logic

Most of the previously described ideas about probability logic had very little in common with the contemporary advances in mathematical logic, i.e., with the proof-theoretical and model-theoretical results of Gödel, Alfred Tarski (1901–1983), Leon Henkin (1921–2006) Abraham Robinson, Saul Kripke, Kenneth Jon Barwise, and other big names in the field [6, 57, 70, 81, 135, 162]. However, this situation started changing with the papers by Gaifman, Hailperin, and Scott and Krauss that appeared in 1960s, and particularly with, a little bit latter, Keisler's work. Also, several authors, e.g., Charles Hamblin, John Burgess, and Krister Segerberg discussed probability in the framework of modal logics.

2.7.1 Probabilities in First-Order Settings

Haim Gaifman in [53] introduced the so-called probability models of the form $\langle U, m\rangle$, where U is a nonempty domain containing all constant symbols from a first-order calculus $\mathfrak{B}$, and m satisfies, for all quantifier free formulas without free variables:

- $m(\phi) \in [0, 1]$,
- if ϕ is a theorem, i.e., $\vdash \phi$, then $m(\phi) = 1$, and
- if $\vdash \neg(\phi \wedge \psi)$, then $m(\phi \vee \psi) = m(\phi) + m(\psi)$.

One can understand $m(\phi)$ as a measure of the set of models of ϕ, or as an extension of a usual assignment of truth values to an assignment of a probabilistic truth values to ϕ. However, since in the above definition m determines $[0, 1]$-values only for a limited set of sentences, it has to be extended to the set of all sentences:

- (Gaifman condition) for every probability model $\langle U, m\rangle$ there is a unique extension m^* of m to the set of all sentences such that $m^*(\exists x\phi(x)) = \sup\{m^*(\phi(a_i) : a_i \in U\}$, the supremum taken over all finite subsets of U.

Geifman analyzed properties of probability models and, for example, proved the existence of certain subclasses of probability models, e.g., of symmetric models that satisfy $m(\phi(a_1, \ldots, a_n) = m(\phi(\pi(a_1), \ldots, \pi(a_n))$ for every formula ϕ and every permutation π.

Motivated by Gaifman's paper [53] Dana Scott (1932) and Peter Krauss wrote (citations from [144]):

> In this paper we wish to investigate how probabilities behave on statements, where to be definite we take the word "statement" to mean "formula of a suitable formalized logical calculus" ...It would be fair to say that our position is midway between that of Carnap and that of Kolmogorov. In fact, we hope that this investigation can eventually make clear the relationships between the two approaches ...The main task we have set ourselves in this paper is to carry over the standard concepts from ordinary logic to what might be called probability logic. Indeed ordinary logic is a special case: the assignment of truth values to formulas can be viewed as assigning probabilities that are either 0 (for false) or 1 (for true).

They used the first-order infinitary language $L_{\omega_1\omega}$, which allows countable conjunctions and disjunctions, and finite quantification, while in semantics they considered strictly positive probabilities[55] defined on Boolean σ-algebras representing values of sentences. As a consequence they obtained countably additive probability models. Scott and Krauss proved the generalization of the Gaifman condition for the infinitary language and introduced several model-theoretic concepts, e.g.:

- a probability assertion Ψ is a tuple

$$\langle\Phi, \phi_1, \ldots \phi_N\rangle$$

---

[55] $m(a) = 0$ iff a is 0 in the Boolean algebra.

where Φ is a formula in the first order language of algebra of real numbers (with N free variables) which speaks about probabilities of the sentences ϕ_i from the initial language,

- a probability system $\langle T, m\rangle$ is a probability model of $\langle \Phi, \phi_1, \dots \phi_N\rangle$ if the n-tuple of reals $\langle m(\phi_1), \dots m(\phi_N)\rangle$ satisfies Φ in $\langle \mathbb{R}, \leq, +, \cdot, 0, +1, 1\rangle$,
- the probability assertion Ψ is a probability consequences of the set of probability assertions Σ if all probability models of Σ are models of Ψ, which introduced probability consequences as a generalization of the consequence relation in classical logic, and
- a probability assertion Ψ is a probability law if Ψ is a probability consequence of the empty set.

It was proved that every countable consistent set of sentences has a countable probability model. Regarding the existence of a method of deductively generating probability consequences Scott and Krauss were pessimists. On the other hand, they proved somehow restricted goal for probability laws:

Theorem 6.7. Let $\langle \Phi, \varphi_0, \dots \varphi_{n-1}\rangle$ be a probability assertion such that the freee variables of Φ are $\lambda_0, \dots, \lambda_{n-1}$; further $\vdash \neg(\varphi_i \wedge \varphi_j)$ if $i \neq j$, and $\vdash \vee_{i<n}\varphi_i$. Let $I = \{i < n :\vdash \neg\varphi_i\}$. Then $\langle \Phi, \varphi_0, \dots \varphi_{n-1}\rangle$ is a probability law iff the sentence

$$\forall\lambda_0 \dots \forall\lambda_{n-1}[[\wedge_{i\in I}\lambda_i = 0 \ \wedge \ \wedge_{i<n}\lambda_i \geq 0 \ \wedge \ \lambda_0 + \cdots + \lambda_{n-1} = 1] \rightarrow \Phi]$$

is a theorem of real algebra.

Using Tarski's result about decidability of theorems of real algebra, this theorem guarantees that for a class of probability assertions it can be decided whether the corresponding sentences in real algebra are theorems.

2.7.2 *Probability Quantifiers*

The most important advancement in probability logic, after work of Leibnitz and Boole, was made by Howard Jerome Keisler (1936). The purpose of his famous paper [76] was to develop, within Robinson's nonstandard infinitesimal analysis [135], model theory appropriate for studying and classifying probability models that arise in applied mathematics, e.g., in analyzing infinitesimal Poisson processes, and Brownian motion. While, Gaifman, Scott and Krauss defined probabilities on sentences, Keisler considered probability distributions on domains of first-order structures. Instead of classical universal and existential quantifiers, he introduced probability quantifiers, for example $Px > r$. The formula $(Px > r)\phi(x)$ means that the probability of the set $\{x : \phi(x)\}$ is greater than r. Keisler studied both finitary ($L_{\omega P}$) and infinitary ($L_{\omega_1 P}$, i.e., with countably infinite conjunctions and disjunctions) languages. Examples of $L_{\omega P}$-valid formulas are:

- nonnegativity: $\models (Px \geq 0)\varphi$
- monotonicity: $\models (Px > r)\varphi \rightarrow (Px > s)\varphi$, for $r > s$,

- additivity: $\models [(Px > r)(\varphi \wedge \psi) \wedge (Px > s)(\varphi \wedge \neg\psi)] \rightarrow (Px > r + s)\varphi$, etc.

Mixing of ordinary $\forall$ and $\exists$ and probability quantifiers in this σ-additive framework is still an open problem [77]. In [76], Keisler focused on model-theoretic approach and proved a number of results analogous to the statements from the standard model theory.

Axiomatizations for Keisler-like logics $L_{\omega P}$, $L_{\omega_1 P}$ and[56] $L_{\mathbb{A}P} = L_{\omega_1 P} \cap \mathbb{A}$, with real-valued σ-additive probabilities were given by Douglas Hoover [71]. In this framework, a probability model is a structure

$$\mathcal{M} = (\mathfrak{A}, \mu_n)_{n<\omega}$$

such that:

- $\mathfrak{A} = \langle A, R_i, f_j, c_k \rangle_{i,j,k}$ is a model in the sense of first-order logic,
- each μ_n, $n < \omega$, is a σ-additive probability measure on A^n, and the sequence of measures $\langle \mu_n : n < \omega \rangle$ satisfies the Fubini property,[57] and
- every set of elements of A^n satisfying an atomic formula with n free variables is measurable with respect to μ_n.

The satisfiability relation fulfills the usual requirements, and additionally:

- $\mathcal{M} \models P(\overline{y} \geq r)\varphi(\overline{x}, \overline{y})[\overline{a}]$ iff $\mu_n\{\overline{b} \in A^n : \mathcal{M} \models \varphi[\overline{a}, \overline{b}]\} \geq r$.

The following examples illustrate expressivity of the mentioned logics:

- there are no singletons of positive probability: $(Px \geq 1)(Py \geq 1)x \neq y$,
- there is a countable set of probability 1: $(Px \geq 1)(Py > 0)x = y$,
- (in $L_{\omega_1 P}$) almost everywhere convergence of a sequence of random variables $X_m \rightarrow X$: $(Px \geq 1)(\wedge_n \vee_m \wedge_{q \in \mathbb{Q}}([X(x) > q] \leftrightarrow [X_m(x) > q - \frac{1}{n}]))$.

Some of the axioms and rules presented in the paper are:

- monotonicity: $(Px \geq 1)(\varphi(x) \rightarrow \psi(x)) \rightarrow ((Px \geq r)\varphi(x) \rightarrow (Px \geq r)\psi(x))$
- necessitation:

$$\frac{\text{from } \psi \rightarrow \varphi(x)}{\text{infer } \psi \rightarrow (Px \geq 1)\varphi(x)}$$

- continuity at 0:

$$\frac{\text{for any } n < \infty, n > 0, \quad \text{from } \varphi \rightarrow (Py \geq \frac{1}{n})(Px \in [r - \frac{1}{m}, r))\psi}{\text{infer} \neg\varphi.}$$

In the infinitary logics, the continuity rule implies a formula expressing the Archimedean property:

---

[56] $\mathbb{A}$ is a countable admissible set [6].

[57] Informally, the Fubini property means that the sequence $\langle \mu_n : n < \omega \rangle$ behaves like a sequence of product measures $\langle \mu^n : n < \omega \rangle$.

$$(Px \geq r)\psi(x) \leftrightarrow \wedge_{n\in\mathbb{N}}\left(Px \geq r - \frac{1}{n}\right)\psi(x)$$

while in the finitary logic it corresponds to the rule

$$\frac{\text{from } \{\psi \to (Px \geq r)(Py \geq s - \frac{1}{n})\varphi : n \in \mathbb{N}\}}{\text{infer } \psi \to (Px \geq r)(Py \geq s)\varphi.}$$

Hoover's completeness proofs consist of two steps. First he proved that every consistent set of formulas T is satisfiable in a weak model, which is a probability model with finitely additive probabilities. Then, by applying the Loeb process, he obtained the strong, i.e., σ-additive, model which satisfies T.

In the following years, Keisler and Hoover made very important contributions in the field. They proved completeness theorems for various kinds of models (probability, graded, analytic, hyperfinite, etc.) and many other model-theoretical theorems. The development of probability model theory has engendered the need for the study of logics with greater expressive power than that of the logic $L_{\mathbb{A}P}$. The logic $L_{\mathbb{A}I}$, introduced in [77] as an equivalent of the logic $L_{\mathbb{A}P}$, allows us to express many properties of random variables in an easier way. In this logic the quantifiers $\int \ldots dx$ are incorporated instead of the quantifiers $Px > r$. Since the logic $L_{\mathbb{A}I}$ is not rich enough to express probabilistic notions involving conditional expectations of random variables with respect to σ-algebras, such as martingale, Markov process, Brownian motion, stopping time, optional stochastic process, etc., [77] also introduced the logics $L_{\mathbb{A}E}$ and L_{ad} that are appropriate for the study of random variables and stochastic processes. Keisler's paper finished with a list of research problems. Miodrag Rašković and his co-authors solved some of them in [124, 126, 130, 131]. For example, [124] presented a new method of using Barwise compactness theorem [6] in proving completeness theorem for absolutely continuous measures, while in [123] a new $L_{\mathbb{A}M}$ logic with $[0, +\infty]$-valued measures was introduced. Some efforts to combine ordinary $\forall$ and $\exists$, and probability quantifiers have been made in [125, 127–129].

Comprehensive overviews of work in the field of logics with probability quantifiers are [49, 130].

2.7.3 *Probabilities in Modal Settings*

Charles Hamblin (1922–1985) combined probability and modal logic [68]. He introduced an additional unary modal operator, denoted P, so that Pp (p is probable) was understood as the probability of p is greater than or equal to x. Hamblin considered a finite language, i.e., a language with a finite number of propositional letters p_1, p_2, ..., p_n, while formulas were restricted to be without nesting of the modal operators. The corresponding models contain 2^n possible worlds, each world characterized by a state description,[58] i.e., a conjunction which contains, for every p_i, either p_i or $\neg p_i$. Probabilities are attached to the possible worlds, probabilities of statements are sums of probabilities of possible worlds so that the corresponding state descriptions satisfy the statements, while formulas of the form Pp are valid if they are true in every model for any $x \in [\frac{1}{2}, 1]$. Hamblin proposed an axiom system containing, besides the standard axioms for the modal system T, for example[59]:

- $P\neg p \to \neg Pp$,
- $\Box p \to Pp$, and
- $\Box(p \to q) \to (Pp \to Pq)$.

Hamblin also considered an alternative system in which he used the unary operator Q to express plausibility. Instead of probabilities, numbers that measure plausibility are associated to sets of possible worlds so that in every model at least one of those numbers must be 1, while, instead of the additivity rule, the plausibility of an union of possible worlds is equal to the largest of the plausibilities of those worlds. In both systems, truth does not imply probability, or plausibility, i.e., $p \to Pp$ and $p \to Qp$ are not theorems. Finally, Hamblin gave an alternative definition of probable, as something more plausible then its negation, and formalized it as:

- $\neg(Q\neg p \to Pp)$.

John Burgess (1948) investigated in [21] the system $S5U$, which is the modal $S5$ extended by the probability operator denoted U. He decided to add the axiom

- $Up \to \Box Up$

i.e., what is probable is necessarily probable, to the probability axioms from [68]. He defined the corresponding class of algebraic models, and proved weak completeness of the axiom system and decidability. To study the notion of numerical probabilities, Burgess used Hamblin's approach interpreting Up to mean probability of p is greater than x, for all rational x from $[\frac{1}{2}, 1)$. He proved that for all such different x and y there is an x-valid formula which is not y-valid. Finally, Burgess studied a relationship between his modal logic and an extended first-order framework in which $\Box$ corresponds to $\forall$, while U corresponds to a new quantifier "for most x".

Krister Segerberg considered qualitative probabilities in [145]. He introduced a binary model operator $\gtrsim$ so that

- $A \gtrsim B$

means A is at least as probable as B. He extended the notion of Kripke models so that every possible world x is associated by a σ-algebra B_x of subsets of accessible worlds and a probability measure M_x defined on B_x. Then, $A \gtrsim B$ is satisfied in a world x if the probability of A in x is not less than the probability of B in x, i.e.:

---

[58] Called atom in Sect. 3.1.1.

[59] Actually, Hamblin, as well as Burgess, used Polish notation, so the first axiom in the list was written as $CPNpNPp$.

- $M_x\{y : y \text{ is accessible from } x, y \models A\} \geqslant M_x\{y : y \text{ is accessible from } x, y \models B\}$.

He provided an axiom system PK containing the modal system K and some axioms characterizing $\gtrsim$, for example:

- $[\Box(A \leftrightarrow A') \wedge \Box(B \leftrightarrow B')] \rightarrow [(A \gtrsim B) \rightarrow (A' \gtrsim B')]$, and
- $\bot \gtrsim \bot$.

Furthermore, if

- $A_1 \dots A_m \mathbb{E} B_1 \dots B_m$ denotes $\Box(C_0 \vee \dots \vee C_m)$, and for every integer $p \in [0, m]$, C_p is the disjunction of all conjunctions

$$\delta_1 A_1 \wedge \dots \wedge \delta_m A_m \wedge \epsilon_1 B_1 \wedge \dots \wedge \epsilon_m B_m$$

such that exactly p of the δ's and p of ϵ's are empty strings, and the rest of them are the negation sign,[60]

another Segerberg's axiom is:

- $(A_1 \dots A_m \mathbb{E} B_1 \dots B_m) \rightarrow [(A_1 \gtrsim B_1 \wedge \dots \wedge A_m \gtrsim B_m) \rightarrow (B_1 \gtrsim A_1 \wedge \dots \wedge B_m \gtrsim A_m)]$, for every $m \geqslant 1$.

Segerberg proved completeness and decidability of the system PK.

2.7.4 *Probabilistic Logical Entailment*

Theodore Hailperin (1916–2014) extended Boole's and Fréchet's results [14, 51], and derived an effective procedure to obtain the best possible bounds for probabilities of propositional formulas $\phi(A_1, \dots, A_n)$, when the probabilities of subformulas, i.e., $P(A_1), \dots, P(A_n)$, are known [61]. His approach, like Boole's, was based on methods of linear programming.

Since the middle of 1980s, the interest in probability logics started growing because of development of many fields of application of representation of and reasoning about uncertain knowledge (in economics, artificial intelligence, computer science, philosophy, etc.) resulting in numerous publications. The first of those papers was Nils Nilsson's [112] inspired by the work on developing the PROSPECTOR expert system in geology. PROSPECTOR used Bayes' rule to calculate probabilities of hypotheses given geological evidence about ore deposits. Nilsson tried to give a semantical generalization of classical logic such that the truth values of sentences are replaced by their probabilities. He analyzed the probabilistic entailment, i.e., how to calculate the probability of a sentence given a set of sentences with the corresponding

[60] In other words, $A_1 \dots A_m \mathbb{E} B_1 \dots B_m$ holds in a world x if in every accessible world y exactly the same number of A_i's and B_j's hold.

probabilities such that the obtained value is independent from any assumptions relative to specific probability models. It turned out that Nilsson's approach was already presented by Hailperin, but the timing of [112] was much better and it caused numerous reactions [113]. For us, particularly interesting between them are papers about proof-theoretic methods that will be analyzed in Sect. 7.

The reduction to linear programming problem in Hailperin's and Nilsson's works means, in fact, that they use modal models, even though they do not explicitly mention it. In that sense, [61, 112] are related to the papers of Hamblin, Burgess and Segerberg.

References

1. Adams, E.W.: The logic of Conditional. Reidel, Dordrecht (1975)
2. Aristotle, R.: Translated by W.R. Roberts, University of Adelaide (2015). https://ebooks.adelaide.edu.au/a/aristotle/a8rh/
3. Arnauld, A., Nicole, P.: La logique, ou l'art de penser (Port-Royal Logic, Logique de Port-Royal, Ars cogitandi). Paris (1662)
4. Arnauld, A., Nicole, P.: Logic; or, the art of thinking: in which besides the common, are contain'd many excellent new rules, very profitable for directing of reason, and acquiring of judgment, in things as well relating to the instruction of a mans self, as of others La logique, ou l'art de penser (Port-Royal Logic, Logique de Port-Royal, Ars cogitandi). English translation of [4] "by several hands". Printed by T. B. for John Taylor, London (1693). https://archive.org/details/logicorartofthin00arnaiala
5. Barone, J., Novikoff, A.: A history of the axiomatic formulation of probability from Borel to Kolmogorov: Part I. Arch. Hist. Exact Sci. **18**(2), 123–190 (1978)
6. Barwise, K.J.: Admissible Sets and Structures. Springer, Berlin (1975)
7. Bayes, T.: An Essay towards solving a Problem in the Doctrine of Chances (1763). http://www.jstor.org/stable/2333180
8. Bernoulli, J.: Ars conjectandi (1713). https://books.google.rs/books?id=kD4PAAAAQAAJ&printsec=frontcover&source=gbs_ge_summary_r&cad=0&q&f=falseBasel
9. Bernoulli, J.: Ars conjectandi. An English translation of Part II. London (1795). http://books.google.rs/books?id=B05LAAAAMAAJ&dq=doctrine+of+permutations&pg=PP9&redir_esc=y&q&f=false
10. Bernoulli, J.: Ars conjectandi. An English translation of Part IV (chapters I to V) by Bing Sung. Technical report No. 2 of the Department of Statistics of Harvard University (1966). http://cerebro.xu.edu/math/Sources/JakobBernoulli/ars_sung/ars_sung.html
11. Bernoulli, J.: Ars conjectandi. An English translation of Part IV (chapters I to V) by Oscar Sheynin (2005). www.sheynin.de/download/bernoulli.pdf
12. Bernstein, S.N.: Teoria Veroiatnosti. Mosva-Leningrad (1927)
13. Bolzano, B.: Wissenschaftslehre (1837)
14. Boole, G.: An investigation of the laws of thought on which are founded the mathematical theories of logic and probabilities. London (1854). http://archive.org/details/investigationofl00boolrich
15. Borel, E.: Leçons sur la théorie des fonctions. Gauthier-Villars, Paris (1898)
16. Borel, E.: Remarques sur certaines questions de probabilité. Bulletin de la Société Mathématique de France, p. 33 (1905), . http://archive.numdam.org/article/BSMF_1905__33__123_1.pdf
17. Borel, E.: Les probabilités dénombrables et leurs applications arithmétiques. Rendiconti del Circolo Matematico di Palermo **27**, 247–270 (1909)

18. Borel, E.: A propos d'un traité de probabilitiés (Apropos of a treatise on probability). Revue philosophique 98, 321 – 326. Also In: Kyburg, H., Smokler, H. (eds.) Studies in Subjective Probability, pp. 46–60. John Wiley and Sons (1964. 1924)
19. Bru, B.: Émile Borel. In: Heyde, C., Seneta, E. (eds.), Statisticians of the Centuries, pp. 287–291. Springer (2001)
20. Burch, R.: Charles Sanders Peirce. In: Edward N.Z. (ed.) The Stanford Encyclopedia of Philosophy (2014). http://plato.stanford.edu/archives/win2014/entries/peirce/
21. Burgess, J.: Probability logic. J. Symb. Logic **34**, 264–274 (1969). http://www.jstor.org/stable/2271103
22. Burton, D.M.: The History of Mathematics: An Introduction, 7th edn. McGraw Hill, New York (2011)
23. Cantelli, F.P.: Sulla probabilitá come limite della frequenza. Rendiconti della R. Accademia dei Lincei. Classe di Scienze Fisiche Matematiche e Naturali, Série 5a, vol. 26, pp. 39–45 (1917)
24. Carnap, R.: On inductive logic. Philos. Sci. **12**(2), 72–97 (1945). http://www.jstor.org/stable/184280
25. Carnap, R.: Logical Foundations of Probability, 2nd edn. The University of Chicago Press, Chicago (1962). (1st edition 1950)
26. Charles, S.P.: Collected Papers of Charles Sanders Peirce. Kinds of reasoning. Principles of philosophy, vol. I. Harvard University Press, Cambridge (1931)
27. Chebyshev, Pafnuty: Démonstration élémentaire d'une proposition générale de la théorie des probabilités. Journal für die reine und angewandte Mathematik (Crelles Journal) **33**, 259–267 (1846)
28. Chebyshev, P.: Sur deux théorémes relatifs aux probabilités (On two theorems of probability). Bulletin physico-mathématique de l'Académie Impériale des Sciences St. Pétersbourg , vol. 55(6) (1887)
29. Churpov, A.A.: Essays on the Theory of Statistics (1909)
30. Coletti, G., Scozzafava, R.: Probabilistic Logic in a Coherent Setting. Kluwer Academic Press, Dordrecht (2002)
31. Cournot, A.: Exposition de la théorie des chances et des probabilités. Paris (1843)
32. Cox, R.: Probability, frequency and reasonable expectation. Am. J. Phys. **14**, 1–13 (1946)
33. Cox, D.: A use of complex probabilities in the theory of stochastic processes. Math. Proc. Camb. Philos. Soc. **51**, 313–319 (1955)
34. Cox, R.: The Algebra of Probable Inference. John Hopkins Press, Baltimore (1961)
35. Dale, A.L.: A History of Inverse Probability From Thomas Bayes to Karl Pearson, 2nd edn. Springer, Berlin (1999)
36. de Finetti, B.: Theory of Probability, vol. 1. Wiley, New York (1974)
37. de Finetti, B.: La Prévision: Ses Lois Logiques, Ses Sources Subjectives (Foresight. Its Logical Laws, Its Subjective Sources). Annales de l'Institut Henri Poincaré, **7**, 1–68 (1937). (Also in: Kyburg, H., Smokler, H. (eds.) Studies in Subjective Probability, pp. 53–118 (1980))
38. de Moivre, A.: The Doctrine of Chances: or a method for calculating the probabilities of events in play, London (1711) (the first edition in Latin), 1718, 1738, 1756 (in English). https://archive.org/details/doctrineofchance00moiv
39. de Montmort, P.-R.: Essay d'analyse sur les jeux de hasard, Paris, 1708 (1713). https://books.google.fr/books?id=DIM_AAAAcAAJ&printsec=frontcover&hl=fr&source=gbs_ge_summary_r&cad=0#v=onepage&q&f=false
40. De Morgan, A.: A Treatise on the Theory of Probabilities. W. Clowes and Sons, London (1837). (Reprinted in Encyclopaedia Metropolitana, vol. 2, 393–490 (1845))
41. De Morgan, A.: Théorie Analytique des Probabilités. Par M. le Marquis de Laplace, 3éme edition. Paris 1820. (Reviewed by AdM). Part I. Dublin. Review **2**, 338–354 (1837)
42. De Morgan, A.: Théorie Analytique des Probabilités. Par M. le Marquis de Laplace, 3éme edition. Paris 1820. (Reviewed by AdM). Part II. Dublin. Review **3**, 237–248 (1838)
43. De Morgan, A.: Formal logic, or, The calculus of inference, necessary and probable. Taylor and Walton, London (1847). https://archive.org/details/formallogicorthe00demouoft

44. Dempster, A.: Upper and lower probabilities induced by a multivalued mapping. Ann. Math. Stat. **38**(2), 325–339 (1967)
45. Devlin, K.: The Unfinished Game: Pascal, Fermat, and the Seventeenth-Century Letter that Made the World Modern. Basic Books, New York (2008)
46. Dirac, P.: The physical interpretation of quantum mechanics. Proc. R. Soc. A: Math. Phys. Eng. Sci. **180**(980), 1–39 (1942)
47. Durand, A., Mazliak. L.: Volterra's Prolusione as a source for Borel's interest in probability (2010). https://www.researchgate.net/publication/47648802_Volterra's_Prolusione_as_a_source_for_Borel's_interest_in_probability
48. Eberhardt, F., Glymour, C.: Hans Reichenbach's probability logic. In: Gabbay, D.M., Hartmann, S., Woods, J. (eds.), Handbook of the History of Logic, vol. 10. pp. 357– 388 (2011)
49. Fajardo, S., Keisler, J.: Model Theory of stochastic processes. Lecture Notes in Logic, Association for Symbolic Logic (2002)
50. Fine, T.: Theories of Probability. Academic Press, Cambridge (1973)
51. Fréchet, M.: Généralizations du théoréme des probabilités totales. Fundamenta mathematicae **25**, 379–387 (1935)
52. Frege, G.: Begriffsschrift: eine der arithmetischen nachgebildete Formelsprache des reinen Denkens. Halle (1879)
53. Gaifman, H.: Concerning measures in first order calculi. Isr. J. Math. **2**, 1–18 (1964)
54. Gaifman, H., Snir, M.: Probabilities over rich languages, testing and randomness. J. Symb. Logic **47**(3), 495–548 (1982)
55. Galavotti, M.C.: The moder epistemic interpretations of probability: logicism and subjectivism. In: Gabbay, D.M., Hartmann, S., Woods, J. (eds.), Handbook of the History of Logic, Inductive logic. vol. 10, p. 153 (2011)
56. Glymour, C, Eberhardt, F., Reichenbach, H., In: Edward N.Z. (ed.), The Stanford Encyclopedia of Philosophy (2014). http://plato.stanford.edu/archives/fall2014/entries/reichenbach/
57. Gödel, K.: Über die Vollstandigkeit des Logikkalkulus. Ph.D. thesis. University of Viena(1929)
58. Hacking, I.: Leibniz and descartes: proof and eternal truths. In: Proceedings of the British Academy, vol. 59, pp. 175–188 (1973). http://www.britac.ac.uk/pubs/proc/files/59p175.pdf
59. Hacking, I.: Infinite analysis. Studia Leibnitiana **6**(1), 126–130 (1974)
60. Hacking, I.: Emergence of Probability. A philosophical study of earlu ideas about probability induction and statistical ineference. Cambridge University Press, Cambridge (2006)
61. Hailperin, T.: Best possible inequalities for the probability of a logical function of events. Am. Math. Mon. **72**(4), 343–359 (1965)
62. Hailperin, T.: Booles Logic and Probability, 2nd edn. North-Holland, Amsterdam (1986)
63. Hailperin, T.: Sentential Probability Logic. Origins, development, current status and technical applications. Lehigh univeristy press, Bethlehem (1996)
64. Hailperin, T.: Logic with a Probability Semantics. Lehigh univeristy press, Bethlehem (2011)
65. Hájek, A.: Probability, Logic, and Probability Logic. In: Goble, L. (ed.), The Blackwell Companion to Logic, pp. 362–384. Blackwell (2001)
66. Hájek, A.: In: Edward N. Zalta (ed.) Interpretations of Probability in The Stanford Encyclopedia of Philosophy (Winter 2012 Edition) (2012). http://plato.stanford.edu/archives/win2012/entries/probability-interpret/
67. Hald, A.: A History of Parametric Statistical Inference from Bernoulli to Fisher, 1713 to 1935. Springer, Berlin (2007)
68. Hamblin, C.: The modal probably. Mind **68**, 234–240 (1959). http://www.jstor.org/stable/2251572
69. Hansen, P., Jaumard, B.: Probabilistic satisfiability. In: Gabbay, D., Smets, P. (eds.) Handbook of Defeasible Reasoning and Uncertainty Management Systems, vol.5, pp. 321–367. Springer (2000)
70. Henkin, L.: The completeness of the first-order functional calculus. J. Symb. Logic **14**, 159–166 (1949)
71. Hoover, D.N.: Probability logic. Ann. Math. logic **14**, 287–313 (1978)

72. Huygens, C.: De ratiociniis in ludo aleae (On Reasoning in Games of Chance). Appendix to the book: Frans van Schooten, Exercitationum Mathematicarum Libri Quinque. Leiden (1657). http://doi.library.cmu.edu/10.1184/OCLC/01297338
73. Huygens, C.: Libellus de ratiociniis in ludo aleae, or, The Value of all chances in games of fortune; Cards, dice, wagers, lotteries, &c. Mathematically Demonstrated. London, Printed by S. Keimer for T. Woodward, near the Inner-Temple-Gate in Fleetstreet (1714)
74. Jaynes, E.T.: Probability Theory: The Logic of Science. Cambridge university press, Cambridge (2003)
75. Jeffreys, H.: The Theory of Probability. Oxford (1939)
76. Keisler, H.J.: Hyperfinite model theory. In: Gandy, R.O., Hyland, J.M.E. (eds.), Logic Colloquim, vol. 76, pp. 5–110. North-Holland (1977)
77. Keisler, H.J.: Probability quantifiers. In: Barwise, J., Feferman, S. (eds.) Model Theoretic Logics, pp. 509–556. Springer, Berlin (1985)
78. Keynes, J.M.: Treatise on Probability. Macmillan & Co, London (1921). http://central.gutenberg.org/articles/a_treatise_on_probability
79. Kolmogorov, A.N.: Grundbegriffe der Wahrscheinlichkeitsrechnung (English edition: Foundations of the Theory of Probability, New York, 1950), Springer, Berlin (1933). https://archive.org/details/foundationsofthe00kolm
80. Kraus, M.: Early Greek Probability Arguments and Common Ground in Dissensus. In: Ontario Society for the Study of Argumentation (OSSA) Proceedings, OSSA Conference Archive (2007). http://scholar.uwindsor.ca/ossaarchive/OSSA7/papersandcommentaries/92
81. Kripke, S.: A completeness theorem in modal logic. J. Symb. Logic **24**, 1–14 (1959)
82. Kulstad, M., Carlin. L.: Leibniz's philosophy of mind. In: Edward N.Z. (ed.), The Stanford Encyclopedia of Philosophy (2013). http://plato.stanford.edu/archives/win2013/entries/leibniz-mind/
83. Lambert, J.H.: Neues organon oder gedanken über die erforschung und bezeichnung des wahren und der unterscheidung von irrtum und schein (Novum Organon) (1764). http://www.deutschestextarchiv.de/book/view/lambert_organon02_1764?p=5
84. Laplace, P.-S.: Mémoire sur la probabilité des causes par les événements (Memoir on the probability of causes of events). Mémoires de l'Académie royale des Sciences de Paris, vol. 6, pp. 621–656 (1774)
85. Laplace, P.-S.: Sur le principe de la gravitation universelle et sur les inégalits séculaires des planets qui en dependent. OC (8), 20178 (1776)
86. Laplace, P.-S.: Essai philosophique sur les probabilités (1814). https://books.google.rs/books?id=rDUJAAAAIAAJ&dq=Essai+philosophique+sur+les+Probabilit%C3%A9s&pg=PP12&redir_esc=y#v=onepage&q&f=false
87. Laplace, P.-S.: Théorie analytique des probabilités (1812, 1814, 1820). https://archive.org/details/theorieanaldepro00laplrich
88. Laplace, P.-S.: Essai philosophique sur les probabilités (An English translation by Truscott, F.W., Emory, F.L.) J. Wiley, Chapman & Hall (1902). https://archive.org/details/philosophicaless00lapliala
89. Lebesgue, H.: Sur une généralisation de l'intégrale définie. Comptes rendus des Séances de l'Académie des Sciences **132**, 1025–1028 (1901)
90. Leibnitz, G.: De Conditionibus. Bachelor thesis. Leipzig (1665)
91. Leibnitz, G.: Specimen certitudinis seu demonstrationum in jure, exhibitum in doctrina conditionum (1665)
92. Leibnitz, G.: Dissertatio de arte combinatoria. Leipzig (1666)
93. Leibnitz, G.: De incerti aestimatione (1678)
94. Leibnitz, G.: Nova Methodus pro Maximis et Minimis, Itemque Tangentibus, qua nec Fractas nec Irrationales Quantitates Moratur, et Singulare pro illi Calculi Genus (New Method for the Greatest and the Least). Acta Eruditorum X, pp. 467–473 (1684). http://www.maa.org/book/export/html/641727
95. Leibnitz, G.: The Art of Discovery (1685)

96. Leibnitz, G.: De geometria recondita et analysi indivisibilium atque infinitorum (On a hidden geometry and analysis of indivisibles and infinites). Acta Eruditorum VI, 292–300. http://www.maa.org/book/export/html/641727, 1686
97. Leibnitz, G.: Die philosophischen Schriften. In: von Leibniz, G.W., Gerhardt, C.I. (eds.) Berlin (1875–1890)
98. Leibnitz, G.: Sädmtliche Schriften und Briefen. Preussischen Akademie der Wissenschaften (1923)
99. Leibnitz, G.: In: Grua, G. (ed.), Textes inédits. Paris (1948)
100. Leibnitz, G.: New Essays on Human Understanding. Cambridge University Press, Cambridge (1981). Translated by Peter Remnant and Jonathan Bennett
101. Lenzen, W.: Leibniz's logic. In: Gabbay, D.M., Woods, J. (eds.) The Rise of Modern Logic: From Leibniz to Frege, pp. 1–83. Elsevier, Amsterdam (2004)
102. Lévy, P.: Calcul des probabilités. Gauthier-Villars, Paris (1925)
103. Lewis, D.: Subjectivists guide to objective chance. In: Jeffrey, R. (ed.), Studies in Inductive Logic and Probability, Vol. II, pp. 263–293. University of California Press (1980)
104. MacColl, H.: Symbolic Logic and its Applications. Longmans. Green and Co., London (1906). https://archive.org/details/symboliclogicand00maccuoft
105. MacColl, H.: If and Imply. Mind **17**, 151–152 (1908)
106. MacColl, H.: If and Imply. Mind **17**, 453–455 (1908)
107. Macfarlane, A.: Principles of the Algebra of Logic, With Examples. Edinburg (1879). https://archive.org/details/cu31924031227386
108. Markov, A.A.: Generalization of the law of large numbers to dependet quantities. Izvestia fiziko-matematičeskogo obschestva pri Kazanskom Universitete, ser. 2, tom 15, pp. 135–156 (1906)
109. Markov, A.A.: Ischislenie Verovatnostei (The calculus of probability) (1913). (other editions: 1900, 1908, 1924)
110. McGrayne, S.B.: The Theory That Would Not Die: How Bayes Rule Cracked the Enigma Code, Hunted Down Russian Submarines, and Emerged Triumphant from Two Centuries of Controversy. Yale University Press, New Haven (2011)
111. Molina, E.C.: The theory of probability: some comments on Laplace's théorie analytique. Bull. Am. Math. Soc. **36**(6), 369–392 (1930)
112. Nilsson, N.: Probabilistic logic. Artif. Intell. **28**, 71–87 (1986)
113. Nilsson, N.: Probabilistic logic revisited. Artif. Intell. **59**, 39–42 (1993)
114. Pascal, B.: Pensées de M. Pascal sur la religion, et sur quelques autres sujets (Thoughts of M. Pascal on religion, and on some other subjects) (1669)
115. Peirce, C.S.: On an improvement in Booles calculus of logic. In: Proceedings of the American Academy of Arts and Sciences, vol. 7, pp. 250–261 (In: Collected Papers of Charles Sanders Peirce, vol. III Exact Logic, Harvard University Press (1933)) (1867)
116. Peirce, C.S.: Notes on the doctrine of chances. In: Collected Papers of Charles Sanders Peirce. Elements of logic, vol. II. Harvard University Press (1932)
117. Pölya, G.: Mathematics and Plausible Reasoning (volumes I and II). Princeton University Press, New York (1954). https://archive.org/details/Patterns_Of_Plausible_Inference_2_
118. Popper, K.: The propensity interpretation of the calculus of probability and the quantum theory. In: Körner, S. (ed.), Observation and Interpretation: A Symposium of Philosophers and Physicists: Proceedings of the Ninth Symposium of the Colston Research Society, pp. 65–70 (1957)
119. Popper, K.: The propensity interpretation of probability. Br. J. Philos. Sci. **10**, 25–42 (1959)
120. Poretskiy, P.S.: Solution to the General Problem of the Theory of Probability by Using Mathematical Logic. Kazan (1887)
121. Rahman, S., Redmond, J.: Hugh MacColl and the birth of logical pluralism. In: Gabbay, D.M., Woods, J. (eds.), Handbook of the History of Logic, vol. 4, pp. 533–604 (2008)
122. Ramsey, F.P.: Truth and probability. In: Braithwaite, R.B., Ramsey, F.P. (eds.) The Foundations of Mathematics and other Logical Essays, Ch. VII, 156–198, 1931, 1999 electronic edition. Harcourt, Brace and Company, London: Kegan, Paul, Trench, Trubner & Co., New York (1926). https://core.ac.uk/download/pdf/7048428.pdf

123. Rašković, M.: Model theory for L_{AM} logic. Publications de L'Institut Mathematique **37**(51), 17–22 (1985). http://elib.mi.sanu.ac.rs/files/journals/publ/57/n051p017.pdf
124. Rašković, M.: Completeness theorem for biprobability models. J. Symb. Logic **51**(3), 586–590 (1986)
125. Rašković, M.: Weak completeness theorem for $L_{AP\forall}$. Zbornik radova PMF Kragujevac **8**, 69–72 (1987)
126. Rašković, M.: Completeness theorem for singular biprobability models. Proc. Am. Math. Soc. **102**, 389–392 (1988)
127. Rašković, M.: Completeness theorem for a monadic logic with both ordinary first-order and probability quantifiers. Publications de L'Institut Mathematique Ns. **47**(61), 1–4 (1990). http://elib.mi.sanu.ac.rs/files/journals/publ/67/n061p001.pdf
128. Rašković, M.: A completeness theorem for an infinitary intuitionistic logic with both ordinary and probability quantifiers. Publications de L'Institut Mathematique Ns **50**(64), 7–13 (1991). http://elib.mi.sanu.ac.rs/files/journals/publ/70/n064p007.pdf
129. Rašković, M., Đorđevic, R.: Second order probability logic. Math. Balkanica **6**, 105–108 (1992)
130. Rašković, M., Dordevic, R.: Probability Quantifiers and Operators. VESTA, Beograd (1996). http://elibrary.matf.bg.ac.rs/bitstream/handle/123456789/189/phdRadosavSDjordjevic.pdf?sequence=1
131. Rašković, M., Živaljević, R.: Barwise completeness theorem for some biprobability logics. Zeitschrift für mathematische Logik und Grundlagen der Mathematik **31**, 133–135 (1986)
132. Reichenbach, H.: The theory of probability. An Inquiry into the Logical and Mathematical Foundations of the Calculus of Probability (the first edition: Reichenbach, H.: Wahrscheinlichkeitslehre. Eine Untersuchung über die Logischen nd Mathematischen Grundlagen der Wahrscheinlichkeitsrechnung, Leiden (1935)). University of California Press (1949)
133. Rényi, A.: Foundations of Probability. Holden-Day, Boca Raton (1970)
134. Rescher, N.: Many-Valued Logic. McGraw-Hill, New York (1969)
135. Robinson, A.: Non-standard Analysis. North-Holland, Amsterdam (1966)
136. Rusnock, P., George, R.: Bolzano as logician. In: Gabbay, D.M., Woods, J. (eds.) The Rise of Modern Logic: From Leibniz to Frege, pp. 3–177. Elsevier, Amsterdam (2004)
137. Savage, L.: Foundations of Statistics. Wiley, New York (1954)
138. Schneider, I.: Christiaan Huygens (version 3). In: StatProb: The Encyclopedia Sponsored by Statistics and Probability Societies. http://statprob.com/encyclopedia/ChristiaanHUYGENS.html
139. Schneider, I.: Jakob Bernoulli (version 3). In: StatProb: The Encyclopedia Sponsored by Statistics and Probability Societies. http://statprob.com/encyclopedia/JakobBERNOULLI.html
140. Schneider, I.: Leibniz on the probable. In: Joseph W. D. (ed.), Mathematical Perspectives. Essays on Mathematics and its Historical Development, pp. 201–219. Elsevier Inc, Academic Press (1981)
141. Schneider, I.: The role of Leibniz and of Jakob Bernoulli for the development of probability theory. LLULL, Boletin de la Sociedad Española de Historia de las. Ciencias **7**, 68–89 (1984)
142. Schneider, I.: Christiaan Huygens' non-probabilistic approach to a calculus of games of chance. De Zeventiende Eeuw **12**(1), 171–185 (1996)
143. Schoenflies, A.: Die Entwickelung der Lehre von den Punktmannigfaltigkeiten. Jahresbericht der Deutschen Mathematiker-Vereinigung **8**, 1–250 (1900). http://gdz.sub.uni-goettingen.de/dms/resolveppn/?PPN=GDZPPN002116928
144. Scott, D., Krauss, P.: Assigning probabilities to logical formulas. In: Hintikka, J., Suppes, P. (eds.), Aspects of Inductive Logic, pp. 219– 264. North-Holland (1966)
145. Segerberg, K.: Qualitative probability in a modal setting. In: Fenstad, E. (ed.) Proceedings 2nd Scandinavian Logic Symposium, pp. 341–352. North-Holland, Amsterdam (1971)
146. Seneta, E.W.: On the history of the strong law of large numbers and Boole's inequality. Historia Mathematica **19**, 24–39 (1992)
147. Seneta, E.W.: A tricentenary history of the law of large numbers. Bernoulli **19**(4), 1088–1121 (2013)

148. Shafer, G.: A Mathematical Theory of Evidence. Princeton University Press, New Jersey (1976)
149. Shafer, G.: Bayes's two arguments for the rule of conditioning. Ann. Stat. **10**(4), 1075–1089 (1982)
150. Shafer, G.: In: Grattan-Guinness, I. (ed.) The Early Development of Mathematical Probability in Encyclopedia of the History and Philosophy of the Mathematical Sciences. Routledge, London (1991)
151. Shafer, G., Vovk, V.: The sources of Kolmogorovs Grundbegriffe. Stat. Sci. **21**(1), 70–98 (2006). www.probabilityandfinance.com/articles/04.pdf
152. Shafer, G.: Non-additive probabilities in the work of Bernoulli and Lambert. Archive for History of Exact Sciences, vol. 19, pp. 309–370 (1978). (Reprinted as Chapter 6 of Classic Works of the Dempster-Shafer Theory of Belief Function, Yager, R., Liping L. (eds.) Springer (2008))
153. Šikić, Z.: What is probability and why does it matter. Eur. J. Anal. Philos. **10**(1), 21–43 (2014)
154. Simons, P.: MacColl and many-valued logic: an exclusive conjunction. Nordic J. Philos. Logic **3**(1), 85–90 (1999). http://www.hf.uio.no/ifikk/forskning/publikasjoner/tidsskrifter/njpl/vol3no1/manyval.pdf
155. Skyrms, B.: Causal Necessity. Yale University Press, New Haven (1980)
156. Solovay, R.M., Strassen, V.: A fast Monte–Carlo test for primality. SIAM J. Comput. **6**, 8–85 (1977)
157. Solovay, R.M., Strassen, V.: Erratum: a fast Monte–Carlo test for primality. SIAM J. Comput. **7**(1), 118 (1978)
158. Stigler, S.M.: Laplace's 1774 memoir on inverse probability. Stat. Sci. **1**(3), 359–363 (1986). http://www.jstor.org/stable/2245475
159. Styazhkin, N.I.: History of Mathematical Logic from Leibnitz to Peano. MIT Press, Cambridge (1969)
160. Suppes, P.: Probabilistic inference and the concept of total evidence. In: Hintikka, J., Suppes , P. (eds.), Aspects of Inductive Logic, pp. 49–65. North-Holland (1966)
161. Sylla, E.: The emergence of mathematical probability from the perspective of the Leibniz–Jacob Bernoulli correspondence. Perspect. Sci. **6**(1 & 2), 41–76 (1998)
162. Tarski, A.: Logic, Semantics, Metamathematics: Papers from 1923 to 1938. Clarendon Press, Oxford (1956)
163. Tartaglia, N.F.: Generale Trattato di numeri, et misure (General Treatise). (1556–1560)
164. Todhunter, I.: A History of the Mathematical Theory of Probability. Macmillan, London (1865)
165. Uspenski, Y.V.: Introduction to Mathematical Probability. McGraw-Hill, New York (1937)
166. Velasco, J.: The Failure of Leibniz's Infinite Analysis view of Contingency. http://philpapers.org/go.pl?id=VELTFO&proxyId=&u=http%3A%2F%2Fjoelvelasco.net%2FPapers%2FVelascoLeibnizcontingency.pdf
167. Venn, J.: The Logic of Chance, 3rd edn. McMillan, London (1888). https://archive.org/details/logicofchance029416mbp (1866, 1876)
168. Venn, J.: The Principles of Empirical or Inductive Logic, 3rd edn. McMillan, London (1889). https://archive.org/details/principlesempir00venngoog
169. von Mises, R.: Grundlagen der Wahrscheinlichkeitsrechnung (The Foundations of Probability). Mathematische Zeitschrift **5**, 52–99 (1919)
170. von Mises, R.: Wahrscheinlichkeit, statistik und wahrheit (Probability, Statistics and Truth), 2nd edn. Dover, New York (1981). Springer, 1928
171. von Plato, J.: Creating Modern Probability. Physics and Philosophy in Historical Perspective. Cambridge University Press, Cambridge (1994). Its Mathematics
172. Whitehead, A.N., Russell, B.: Principia Mathematica. Cambridge University Press, Cambridge (1910,1912,1913). (in 3 volumes)
173. Wigner, E.: On the quantum correction for thermodynamic equilibrium. Phys. Rev. **40**(5), 749–759 (1932)

174. Wiman, A.: Über eine Wahrscheinlichkeitsaufgabe bei Kettenbruchentwicklungen. Öfversigt af Kongliga Vetenskaps-Akademiens Förhandlingar. Femtiondesjunde Årgången **57**(7), 829–841 (1900)
175. Zabell, S.: De Morgan and Laplace: A tale of two cities. Electron. J. Hist. Probab. Stat. **8** (2012). www.jehps.net

Chapter 3
LPP_2, a Propositional Probability Logic Without Iterations of Probability Operators

Abstract The probability logic denoted LPP_2 is described with the aim to give a clear, step-by-step introduction to the field and the main proof techniques that will be used elsewhere in the book. The logic enriches propositional calculus with probabilistic operators of the form $P_{\geq s}$ with the intended meaning "probability is at least s". In LPP_2 the operators are applied to propositional formulas, while iterations of probability operators are not allowed. Possible world semantics with a finitely additive probability measure on sets of worlds definable by formulas is defined, so that formulas remain true or false. The corresponding axiomatization is provided. The axiom system is infinitary. It contains an infinitary rule with countable many premisses and one conclusion. The rule is related to the Archimedean property of real numbers. The logic LPP_2 is not compact: there are unsatisfiable sets of formulas that are finitely satisfiable. Some of the consequences of non-compactness are described. Then, soundness and strong completeness of the logic is proved with respect to several classes of probability models. This is followed by a proof of decidability of PSAT, the satisfiability problem for LPP_2, which is NP-complete. Finally, a heuristic approach to PSAT is presented. This Chapter covers some results from Ikodinović et al., Int J Approx Reason, (55):1830–1842, 2014, [3], Jovanović et al., Variable neighborhood search for the probabilistic satisfiability problem, 2007, [4], Kokkinis et al., Logic J. IGPL, (23):662–687, 2015, [5], Ognjanović, J. Logic Comput, (9):181–195, 1999, [6], Ognjanović et al., Theor Comput Sci, (247):191–212, 2000, [7], Ognjanović et al., A genetic algorithm for satisfiability problem in a probabilistic logic, 2001, [8],Ognjanović et al., A Genetic Algorithm for Probabilistic SAT Problem, 2004, [9], Ognjanović et al., A Hybrid Genetic and Variable Neighborhood Descent for Probabilistic SAT Problem, 2005, [10], Ognjanović et al., Zbornik Radova, Subseries Logic in Computer Science, 2009, [11], Rašković and Ognjanović, Some propositional probabilistic logics, 1996, [12], Rašković and Ognjanović, A first order probability logic, LP_Q, 1999, [13], Stojanović et al., Appl Soft Comput, (31):339–347, 2015, [14].

Z. Ognjanović et al., *Probability Logics*,
DOI 10.1007/978-3-319-47012-2_3

3.1 Syntax and Semantics

3.1.1 Syntax

Let $[0, 1]_{\mathbb{Q}}$ be the set of all rational numbers from the real unit interval $[0, 1]$. The symbols of the language of LPP_2 are

- primitive propositions from the denumerable set $\phi = \{p, q, r, \ldots\}$,
- classical propositional connectives $\neg$, and $\wedge$, and
- a list of probability operators $P_{\geq s}$ for every $s \in [0, 1]_{\mathbb{Q}}$.

The set For_C of all classical propositional formulas over the set ϕ is the smallest set

- containing all primitive propositions, i.e., $p \in \mathrm{For}_C$, for every $p \in \phi$, and
- closed under the formation rules: if $\alpha, \beta \in \mathrm{For}_C$, then $\neg\alpha, \alpha \wedge \beta \in \mathrm{For}_C$.

The formulas from the set For_C will be denoted by α, β, ..., indexed if necessary. Probability formulas are defined as follows:

Definition 3.1 If $\alpha \in \mathrm{For}_C$ and $s \in [0, 1]_{\mathbb{Q}}$, then

$$P_{\geq s}\alpha$$

is a *basic probability formula*.

The set For_P of all probability formulas is the smallest set

- containing all basic probability formulas, and
- closed under the formation rules: if $A, B \in \mathrm{For}_P$, then $\neg A, A \wedge B \in \mathrm{For}_P$.

The set of all LPP_2-formulas is $\mathrm{For}_{LPP_2} = \mathrm{For}_C \cup \mathrm{For}_P$. ■

The intended meaning of $P_{\geq s}\alpha$ is "the probability of α is at least s." The formulas from the sets For_P and For_{LPP_2} will be denoted by $A, B, \ldots$, and $\varphi, \psi, \ldots$, respectively, and indexed if necessary.

We use the usual abbreviations for the other classical connectives, and also denote

- $\neg P_{\geq s}\alpha$ by $P_{<s}\alpha$,
- $P_{\geq 1-s}\neg\alpha$ by $P_{\leq s}\alpha$,
- $\neg P_{\leq s}\alpha$ by $P_{>s}\alpha$,
- $P_{\geq s}\alpha \wedge P_{\leq s}\alpha$ by $P_{=s}\alpha$, and
- both $\alpha \wedge \neg\alpha$ and $A \wedge \neg A$ by $\bot$, letting the context determine the meaning.

As it can be seen from Definition 3.1, neither mixing of pure propositional formulas and probability formulas, nor nested probability operators are allowed in LPP_2. Thus

- $\alpha \wedge P_{\geq s}\beta$ and
- $P_{\geq s}P_{\geq r}\alpha$

do not belong to For_{LPP_2}. It means that in LPP_2 we can formally reason about probabilities of formulas, but cannot express higher order probabilities, i.e., probabilities of probabilities of formulas.

3.1.2 Semantics

The semantics for For_{LPP_2} will be based on the possible world approach.

Definition 3.2 An *LPP_2-model* is a structure $\mathbf{M} = \langle W, H, \mu, v\rangle$ where:

- W is a nonempty set of objects called worlds,
- H is an algebra of subsets of W,
- μ is a finitely additive probability measure, $\mu : H \rightarrow [0, 1]$, and
- $v : W \times \phi \rightarrow \{\mathit{true}, \mathit{false}\}$ provides for each world $w \in W$ a two-valued valuation of the primitive proposition. ■

For each world $w \in W$, the truth valuation $v(w, \cdot)$ is extended to all classical propositional formulas from For_C as usual.

If $\mathbf{M}$ is an LPP_2-model and $\alpha \in \text{For}_C$, the set of all worlds in which α is true, $\{w : v(w, \alpha) = \mathit{true}\}$, is denoted by $[\alpha]_{\mathbf{M}}$. We will omit the subscript $\mathbf{M}$ from $[\alpha]_{\mathbf{M}}$ and write $[\alpha]$ if $\mathbf{M}$ is clear from the context. Instead of $v(w, \alpha) = \mathit{true}$ and $v(w, \alpha) = \mathit{false}$, we will occasionally write $w \models \alpha$, and $w \not\models \alpha$, respectively.

Definition 3.3 An LPP_2-model $\mathbf{M} = \langle W, H, \mu, v\rangle$ is *measurable* if $[\alpha]_{\mathbf{M}} \in H$ for every formula $\alpha \in \text{For}_C$. The class of all measurable LPP_2-models is denoted by $LPP_{2,\text{Meas}}$. ■

In this book we focus on the class of measurable models.

Example 3.1 Let us consider the a finite set of primitive propositions $\phi = \{p, q, r\}$ and the following structure $\mathbf{M} = \langle W, H, \mu, v\rangle$:

- $W = \{u, t, w\}$,
- H is the power set $\mathbb{P}(W)$,
- $\mu(\emptyset) = 0$, $\mu(u) = \mu(t) = \frac{2}{5}$, $\mu(t) = \frac{1}{5}$, $\mu(\{u, t\}) = \frac{4}{5}$, $\mu(\{u, w\}) = \mu(\{t, w\}) = \frac{3}{5}$, and $\mu(W) = 1$, and
- $v(u, p) = v(u, q) = v(u, \neg r) = \mathit{true}$, $v(t, p) = v(t, \neg q) = v(t, r) = \mathit{true}$, and $v(w, p) = v(w, q) = v(w, r) = \mathit{true}$.

The reader can easily check that $\mathbf{M}$ is an LPP_2-model. Moreover, since μ is defined for all subsets of W, for every $\alpha \in \text{For}_C$, the corresponding set $[\alpha]$ is measurable, so $\mathbf{M}$ is an $LPP_{2,\text{Meas}}$-model. ■

Definition 3.4 *The satisfiability relation* $\models \subseteq LPP_{2,\text{Meas}} \times \text{For}_{LPP_2}$ fulfills the following conditions for every $LPP_{2,\text{Meas}}$-model $\mathbf{M} = \langle W, H, \mu, v\rangle$:

- if $\alpha \in \text{For}_C$, $\mathbf{M} \models \alpha$ iff for every $w \in W$, $v(w, \alpha) = \mathit{true}$,
- if $\mathbf{M} \models P_{\geq s}\alpha$ iff $\mu([\alpha]) \geq s$,
- if $A \in \text{For}_P$, $\mathbf{M} \models \neg A$ iff $\mathbf{M} \not\models A$,
- if $A, B \in \text{For}_P$, $\mathbf{M} \models A \wedge B$ iff $\mathbf{M} \models A$ and $\mathbf{M} \models B$. ■

Concerning expressiveness of our formal language, note that it is not possible to directly say that the probabilities of two formulas are equal. If the probabilities of those formulas is the same rational number, we can write for example $P_{=s}\alpha \leftrightarrow P_{=s}\beta$, but, since our formal language is countable, it is not possible to do that for irrational probabilities. However, those equal probabilities can be described using a set of formulas

$$\{P_{\geq s}\alpha \leftrightarrow P_{\geq s}\beta, P_{\geq s}\beta \leftrightarrow P_{\geq s}\alpha : s \in [0,1]_{\mathbb{Q}}\}.$$

Similarly, we can say that the probability of a formula is not lesser, or not greater, than the probability of another formula. On the other hand, our strong hypothesis is that the predicate $Pr(\alpha) < Pr(\beta)$ cannot be represented as an LPP_2-theory.

Now, we introduce the notions of satisfiable and valid formulas, and satisfiable sets of formulas.

Definition 3.5 A formula $\varphi \in \mathrm{For}_{LPP_2}$ is *satisfiable* if there is an $LPP_{2,\mathrm{Meas}}$-model $\mathbf{M}$ such that $\mathbf{M} \models \varphi$; φ is *valid* (denoted $\models \varphi$) if for every $LPP_{2,\mathrm{Meas}}$-model $\mathbf{M}$, $\mathbf{M} \models \varphi$.

A set T of For_{LPP_2}-formulas is satisfiable if there is an $LPP_{2,\mathrm{Meas}}$-model $\mathbf{M}$ such that $\mathbf{M} \models \varphi$ for every $\varphi \in T$ (denoted $\mathbf{M} \models T$). ■

If $\mathbf{M} \models \varphi$, we say that $\mathbf{M}$ is a model of φ.

Example 3.2 Let us consider the model $\mathbf{M}$ described in Example 3.1. We can see, for example, that the following holds:

- $u \models p$ and $u \models \neg r$, so $u \models p \wedge \neg r$,
  $t \models p$ and $t \models r$, so $t \models p \wedge r$ and $t \not\models p \wedge \neg r$,
  since $u \models p \wedge \neg r$, and $t \not\models p \wedge \neg r$, we have that $M \not\models p \wedge \neg r$.
- Thus, $M \models P_{\geq 1}p$, and $M \not\models P_{\geq 1}(p \wedge \neg r)$.

The reader can also check that, for example, since $[p \wedge r] = \{t, w\}$, we have that $M \models P_{=\frac{3}{5}}(p \wedge r)$. Similarly, since $[q] = \{u, w\}$, and $[p \wedge \neg q] = \{t\}$:

- $M \models P_{<\frac{4}{5}}q$
- $M \not\models P_{>\frac{2}{5}}(p \wedge \neg q)$, etc. ■

Note that the classical formul as do not behave in the usual way.

- For some $\alpha, \beta \in \mathrm{For}_C$ and an $LPP_{2,\mathrm{Meas}}$-model $\mathbf{M}$ it can be $\mathbf{M} \models \alpha \vee \beta$, but that neither $\mathbf{M} \models \alpha$, nor $\mathbf{M} \models \beta$.
- It can be simultaneously $\mathbf{M} \not\models \alpha$ and $\mathbf{M} \not\models \neg\alpha$.

Nevertheless, the set of all classical formulas that are valid with respect to the above given semantics and the set of all classical valid formulas coincide, because every world from an arbitrary $LPP_{2,\mathrm{Meas}}$-model can be seen as a classical propositional interpretation.

Example 3.3 Let us consider the model $\mathbf{M} = \langle\{u, w\}, \{\emptyset, \{u\}, \{w\}, \{u, w\}\}, \mu, v\rangle$ such that

- $\mu(\{u\}) = \mu(\{w\}) = \frac{1}{2}$, and
- $u \models p$, and $w \models \neg p$, for a $p \in \phi$.

Then, it is easy to see that: $M \not\models p$ (since $w \models \neg p$), and $M \not\models \neg p$ (since $u \models p$), while obviously $M \models p \vee \neg p$. ■

Definition 3.6 A formula $\varphi \in \mathrm{For}_{LPP_2}$ is a *(semantical) consequence* of a set T of For_{LPP_2}-formulas, denoted

$$T \models \varphi$$

if every model of T is a model of φ. ■

In the sequel we will also consider the following classes of LPP_2-models

- $LPP_{2,\mathrm{Meas},\mathrm{All}}$,
- $LPP_{2,\mathrm{Meas},\sigma}$ and
- $LPP_{2,\mathrm{Meas},\mathrm{Neat}}$.

A model $\mathbf{M} = \langle W, H, \mu, v\rangle$ belongs to the first class if $H = \mathbb{P}(W)$, i.e., if every subset of W is μ-measurable. A model $\mathbf{M}$ belongs to the second class if it is a σ-additive measurable model, i.e., if μ is a σ-additive probability measure. Finally, a model $\mathbf{M}$ belongs to the third class if it is a measurable model such that $\mu(H_1) = 0$ iff $H_1 = \emptyset$, i.e., if only the empty set has the zero probability.

3.1.3 Atoms

Let $\varphi \in \mathrm{For}_{LPP_2}$ be a formula and $\{p_1, \ldots, p_n\}$ be the set of all primitive propositions that appear in φ. An *atom* a of φ is a conjunction $\pm p_1 \wedge \ldots \wedge \pm p_n$, where $\pm p_i$ is either p_i, or $\neg p_i$, i.e., an atom contains, for every p_i, either p_i itself or its negation. The set of all atoms of φ is denoted by $\mathrm{Atoms}(\varphi)$. Note that

- $|\mathrm{Atoms}(\varphi)| = 2^n \leq 2^{\mathrm{len}(\varphi)}$.
- Different atoms are mutually exclusive, i.e., for $a, b \in \mathrm{Atoms}(\varphi)$, if a and b are different, then $\models \neg(a \wedge b)$.
- Every classical propositional formula $\alpha \in \mathrm{For}_C$ is equivalent to its complete disjunctive normal form $\mathrm{CDNF}(\alpha)$, i.e., to a disjunction of some atoms from $\mathrm{Atoms}(\alpha)$.

When we analyze a classical propositional formula $\alpha \in \mathrm{For}_C$ in a world w of a model $\mathbf{M} = \langle W, H, \mu, v\rangle$, w is completely specified by an atom $a_w \in \mathrm{Atoms}(\alpha)$ which contains all p_i's from α such that $w \models p_i$, and all $\neg p_j$'s such that $w \not\models p_j$. Then, $w \models \alpha$ iff $a_w \in \mathrm{CDNF}(\alpha)$. In other words:

- $[\alpha] = \cup_{a \in \mathrm{CDNF}(\alpha)}[a]$,
- if $a, b \in \mathrm{Atoms}(\alpha)$ are different, then $[a] \cap [b] = \emptyset$, and
- $\mu([\alpha]) = \sum_{a \in \mathrm{CDNF}(\alpha)} \mu([a])$.

Later in this text we will describe probability models by specifying probabilities of the sets of the form $[a]$, where a is an atom, with the condition that the sum of those probabilities is 1.

3.2 Complete Axiomatization

The set of all $LPP_{2,\text{Meas}}$-valid formulas can be characterized by the following set of axiom schemata:

1. all For_C-substitutional instances and all For_P-substitutional instances of the classical propositional tautologies
2. $P_{\geq 0}\alpha$
3. $P_{\leq r}\alpha \rightarrow P_{<s}\alpha$, $s > r$
4. $P_{<s}\alpha \rightarrow P_{\leq s}\alpha$
5. $(P_{\geq r}\alpha \wedge P_{\geq s}\beta \wedge P_{\geq 1}(\neg(\alpha \wedge \beta))) \rightarrow P_{\geq \min(1,r+s)}(\alpha \vee \beta)$
6. $(P_{\leq r}\alpha \wedge P_{<s}\beta) \rightarrow P_{<r+s}(\alpha \vee \beta)$, $r + s \leq 1$

and inference rules:

1. From φ and $\varphi \rightarrow \psi$ infer ψ.
2. From α infer $P_{\geq 1}\alpha$.
3. From $A \rightarrow P_{\geq s-\frac{1}{k}}\alpha$, for every integer $k \geq \frac{1}{s}$, and $s > 0$, infer $A \rightarrow P_{\geq s}\alpha$.

We denote this axiom system by Ax_{LPP_2}.

Let us now discuss the above axioms and rules. First note that, by Axiom 1, the classical propositional logic is a sublogic of LPP_2, i.e., all classical propositional tautologies are $LPP_{2,\text{Meas}}$-valid.

The axioms 2–6 concern the probabilistic aspect of LPP_2. Axiom 2 announces that every formula is satisfied by a set of worlds of the measure at least 0. By substituting $\neg\alpha$ for α in the axiom, the formula $P_{\geq 0}\neg\alpha$ is obtained. According to our definition of the operator $P_{\leq 1}$, we have the following instance of Axiom 2:
$2'.P_{\leq 1}\alpha$ $(= P_{\geq 1-s}\neg\alpha$, for $s = 1)$.
It forces that every formula is satisfied by a set of worlds of the measure at most 1, and gives the upper bound for probabilities of formulas in $LPP_{2,\text{Meas}}$-models. In a similar way, the axioms 3 and 4 are equivalent to
$3'.P_{\geq t}\alpha \rightarrow P_{>s}\alpha$, $t > s$
$4'.P_{>s}\alpha \rightarrow P_{\geq s}\alpha$
respectively. The axioms 5 and 6 correspond to the additivity of measures. Suppose that α and β are disjoint. By Axiom 5, the lower bound of $\mu[\alpha \vee \beta]$ cannot be lesser than $\mu[\alpha] + \mu[\beta]$, while by Axiom 6, the upper bound of $\mu[\alpha \vee \beta]$ cannot be greater than $\mu[\alpha] + \mu[\beta]$.

Rule 1 is classical Modus Ponens. Rule 2 can be considered as the rule of necessitation in modal logics, but, since iterations of probability operators are not allowed, it can be applied on the classical propositional formulas only.

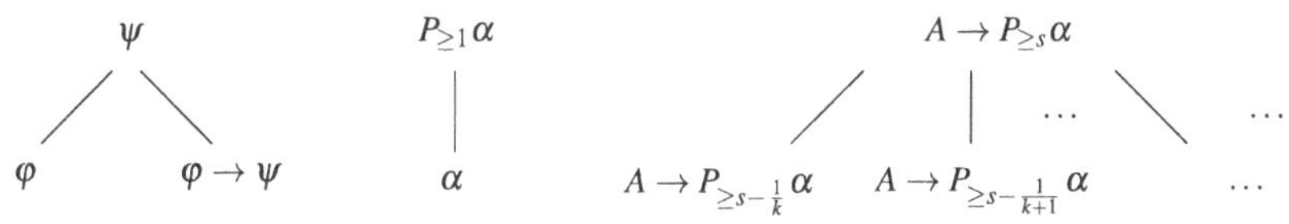

Fig. 3.1 Tree-like representations of the inference rules from Ax_{LPP_2}

Rule 3 is the only infinitary inference rule in the system, i.e., it has a countable set of assumptions and one conclusion. It corresponds to the Archimedean axiom for real numbers and intuitively says that if the probability is arbitrary close to s, then it is at least s. An equivalent form of Rule 3 is

3′. From $A \to P_{\leq s+\frac{1}{k}}\alpha$, for every integer $k \geq \frac{1}{1-s}$, infer $A \to P_{\leq s}\alpha$.

The axioms and rules about probability are similar to those given by Keisler and Hoover (see Sect. 2.7.2), but instead of first order probability quantifiers, we use probability operators, so the corresponding proofs are quite different: while in Keisler–Hoover's approach first order models are considered, our framework relies on Kripke-like modal models.

Since there is an infinitary inference rule in Ax_{LPP_2}, the classical notion of deduction should be modified accordingly.

Definition 3.7 A formula φ is *deducible from a set* T of formulas[1] (denoted by $T \vdash \varphi$) if there is a sequence $\varphi_0, \varphi_1, \ldots, \varphi_{\lambda+1}$ (λ is a finite or countable ordinal[2]) of For_{LPP_2}-formulas, such that

- $\varphi_{\lambda+1} = \varphi$, and
- every φ_i, $i \leq \lambda+1$, is an axiom-instance, or $\varphi_i \in T$, or φ_i is derived by an inference rule applied on some previous members of the sequence.

A *proof* for φ from T is the corresponding sequence of formulas. A formula φ is a *theorem* (denoted by $\vdash \varphi$) if it is deducible from the empty set. ■

It is easy to see that every LPP_2-proof consists of two parts (one of them may be empty). In the first one only classical formulas are involved, while the second one uses formulas from For_P. Two parts are separated by some applications of Rule 2. There is no inverse rule, so we can pass from the classical to the probability level, but we cannot come back. It follows that LPP_2-logic is a conservative extension of the classical propositional logic.

Definition 3.7 introduces proofs as (possibly countable) linear sequences of formulas. Alternatively, we can represent a proof as a (possibly countable) tree where predecessors of nodes imply by inference rules their successors. Since we have an infinitary rule, some nodes can have countably many predecessors. This is illustrated in the Figs. 3.1 and 5.1.

[1] φ is a syntactical consequence of T.

[2] In other words, the length of a proof is an at most countable successor ordinal.

Definition 3.8 A set T of formulas is *consistent* if there are at least a formula from For_C, and at least a formula from For_P that are not deducible from T, otherwise T is *inconsistent*.

A consistent set T of formulas is said to be *maximal consistent* if the following holds:

- for every $\alpha \in \mathrm{For}_C$, if $T \vdash \alpha$, then $\alpha \in T$ and $P_{\geq 1}\alpha \in T$, and
- for every $A \in \mathrm{For}_P$, either $A \in T$ or $\neg A \in T$.

A set T of formulas is *deductively closed* if for every $\varphi \in \mathrm{For}_{LPP_2}$, if $T \vdash \varphi$, then $\varphi \in T$. ∎

Alternatively, we can say that T is inconsistent iff $T \vdash \bot$. Also, note that classical and probability formulas are handled in different ways in Definition 3.8: it is not required that for every classical formula α, either α or $\neg\alpha$ belongs to a maximal consistent set, as it is done for formulas from For_P.

3.3 Non-compactness

Let T be the set

$$\{\neg P_{=0}p\} \cup \{P_{<1/n}p : n \in \mathbb{N}\}$$

for a primitive proposition p. In every finite subset T' of T, there is the largest $k \in \mathbb{N}$ such that $P_{<1/k}p \in T'$. It is easy to see that there is an $LPP_{2,\mathrm{Meas}}$-model $\mathbf{M}_{T'}$ such that $\mu[p] = \frac{1}{k+1} > 0$, and that all formulas from T' are satisfied in $\mathbf{M}_{T'}$. However, there is no $LPP_{2,\mathrm{Meas}}$-model $\mathbf{M}$ which satisfies all formulas from T, since for every $c > 0$, if $\mu[p] = c$, there is a $k \in \mathbb{N}$, such that $\frac{1}{k} < c$, and $\mathbf{M} \not\models P_{<1/k}p$. If $\mu[p] = 0$, then $\mathbf{M} \not\models \neg P_{=0}p$. Thus:

- Although every finite subset of T is $LPP_{2,\mathrm{Meas}}$-satisfiable, the set T itself is not

So, the compactness theorem

Theorem 3.1 *A set of formulas is satisfiable iff every finite subset of it is satisfiable. Does not hold for* $LPP_{2,\mathrm{Meas}}$.

If we have a finitary axiomatization, and if compactness holds, the strong completeness is a consequence of the weak completeness. Let us suppose that a set T of formulas is unsatisfiable. By the compactness theorem, there is a finite subset $T' \subset T$ which is also unsatisfiable. Since T' is finite, by the weak completeness we have that $T' \vdash \bot$, which implies $T \vdash \bot$. Hence, any unsatisfiable set of formulas is inconsistent, or equivalently: every consistent set of formulas is satisfiable. To conclude

- Starting with a finitary axiom system, we cannot hope for the strong completeness theorem for $LPP_{2,\mathrm{Meas}}$.

It means that for every finitary, weakly complete, axiomatization there are consistent $LPP_{2,\mathrm{Meas}}$-unsatisfiable sets of formulas. If we carefully examine the above introduced set $T = \{\neg P_{=0}p\} \cup \{P_{<1/n}p : n \in \mathbb{N}\}$ which illustrated non-compactness, we can notice

- No finitary proof (in the context $LPP_{2,\mathrm{Meas}}$) can prove inconsistency of T, since only finite number of members of T can be used in such a proof, while every finite subset of T is satisfiable and consistent.
- The set T is satisfiable if infinitesimals belong to the range of probability functions. Let ε be a positive infinitesimal, and let $\mu[p] = \varepsilon$ in a model $\mathbf{M}$. Then, it is easy to verify that $\mathbf{M} \models T$.
- The set T is unsatisfiable if the range of probability functions is a finite set of the form $\{0, \frac{1}{k}, \frac{2}{k}, \ldots, \frac{k-1}{k}, 1\}$, but in that case it is possible to give a finitary strongly complete axiomatization such that T is inconsistent.

While logics with finite ranges and with non-standard valued probabilities will be discussed later in Chap. 5, in this chapter we will address the first issue, where Rule 3 guarantees that T is inconsistent:

1. $T \vdash \neg P_{=0}p$, since $\neg P_{=0}p \in T$
2. $T \vdash P_{<1/n}p$, since $P_{<1/n}p \in T$, for every $n \in \mathbb{N}$
3. $T \vdash P_{\leq 1/n}p$, by Axiom 4 and Modus Ponens, for every $n \in \mathbb{N}$
4. $T \vdash P_{\leq 0}p$, by Rule 3 from (3)
5. $T \vdash P_{=0}p$, by Axiom 2 and definition of $P_{=0}$
6. $T \vdash \bot$ from (1) and (5).

In this way Rule 3 eliminates non-Archimedean-valued probability functions.

The reader can observe that the range of probability functions in $LPP_{2,\mathrm{Meas}}$, the unit interval $[0, 1]$ of reals, is a unique, concrete set. Usually, when we have to characterize such a concrete set, we can expect troubles. In terminology of model theory, $[0, 1]$ is not saturated with respect to the considered logical language, i.e., we can define in $[0, 1]$ a non-isolated type.[3] To achieve strong completeness, such a non-isolated type should be omitted, i.e., we have to provide logical instruments which allow us to prove inconsistency of non-isolated types. In the logic LPP_2 that aim is achieved by Rule 3. However, the similar situation appears in other cases, too. In next chapters we will consider other ranges of probability functions: for example $[0, 1]_{\mathbb{Q}}$, the unit intervals of $\mathbb{Q}(\varepsilon)$ or some other countable sets, etc. In those cases, other rules will be given instead of Rule 3 to overcome the problem. To summarize, we will consider ranges of probability functions that are

- uncountable, with everywhere dense subsets (e.g., $[0, 1]$ and $[0, 1]_{\mathbb{Q}}$), in which case rules similar to Rule 3 are used to prove inconsistency of non-Archimedean[4] types,

[3] A set of formulas, a theory, which is finitely satisfiable, but not realized (satisfiable).

[4] Which in the case of LPP_2 coincides with non-isolated types.

- countable (e.g., $[0, 1]_{\mathbb{Q}}$ or the $[0, 1]_{\mathbb{Q}(\varepsilon)}$), in which cases we represent the ranges in syntax, i.e., for every element s from a range, $P_{=s}$ is a probability operator, and formulate infinitary rules of the form: if $A \rightarrow P_{\neq s}\alpha$, for every s in the range, then infer $A \rightarrow \bot$, that guarantee that probabilities of formulas are in the specified range, and
- finite, in which cases we provide infinitary strongly complete axiomatizations.

As we stated in the introduction, we will try to limit the use of infinitary means. While proofs can be infinite, we keep them countable. In our approach infinitary formulas are not necessary to obtain completeness, for example infinite conjunctions are represented by sets of formulas. Wherever appropriate, we use recursive object languages and finite formulas, so that decidability could be achieved.

Finally, we note that compactness is obviously related to classes of models that can be connected via completeness to axiom systems. Is possible that a system axiomatizes valid formulas in two distinct classes of models, and that for one of those classes compactness holds, while it fails for the second class. So it is not impossible that, for an axiom system strong completeness for the former class is proved, while only weak completeness can be shown for the latter.

3.4 Soundness and Completeness

3.4.1 Soundness

Soundness of our system follows from the soundness of classical propositional logic, as well as from the properties of probabilistic measures, so we give only a sketch of a straightforward but tedious proof.

Theorem 3.2 (Soundness) *The axiom system Ax_{LPP_2} is sound with respect to the class of $LPP_{2,\mathrm{Meas}}$-models.*

Proof We can show that every instance of axiom schemata holds in every model, while the inference rules preserve the validity. For example, let us consider Axiom 5. Suppose that $P_{\geq r}\alpha$, $P_{\geq s}\beta$, and $P_{\geq 1}\neg(\alpha \vee \beta)$ hold in a model $\mathbf{M} = \langle W, H, \mu, v\rangle$. It means that $\mu([\alpha]) \geq r$, $\mu([\beta]) \geq s$, and that $[\alpha]$ and $[\beta]$ are disjoint sets. By the definition of finitely additive measures, the measure of $[\alpha] \cup [\beta]$ (which is $[\alpha \vee \beta]$) is $\mu([\alpha]) + \mu([\beta])$. Hence, $\mathbf{M} \models P_{\geq \min(1, r+s)}(\alpha \vee \beta)$, and Axiom 5 holds in M. The other axioms can be proved to be valid in a similar way.

Rule 1 is validity-preserving for the same reason as in classical logic. Consider Rule 2 and suppose that a formula $\alpha \in \mathrm{For}_C$ is valid. Then, for every model $\mathbf{M} = \langle W, H, \mu, v\rangle$, $[\alpha] = W$, and $\mu([\alpha]) = 1$. Hence, $P_{\geq 1}\alpha$ is valid too. Rule 3 preserves validity because of the properties of the set of real numbers. ■

3.4.2 Completeness

In the proof of the completeness theorem we adapt the Henkin style procedure and the strategy developed for proving completeness for modal logics. So, the proof consists of the following main steps:

- We start with a form of Deduction theorem (Theorem 3.3) and some other auxiliary statements (the Lemmas 3.1, 3.2 and 3.3).
- Then, we prove Lindenbaum's theorem, i.e., we show how to extend a consistent set T of formulas to a maximal consistent set T^* (Theorem 3.4).
- Finally, the canonical model $\mathbf{M}_T$ is constructed using the set T^* (Theorem 3.5) such that $\mathbf{M}_T \models \varphi$ iff $\varphi \in T^*$ (Theorem 3.6).

The next formulation of Deduction theorem takes into account that the formal language is restricted so that iterations of probability operators are not allowed.

Theorem 3.3 (Deduction theorem) *If T is a set of formulas and $\varphi, \psi \in \mathrm{For}_C$ or $\varphi, \psi \in \mathrm{For}_P$, then*

$$T \cup \{\varphi\} \vdash \psi \textit{ iff } T \vdash \varphi \rightarrow \psi.$$

Proof The implication from right to left can prove exactly in the same way as in the classical propositional case. For the other direction we use the transfinite induction on the length of the proof of ψ from $T \cup \{\varphi\}$. The cases when either $\vdash \psi$ or $\varphi = \psi$ or ψ is obtained by application of Modus Ponens (Rule 1) are standard.

Thus, let us consider the case where $\psi = P_{\geq 1}\alpha$ is obtained from $T \cup \{\varphi\}$ by an application of Rule 2, and $\varphi \in \mathrm{For}_P$. In that case

$T, \varphi \vdash \alpha$
$T, \varphi \vdash P_{\geq 1}\alpha$ by Rule 2.

However, since $\alpha \in \mathrm{For}_C$, and $\varphi \in \mathrm{For}_P$, φ does not affect the proof of α from $T \cup \{\varphi\}$, and we have

$T \vdash \alpha$
$T \vdash P_{\geq 1}\alpha$ by Rule 2
$T \vdash P_{\geq 1}\alpha \rightarrow (\varphi \rightarrow P_{\geq 1}\alpha)$
$T \vdash \varphi \rightarrow P_{\geq 1}\alpha$ by Rule 1.

Next, let us consider the case where $\psi = A \rightarrow P_{\geq s}\alpha$ is obtained from $T \cup \{\varphi\}$ by an application of Rule 3, and $\varphi \in \mathrm{For}_P$. Then

1. $T, \varphi \vdash A \rightarrow P_{\geq s - \frac{1}{k}}\alpha$, for every integer $k \geq \frac{1}{s}$
2. $T \vdash \varphi \rightarrow (A \rightarrow P_{\geq s - \frac{1}{k}}\alpha)$, for $k \geq \frac{1}{s}$, by the induction hypothesis
3. $T \vdash (\varphi \wedge A) \rightarrow P_{\geq s - \frac{1}{k}}\alpha$, for $k \geq \frac{1}{s}$
4. $T \vdash (\varphi \wedge A) \rightarrow P_{\geq s}\alpha$, from (3) by Rule 3
5. $T \vdash \varphi \rightarrow \psi$. ■

The next statement Lemma 3.1(1)) is a generalization of the well known modal axiom K: $\Box(\alpha \rightarrow \beta) \rightarrow (\Box\alpha \rightarrow \Box\beta)$.

Lemma 3.1

1. $\vdash P_{\geq 1}(\alpha \rightarrow \beta) \rightarrow (P_{\geq s}\alpha \rightarrow P_{\geq s}\beta)$,
2. *if* $\vdash \alpha \leftrightarrow \beta$, *then* $\vdash P_{\geq s}\alpha \leftrightarrow P_{\geq s}\beta$,
3. $\vdash P_{\geq s}\alpha \rightarrow P_{\geq r}\alpha$, *for* $s \geq r$,
4. $\vdash P_{\leq r}\alpha \rightarrow P_{\leq s}\alpha$, $s \geq r$.

Proof (1) First note that using Rule 2, from $\vdash \neg\alpha \vee \neg\bot$, we obtain

$$\vdash P_{\geq 1}(\neg\alpha \vee \neg\bot), \tag{3.1}$$

and similarly, from $\vdash (\neg\alpha \wedge \neg\bot) \vee \neg\neg\alpha$ we have

$$\vdash P_{\geq 1}((\neg\alpha \wedge \neg\bot) \vee \neg\neg\alpha). \tag{3.2}$$

By Axiom 5, we have $\vdash (P_{\geq s}\alpha \wedge P_{\geq 0}\bot \wedge P_{\geq 1}(\neg\alpha \vee \neg\bot)) \rightarrow P_{\geq s}(\alpha \vee \bot)$. Since $\vdash P_{\geq 0}\bot$ by Axiom 2, from (3.1) it follows that:

$$\vdash P_{\geq s}\alpha \rightarrow P_{\geq s}(\alpha \vee \bot). \tag{3.3}$$

The expression $P_{\geq s}(\alpha \vee \bot)$ denotes $P_{\geq s}\neg(\neg\alpha \wedge \neg\bot)$, $P_{\geq 1-(1-s)}\neg(\neg\alpha \wedge \neg\bot)$, and $P_{\leq 1-s}(\neg\alpha \wedge \neg\bot)$. Similarly, $\neg P_{\geq s}\neg\neg\alpha$ denotes $P_{<s}\neg\neg\alpha$. By Axiom 6, we have

$$\vdash (P_{\leq 1-s}(\neg\alpha \wedge \neg\bot) \wedge P_{<s}\neg\neg\alpha) \rightarrow P_{<1}((\neg\alpha \wedge \neg\bot) \vee \neg\neg\alpha). \tag{3.4}$$

Since $P_{\geq 1}((\neg\alpha \wedge \neg\bot) \vee \neg\neg\alpha)$ denotes $\neg P_{<1}((\neg\alpha \wedge \neg\bot) \vee \neg\neg\alpha)$, from (3.2) we obtain that

$$\begin{aligned}\vdash (P_{\leq 1-s}(\neg\alpha \wedge \neg\bot) \wedge P_{<s}\neg\neg\alpha) \rightarrow \\ (P_{<1}((\neg\alpha \wedge \neg\bot) \vee \neg\neg\alpha) \ \wedge\ \neg P_{<1}((\neg\alpha \wedge \neg\bot) \vee \neg\neg\alpha).\end{aligned} \tag{3.5}$$

It follows that $\vdash P_{\leq 1-s}(\neg\alpha \wedge \neg\bot) \rightarrow \neg P_{<s}\neg\neg\alpha$, i.e.,

$$\vdash P_{\geq s}(\alpha \vee \bot) \rightarrow P_{\geq s}\neg\neg\alpha. \tag{3.6}$$

From (3.3) and (3.6) we obtain

$$\vdash P_{\geq s}\alpha \rightarrow P_{\geq s}\neg\neg\alpha. \tag{3.7}$$

The negation of the formula $P_{\geq 1}(\alpha \rightarrow \beta) \rightarrow (P_{\geq s}\alpha \rightarrow P_{\geq s}\beta)$ is equivalent to

$$P_{\geq 1}(\neg\alpha \vee \beta) \wedge P_{\geq s}\alpha \wedge P_{<s}\beta. \tag{3.8}$$

Since $\vdash P_{\geq s}\alpha \to P_{\geq s}\neg\neg\alpha$, this formula implies

$$P_{\geq 1}(\neg\alpha \vee \beta) \wedge P_{\geq s}\neg\neg\alpha \wedge P_{<s}\beta \tag{3.9}$$

which can be rewritten as

$$P_{\geq 1}(\neg\alpha \vee \beta) \wedge P_{\leq 1-s}\neg\alpha \wedge P_{<s}\beta. \tag{3.10}$$

Finally, from

Axiom 6, $P_{\leq 1-s}\neg\alpha \wedge P_{<s}\beta \to P_{<1}(\neg\alpha \vee \beta)$, and
$P_{<1}\alpha = \neg P_{\geq 1}\alpha$,

we have

$$\vdash \neg(P_{\geq 1}(\alpha \to \beta) \to (P_{\geq s}\alpha \to P_{\geq s}\beta)) \to P_{\geq 1}(\neg\alpha \vee \beta) \wedge \neg P_{\geq 1}(\neg\alpha \vee \beta), \tag{3.11}$$

a contradiction. It follows that

$$\vdash P_{\geq 1}(\alpha \to \beta) \to (P_{\geq s}\alpha \to P_{\geq s}\beta). \tag{3.12}$$

(2) It is an easy consequence of Lemma 3.1(1).
(3) This formula expresses monotonicity of probabilities. From

Axiom 3' $P_{\geq s}\alpha \to P_{>r}\alpha$, $s > r$, and
Axiom 4' $P_{>r}\alpha \to P_{\geq r}\alpha$,

we obtain $\vdash P_{\geq s}\alpha \to P_{\geq r}\alpha$ for $s > r$. If $s = r$, the formula is trivially a theorem of the form $\vdash \varphi \to \varphi$.
(4) Similarly as (3). ■

Lemma 3.2 *Let T be a consistent set of formulas.*

1. *For any formula $A \in \mathrm{For}_P$, either $T \cup \{A\}$ is consistent or $T \cup \{\neg A\}$ is consistent.*
2. *If $\neg(\alpha \to P_{\geq s}\beta) \in T$, then there is some $n > \frac{1}{s}$ such that $T \cup \{\alpha \to \neg P_{\geq s-\frac{1}{n}}\beta\}$ is consistent.*

Proof (1) The proof is standard: if $T \cup \{A\} \vdash \bot$, and $T \cup \{\neg A\} \vdash \bot$, by Deduction Theorem we have $T \vdash \bot$.
(2) Suppose that for every $n > \frac{1}{s}$

$$T, \alpha \to \neg P_{\geq s-\frac{1}{n}}\beta \vdash \bot.$$

By Deduction Theorem, and manipulation at the propositional level, we have

$$T \vdash \alpha \to P_{\geq s-\frac{1}{n}}\beta,$$

for every $n > \frac{1}{s}$. By application of Rule 3 we obtain

$$T \vdash \alpha \rightarrow P_{\geq s}\beta,$$

a contradiction with the fact that $\neg(\alpha \rightarrow P_{\geq s}\beta) \in T$. ■

Lemma 3.3 *Let T be a maximal consistent set of formulas. Then,*

1. *for any formula $A \in \mathrm{For}_P$, exactly one member of $\{A, \neg A\}$ is in T,*
2. *for all formulas $A, B \in \mathrm{For}_P$, $A \vee B \in T$ iff $A \in T$ or $B \in T$,*
3. *for all formulas φ, ψ, where either $\varphi, \psi \in \mathrm{For}_C$ or $\varphi, \psi \in \mathrm{For}_P$, $\varphi \wedge \psi \in T$ iff $\{\varphi, \psi\} \subset T$,*
4. *for every $\varphi \in \mathrm{For}_{LPP_2}$, if $T \vdash \varphi$, then $\varphi \in T$,*
5. *for all formulas φ, ψ, where either $\varphi, \psi \in \mathrm{For}_C$ or $\varphi, \psi \in \mathrm{For}_P$, if $\{\varphi, \varphi \rightarrow \psi\} \subset T$, then $\psi \in T$*
6. *for all formulas φ, ψ, where either $\varphi, \psi \in \mathrm{For}_C$ or $\varphi, \psi \in \mathrm{For}_P$, if $\varphi \in T$ and $\vdash \varphi \rightarrow \psi$, then $\psi \in T$,*
7. *for any formula α, if $t = \sup_s\{P_{\geq s}\alpha \in T\}$, and $t \in \mathbb{Q}$, then $P_{\geq t}\alpha \in T$.*

Proof Proofs (1)–(6) are standard.
(7) Let $t = \sup_s\{P_{\geq s}\alpha \in T\} \in \mathbb{Q}$. By the monotonicity of the measure (Lemma 3.1(3)), for every $s \in \mathbb{Q}$, $s < t$, $T \vdash P_{\geq s}\alpha$. Using Rule 3 we have $T \vdash P_{\geq t}\alpha$. Since T is a maximal consistent set, it follows from Lemma 3.3(4) that $P_{\geq t}\alpha \in T$. ■

Theorem 3.4 (Lindenbaum's theorem) *Every consistent set of formulas can be extended to a maximal consistent set.*

Proof Let T be a consistent set, $Cn_C(T)$ the set of all classical formulas that are consequences of T, and $A_0, A_1, \ldots$ an enumeration of all formulas from For_P. We define a sequence of sets T_i, $i = 0, 1, 2, \ldots$ such that

1. $T_0 = T \cup Cn_C(T) \cup \{P_{\geq 1}\alpha : \alpha \in Cn_C(T)\}$
2. for every $i \geq 0$,
   a. if $T_i \cup \{A_i\}$ is consistent, then $T_{i+1} = T_i \cup \{A_i\}$, otherwise
   b. if A_i is of the form $\beta \rightarrow P_{\geq s}\gamma$, then $T_{i+1} = T_i \cup \{\neg A_i, \beta \rightarrow \neg P_{\geq s - \frac{1}{n}}\gamma\}$, for some positive $n \in \mathbb{N}$, so that T_{i+1} is consistent, otherwise
   c. $T_{i+1} = T_i \cup \{\neg A_i\}$.
3. $\mathscr{T} = \cup_{i=0}^{\infty} T_i$.

The set T_0 is consistent since it is contains consequences of an consistent set, and similarly for the other members of the family of sets, by Lemma 3.3 each T_i, $i > 0$, is consistent.

It remains to show that $\mathscr{T}$ is maximal and consistent. The steps 1 and 2 of the above construction fulfill all requirements from Definition 3.8 which guarantees that $\mathscr{T}$ is maximal. We continue by showing that $\mathscr{T}$ is a deductively closed set which does not contain all formulas, and, as a consequence, that $\mathscr{T}$ is consistent.

First of all, $\mathcal{T}$ does not contain all formulas. If $\alpha \in \mathrm{For}_C$, by the construction of T_0, α and $\neg\alpha$ cannot be simultaneously in T_0. For a formula $A \in \mathrm{For}_P$ the set $\mathcal{T}$ does not contain both $A = A_i$ and $\neg A = A_j$, because $T_{\max(i,j)+1}$ is consistent.

I remains to show that $\mathcal{T}$ is deductively closed. If a formula $\alpha \in \mathrm{For}_C$ and $\mathcal{T} \vdash \alpha$, then by the construction of T_0, $\alpha \in \mathcal{T}$ and $P_{\geq 1}\alpha \in \mathcal{T}$.

Let $A \in \mathrm{For}_P$. It can be proved by the induction on the length of the inference that if $\mathcal{T} \vdash A$, then $A \in \mathcal{T}$. Note that if $A = A_j$ and $T_i \vdash A$, it must be $A \in \mathcal{T}$ because $T_{\max(i,j)+1}$ is consistent.

Suppose that the sequence $\varphi_1, \varphi_2, \ldots, A$ forms the proof of A from $\mathcal{T}$. If the sequence is finite, there must be a set T_i such that $T_i \vdash A$, and $A \in \mathcal{T}$.

Thus, suppose that the sequence is countably infinite. We can show that for every i, if φ_i is obtained by application of an inference rule, and all premisses belong to $\mathcal{T}$, then it must be $\varphi_i \in \mathcal{T}$. If the rule is a finitary one, then there must be a set T_j which contains all premisses and $T_j \vdash \varphi_i$. Reasoning as above, we conclude $\varphi_i \in \mathcal{T}$. Next, we consider the only infinitary rule 3. Let $\varphi_i = B \to P_{\geq s}\alpha$ be obtained from the set of premisses $\{\varphi_i^k = B \to P_{\geq s_k}\gamma : s_k \in \mathbb{Q}\}$. By the induction hypothesis, $\varphi_i^k \in \mathcal{T}$ for every k. If $\varphi_i \notin \mathcal{T}$, by the step 2b of the construction, there are some l and j such that $\neg(B \to P_{\geq s}\alpha), B \to \neg P_{\geq s-\frac{1}{l}}\gamma \in T_j$. It means that for some $j' \geq j$

- $B \wedge \neg P_{\geq s}\alpha \in T_{j'}$,
- $B \in T_{j'}$,
- $\neg P_{\geq s-\frac{1}{l}}\gamma, P_{\geq s-\frac{1}{l}}\gamma \in T_{j'}$,

which is in contradiction with consistency of $T_{j'}$. Hence, $\varphi_i \in \mathcal{T}$ which finally means that $\mathcal{T}$ is deductively closed. ■

The set $\mathcal{T}$ is used to define a tuple $\mathbf{M}_T = \langle W, H, \mu, v\rangle$, where

- $W = \{w \models Cn_C(T)\}$ contains all classical propositional interpretations that satisfy the set $Cn_C(T)$ of all classical consequences of the set T,
- $[\alpha] = \{w \in W : w \models \alpha\}$ and $H = \{[\alpha] : \alpha \in \mathrm{For}_C\}$,
- $\mu : H \to [0, 1]$ such that $\mu([\alpha]) = \sup_s\{P_{\geq s}\alpha \in \mathcal{T}\}$, and
- for every world w and every primitive proposition $p \in \phi$, $v(w, p) = true$ iff $w \models p$.

The next theorem states that $\mathbf{M}_T$ is an $LPP_{2,\mathrm{Meas}}$-model, called the canonical model of T.

Theorem 3.5 *Let* $\mathbf{M}_T = \langle W, H, \mu, v\rangle$ *be defined as above and* $\alpha, \beta \in \mathrm{For}_C$. *Then, the following hold*

1. *H is an algebra of subsets of W,*
2. *If $[\alpha] = [\beta]$, then $\mu([\alpha]) = \mu([\beta])$,*
3. *$\mu([\alpha]) \geq 0$.*
4. *$\mu(W) = 1$ and $\mu(\emptyset) = 0$.*
5. *$\mu([\alpha]) = 1 - \mu([\neg\alpha])$.*
6. *$\mu([\alpha] \cup [\beta]) = \mu([\alpha]) + \mu([\beta])$, for all disjoint $[\alpha]$ and $[\beta]$.*

Proof (1) Let $\alpha, \alpha_1, \alpha_2, \ldots \alpha_n$ be formulas from For_C. It is not hard to see that the following hold:

- $W = [\alpha \vee \neg\alpha]$, and $W \in H$,
- if $[\alpha] \in H$, then its complement $[\neg\alpha]$ belongs to H, and
- if $[\alpha_1], \ldots, [\alpha_n] \in H$, then the union $[\alpha_1] \cup \ldots \cup [\alpha_n] \in H$ because $[\alpha_1] \cup \ldots \cup [\alpha_n] = [\alpha_1 \vee \ldots \vee \alpha_n]$.

Thus, H is an algebra of subsets of W.
(2) It is enough to prove that $[\alpha] \subset [\beta]$ implies $\mu([\alpha]) \leq \mu([\beta])$. By the completeness of the propositional logic, $[\alpha] \subset [\beta]$ means that $\alpha \to \beta \in Cn_C(T)$ and $P_{\geq 1}(\alpha \to \beta) \in \mathcal{T}$. By Lemma 3.1(1) we have that for every $s \in \mathbb{Q}$, $P_{\geq s}\alpha \to P_{\geq s}\beta \in \mathcal{T}$. Thus, $\mu([\alpha]) \leq \mu([\beta])$.
(3) Since $P_{\geq 0}\alpha$ is an axiom, $\mu([\alpha]) \geq 0$.
(4) Since $p \vee \neg p \in Cn_C(T)$ and $P_{\geq 1}(p \vee \neg p) \in \mathcal{T}$ for every $p \in \phi$, we have $W = [p \vee \neg p]$ and $\mu(W) = 1$.

On the other hand, obviously, $\mu(\emptyset) \geq 0$. Since $P_{\geq 1}(p \vee \neg p) = P_{\geq 1-0}(p \vee \neg p) = P_{\leq 0}\neg(p \vee \neg p) = P_{\leq 0}(p \wedge \neg p) = \neg P_{>0}(p \wedge \neg p)$, by Axiom 3', $\sup_s\{P_{\geq s}(p \wedge \neg p) \in \mathcal{T}\} = 0$, and $\mu(\emptyset) = 0$.
(5) Let $r = \mu([\alpha]) = \sup_s\{P_{\geq s}\alpha \in \mathcal{T}\}$. Suppose that $r = 1$. By Lemma 3.3(7), $P_{\geq 1}\alpha) \in \mathcal{T}$. Thus, $\neg P_{>0}\neg\alpha (= P_{\leq 0}\neg\alpha = P_{\geq 1}\alpha)$ belongs to $\mathcal{T}$. If for some $s > 0$, $P_{\geq s}\neg\alpha \in \mathcal{T}$, by Axiom 3' it must be $P_{>0}\neg\alpha \in w$, a contradiction. It follows that $\mu([\neg\alpha]) = 1$.

Next, suppose that $r < 1$. Then, for every rational number $r' \in (r, 1]$, $\neg P_{\geq r'}\alpha = P_{<r'}\alpha$, and $P_{<r'}\alpha \in \mathcal{T}$. By Axiom 4, $P_{\leq r'}\alpha$ and $P_{\geq 1-r'}\neg\alpha$ belong to $\mathcal{T}$. On the other hand, if there is a rational number $r'' \in [0, r)$ such that $P_{\geq 1-r''}\neg\alpha \in \mathcal{T}$, then $\neg P_{>r''}\alpha \in \mathcal{T}$, a contradiction.

Hence, $\sup_s\{P_{\geq s}(\neg\alpha) \in \mathcal{T}\} = 1 - \sup_s\{P_{\geq s}\alpha \in \mathcal{T}\}$, i.e., $\mu([\alpha]) = 1 - \mu([\neg\alpha])$.
(6) Let $[\alpha] \cap [\beta] = \emptyset$, $\mu([\alpha]) = r$ and $\mu([\beta]) = s$. Since $[\beta] \subset [\neg\alpha]$, by the above steps (2) and (5), we have $r + s \leq r + (1 - r) = 1$.

Suppose that $r > 0$, and $s > 0$. By the well known properties of the supremum, for every rational number $r' \in [0, r)$, and every rational number $s' \in [0, s)$, we have $P_{\geq r'}\alpha$, $P_{\geq s'}\beta \in \mathcal{T}$. It follows by the axiom 5 that $P_{\geq r'+s'}(\alpha \vee \beta) \in \mathcal{T}$. Hence, $r + s \leq t_0 = \sup_t\{P_{\geq t}(\alpha \vee \beta) \in \mathcal{T}\}$.

If $r + s = 1$, then the statement trivially holds.

Suppose $r + s < 1$. If $r + s < t_0$, then for every rational number $t' \in (r + s, t_0)$ we have $P_{\geq t'}(\alpha \vee \beta) \in \mathcal{T}$. We can choose rational numbers $r'' > r$ and $s'' > s$ such that

- $\neg P_{\geq r''}\alpha$, $P_{<r''}\alpha \in \mathcal{T}$,
- $\neg P_{\geq s''}\beta$, $P_{<s''}(\beta) \in \mathcal{T}$ and
- $r'' + s'' = t' \leq 1$.

By Axiom 4, $P_{\leq r''}\alpha \in \mathscr{T}$. Using Axiom 6 we have

- $P_{<r''+s''}(\alpha \vee \beta) \in \mathscr{T}$,
- $\neg P_{\geq r''+s''}(\alpha \vee \beta) \in \mathscr{T}$ and
- $\neg P_{\geq t'}(\alpha \vee \beta) \in \mathscr{T}$,

a contradiction. Hence, $r + s = t_0$ and $\mu([\alpha] \cup [\beta]) = \mu([\alpha]) + \mu([\beta])$.

Finally suppose that $r = 0$ or $s = 0$. Then we can reason as above, with the only exception that $r' = 0$ or $s' = 0$. ■

Finally, we can summarize the previous statements to prove the strong completeness for $LPP_{2,\mathrm{Meas}}$.

Theorem 3.6 (Strong completeness theorem for $LPP_{2,\mathrm{Meas}}$) *A set T of formulas is Ax_{LPP_2}-consistent iff it is $LPP_{2,\mathrm{Meas}}$-satisfiable.*

Proof The ($\Leftarrow$)-direction follows from the soundness of the above axiomatic system. In order to prove the ($\Rightarrow$)-direction we can construct $\mathbf{M}_T$, the $LPP_{2,\mathrm{Meas}}$-canonical model of T, and show that for every $\varphi \in \mathrm{For}_{LPP_2}$, $\mathbf{M}_T \models \varphi$ iff $\varphi \in \mathscr{T}$.

To begin the induction, let $\varphi = \alpha \in \mathrm{For}_C$. If $\alpha \in Cn_C(T)$, then by the definition of $\mathbf{M}_T$, $\mathbf{M}_T \models \alpha$. Conversely, if $\mathbf{M}_T \models \alpha$, by the completeness of classical propositional logic, $\alpha \in Cn_C(T)$.

Next, let $\varphi = P_{\geq s}\alpha$. If $P_{\geq s}\alpha \in \mathscr{T}$, then $\sup_r\{P_{\geq r}(\alpha) \in \mathscr{T}\} = \mu([\alpha]) \geq s$, and $\mathbf{M}_T \models P_{\geq s}\alpha$. For the other direction, suppose that $\mathbf{M}_T \models P_{\geq s}\alpha$, i.e., that $\sup_r\{P_{\geq r}(\alpha) \in \mathscr{T}\} \geq s$. If $\mu([\alpha]) > s$, then, by the well known property of supremum and monotonicity of μ, $P_{\geq s}\alpha \in \mathscr{T}$. If $\mu([\alpha]) = s$, then by Lemma 3.3(7), $P_{\geq s}\alpha \in \mathscr{T}$.

Let $\varphi = \neg A \in \mathrm{For}_P$. Then $\mathbf{M}_T \models \neg A$ iff $\mathbf{M}_T \not\models A$ iff $A \notin \mathscr{T}$ iff (by Lemma 3.3(1)) $\neg A \in \mathscr{T}$.

Finally, let $\varphi = A \wedge B \in \mathrm{For}_P$. $\mathbf{M}_T \models A \wedge B$ iff $\mathbf{M}_T \models A$ and $\mathbf{M}_T \models B$ iff A, $B \in \mathscr{T}$ iff (by Lemma 3.3(3)) $A \wedge B \in \mathscr{T}$. ■

Using the notion of consequences, the same can be formulated as

Theorem 3.7 *Let T be a set of formulas, and φ a formula. Then:*

$$T \models \varphi \quad \textit{iff} \quad T \vdash \varphi.$$

Proof ($\Leftarrow$) Let $T \models \varphi$. It means that $T \cup \{\varphi\}$ is not satisfiable. By Theorem 3.6, $T \cup \{\neg\varphi\}$ is inconsistent, i.e., $T \cup \{\neg\varphi\} \vdash \bot$. By Deduction theorem, $T \vdash \neg\varphi \rightarrow \bot$. Thus, $T \vdash \varphi$.
($\Rightarrow$) Let $T \vdash \varphi$. We use the induction on the length of the proof of φ from T. If φ is an axiom or belongs to T, the statement trivially holds. So, let φ is obtained by an application of a inference rule.

In the cases of Modus Ponens (Rule 1) it means that we have

$T \vdash \psi \to \varphi$
$T \vdash \psi$
$T \vdash \varphi$ (by Rule 1),

and that by the induction hypothesis

- $T \models \psi \to \varphi$, and
- $T \models \psi$.

Then, by classical reasoning we obtain that $T \models \varphi$.

If φ is obtained by an application of Rule 3, we have

- $T \models \beta \to P_{\geq s-1/k}\alpha$, for every $k \geq \frac{1}{s}$.

By the properties of real-valued probability functions, it follows that $T \models P_{\geq s}\alpha$.

Finally, let φ is obtained by an application of Rule 2. It means that $\varphi = P_{\geq 1}\alpha$, and that $T \models \alpha$. Thus, for every **M** which is a model of T it is also $\mathbf{M} \models \alpha$, and in every world w from **M**, $w \models \alpha$. Furthermore, $\mathbf{M} \models P_{\geq 1}\alpha$, i.e., $\mathbf{M} \models \varphi$. ∎

3.4.3 The Role of the Infinitary Rule

Let us note a few things that could help in understanding Rule 3. The infinitary rule

- From $A \to P_{\geq s-\frac{1}{k}}\alpha$, for every integer $k \geq \frac{1}{s}$, and $s > 0$ infer $A \to P_{\geq s}\alpha$

has the role to ensure that some infinitary sets are inconsistent. That is the reason that Rule 3 is infinite—all formulas from an infinite set should be taken as premisses in the corresponding application of the rule.

In the proof of Lindenbaum's Theorem 3.4, the step of the construction of a maximal consistent extension $\mathcal{T}$ of a consistent set T which is devoted to Rule 3

- if A_i is of the form $\beta \to P_{\geq s}\gamma$, and $T_i \cup \{A_i\}$ is not consistent, then $T_{i+1} = T_i \cup \{\neg A_i, \beta \to \neg P_{\geq s-\frac{1}{n}}\gamma\}$, for some positive $n \in \mathbb{N}$, so that T_{i+1} is consistent,

guarantees that it is not simultaneously possible that

- $\mathcal{T} \cup \{\beta \to P_{\geq s}\gamma\}$ is inconsistent, and
- $\beta \to P_{\geq s-\frac{1}{k}}\gamma \in \mathcal{T}$, for all $k \geq \frac{1}{s}$.

In that sense, the chosen formula $\beta \to \neg P_{\geq s-\frac{1}{n}}\gamma$ is a witness that

$$\mathcal{T} \nvdash \beta \to P_{\geq s}\gamma.$$

Note that this step in the construction of a maximal consistent extension is a propositional counterpart of the corresponding step in the Henkin construction of saturated sets (see Sect. 4.4).

Finally, Rule 3 is given in the implicative form to allow a straightforward proof of Deduction Theorem 3.3.

3.4.4 Completeness for Other Classes of Models

The canonical model $\mathbf{M}_T$ from Theorem 3.6 will be a tool in proving completeness with respect to the classes

- $LPP_{2,\text{Meas},\text{All}}$,
- $LPP_{2,\text{Meas},\sigma}$, and
- $LPP_{2,\text{Meas},\text{Neat}}$.

Theorem 3.8 (Strong completeness theorem for $LPP_{2,\text{Meas},\text{All}}$) *A set T of formulas is Ax_{LPP_2}-consistent iff it is $LPP_{2,\text{Meas},\text{All}}$-satisfiable.*

Proof The proof can be obtained by applying the extension theorem for additive measure[5] on the measure μ from the weak canonical model $\mathbf{M}_T$. Thus, there is a finitely additive measure $\overline{\mu}$ defined on the power set of W that is an extension of the measure μ. It is easy to verify that T is satisfied in that extension of $\mathbf{M}_T$. ∎

In Theorem 3.9 the canonical model $\mathbf{M}_T$ will be used as a base for application of the Carathéodory theorem[6] to show the extended completeness theorem for σ-additive models.

Theorem 3.9 (Strong completeness theorem for $LPP_{2,\text{Meas},\sigma}$) *A set T of formulas is Ax_{LPP_2}-consistent iff it is $LPP_{2,\text{Meas},\sigma}$-satisfiable.*

Proof Let $\mathbf{M}_T = \langle W, H, \mu, v\rangle$ be the canonical model of T. We have proved in Theorem 3.5 that μ is finitely additive. Note that, if $\mathbf{M}_T$ is finite, μ is trivially σ-additive on H, and $\mathbf{M}_T$ is a σ-additive of T.

So, let $\mathbf{M}_T$ be infinite, in which case we will use the Carathéodory theorem.

Let $\beta_0, \beta_1, \ldots, \beta_n, \ldots$ be an infinite sequence of For_C-formulas and let $\bigcup_{n\in\mathbb{N}}[\beta_n] \in H$. Then, by the above construction of $\mathbf{M}_T$, there is $\alpha \in \text{For}_C$ such that $\bigcup_{n\in\mathbb{N}}[\beta_n] = [\alpha]$. Note that:

- for every $k \in \mathbb{N}$, $\bigcup_{n\leq k}[\beta_n] \subseteq [\alpha]$,
- if $\bigcup_{n\leq k}[\beta_n] \neq [\alpha]$, the For_C formula $\neg\beta_0 \wedge \ldots \wedge \neg\beta_k \wedge \alpha$ is satisfiable, and
- by the compactness theorem for classical logic the set $S = \{\alpha, \neg\beta_0, \neg\beta_1, \ldots, \neg\beta_n, \ldots\}$ is satisfiable.

---

[5]**Theorem 3.2.10 from** [1]. Let C be an algebra of subsets of a set Ω and $\mu(w)$ a positive bounded charge—a finitely additive measure—on C. Let F be an algebra on Ω containing C. Then there exists a positive bounded charge $\overline{\mu(w)}$ on F such that $\overline{\mu(w)}$ is an extension of $\mu(w)$ from C to F and that the range of $\overline{\mu(w)}$ is a subset of the closure of the range of $\mu(w)$ on C.

[6]**Carathéodory theorem 1.3.10** Let μ be a measure on the algebra H, and assume that μ is σ-finite, i.e.:

- if $\mathscr{F}_i \in H$, for $i \in \mathbb{N}$, and $\bigcup_i \mathscr{F}_i \in H$, then
- $\mu(\bigcup_i \mathscr{F}_i) = \lim_i \mu(\mathscr{F}_0 \cup \ldots \cup \mathscr{F}_i)$.

Then μ has a unique extension to a measure on the minimal σ-algebra $\overline{H}$ over H.

It means that $\bigcup_{n\in\mathbb{N}}[\beta_n] \neq [\alpha]$, which contradicts the assumption that $\bigcup_{n\in\mathbb{N}}[\beta_n] = [\alpha]$.

Hence, there must be $k \in \mathbb{N}$, such that $\beta_0 \cup \ldots \cup \beta_k = \bigcup_{n\in\mathbb{N}}[\beta_n] = [\alpha]$, and we conclude that for every countably infinite union $\bigcup_{n\in\mathbb{N}}[\beta_n]$ which belongs to H, $\mu(bigcup_{n\in\mathbb{N}}[\beta_n])$ is defined, i.e., it is equal to $\mu(\beta_0 \cup \ldots \cup \beta_k)$. Thus, μ is σ-finite.

Now, according to the Carathéodory theorem, we can transform $\mathbf{M}_T$ to an $LPP_{2,\text{Meas},\sigma}$-model $\overline{\mathbf{M}_T} = \langle W, \overline{H}, \overline{\mu}, v\rangle$ such that

- $\overline{H}$ is the minimal σ-algebra over H, and
- $\overline{\mu}$ is a σ-additive extension of the measure μ such that for every $\varphi \in \text{For}_{LPP_2}$, $\overline{\mathbf{M}_T} \models \varphi$ iff $\varphi \in \mathscr{T}$.

Thus, T is $LPP_{2,\text{Meas},\sigma}$-satisfiable. ∎

Theorem 3.10 (Strong completeness theorem for $LPP_{2,\text{Meas},\text{Neat}}$) *A set T of formulas is Ax_{LPP_2}-consistent iff it is $LPP_{2,\text{Meas},\text{Neat}}$-satisfiable.*

Proof In this proof we use a slightly changed construction of the set $\mathscr{T}$ from Theorem 3.4. Using the same notation as above, the sequence of sets T_i, $i = 0, 1, 2, \ldots$ is now defined in the following way:

1. $T_0 = T \cup Cn_C(T) \cup \{P_{\geq 1}\alpha : \alpha \in Cn_C(T)\}$
2. for every $i \geq 0$,

   a. if $T_i \cup \{A_i\}$ is consistent, then $T_{i+1} = T_i \cup \{A_i\}$, otherwise
   b. if A_i is of the form $\beta \to P_{\geq s}\gamma$, then $T_{i+1} = T_i \cup \{\neg A_i, \beta \to \neg P_{\geq s-\frac{1}{n}}\gamma\}$, for some positive integer n, so that T_{i+1} is consistent, otherwise
   c. $T_{i+1} = T_i \cup \{\neg A_i\}$.
   d. if T_i is enlarged by a formula of the form $P_{=0}\alpha$, add $\neg\alpha$ to T_{i+1} as well.

3. $\mathscr{T} = \cup_{i=0}^{\infty} T_i$.

As it can be seen, the only new step is 2d. We can show that it produces consistent sets, too. So, suppose that for some $\alpha \in \text{For}_C$, $(T_i \cup \{P_{=0}\alpha\}) \cup \{\neg\alpha\} \vdash \bot$. By Deduction theorem, we have that $T_i \cup \{P_{=0}\alpha\} \vdash \alpha$. Since $\alpha \in \text{For}_C$, α belongs to $Cn_C(T)$, and by the construction, we have that $P_{\geq 1}\alpha \in T_0$ which leads to inconsistency of $T_i \cup \{P_{=0}\alpha\}$ since

1. $T_i, P_{=0}\alpha \vdash P_{\geq 1}\alpha$, since $P_{\geq 1}\alpha \in T_i$,
2. $T_i, P_{=0}\alpha \vdash P_{\leq 0}\alpha$, by the definition of $P_{=0}$,
3. $T_i, P_{=0}\alpha \vdash P_{<1}\alpha$, by Axiom 3

and $P_{<1}\alpha = \neg P_{\geq 1}\alpha$. The rest of the completeness proof is the same as in Theorem 3.9. ∎

The situation that the axiomatic system Ax_{LPP_2} is sound and complete with respect to three different classes of models is similar to the one from the modal framework where, for example, the modal system K is characterized by the class of all models, but also by the class of all irreflexive models. In other words, LPP_2-formulas cannot express the differences between the mentioned classes of probability models.

3.5 Decidability and Complexity

In this subsection we will consider the problem of satisfiability of For_{LPP_2} formulas. Since there is a procedure for deciding satisfiability and validity for classical propositional formulas, we will consider For_P-formulas only.

So, let $A \in \mathrm{For}_P$. It is easy, using propositional reasoning and Lemma 3.1(2), to show that A is equivalent to a formula

$$\mathrm{DNF}(A) = \bigvee_{i=1}^{m} \bigwedge_{j=1}^{k_i} X^{i,j}(p_1, \ldots, p_n)$$

called a disjunctive normal form of A, where

- $X^{i,j}$ is a probability operator from the set $\{P_{\geq s_{i,j}}, P_{<s_{i,j}}\}$, and
- $X^{i,j}(p_1, \ldots, p_n)$ denotes that the propositional formula which is in the scope of the probability operator $X^{i,j}$ is in the complete disjunctive normal form, i.e., the propositional formula is a disjunction of the atoms of A.

Theorem 3.11 (Decidability theorem) *The logic LPP_2 is decidable.*

Proof Every For_P-formula A is equivalent to $\mathrm{DNF}(A) = \bigvee_{i=1}^{m} \bigwedge_{j=1}^{k_i} X^{i,j}(p_1, \ldots, p_n)$. A is satisfiable iff at least one disjunct from $\mathrm{DNF}(A)$ is satisfiable. Let the measure of the atom a_i be denoted by y_i. We use an expression of the form

$$a_t \in X(p_1, \ldots, p_n)$$

to denote that the atom a_t appears in the propositional part of $X(p_1, \ldots, p_n)$.

A disjunct $D = \bigwedge_{j=1}^{k} X^j(p_1, \ldots, p_n)$ from $\mathrm{DNF}(A)$ is satisfiable iff the following system of linear equalities and inequalities is satisfiable

$$\begin{array}{r}
\sum_{i=1}^{2^n} y_i = 1 \\
y_i \geq 0 \text{ , for } i = 1, \ldots, 2^n \\
\sum_{a_t \in X^1(p_1,\ldots,p_n) \in D} y_t \begin{cases} \geq s_1 \text{ if } X^1 = P_{\geq s_1} \\ < s_1 \text{ if } X^1 = P_{<s_1} \end{cases} \\
\ldots \\
\sum_{a_t \in X^k(p_1,\ldots,p_n) \in D} y_t \begin{cases} \geq s_k \text{ if } X^k = P_{\geq s_k} \\ < s_k \text{ if } X^k = P_{<s_k} \end{cases}
\end{array} \qquad (3.13)$$

Since the problem of $LPP_{2,\mathrm{Meas}}$-satisfiability of A is reduced to the linear systems solving problem, the satisfiability problem for LPP_2-logic is decidable. Finally, since A is $LPP_{2,\mathrm{Meas}}$-valid iff $\neg A$ is not $LPP_{2,\mathrm{Meas}}$-satisfiable, the validity problem is also decidable. ■

We can show that the $LPP_{2,\mathrm{Meas}}$-satisfiability problem is NP-complete.

Theorem 3.12 *The $LPP_{2,\mathrm{Meas}}$-satisfiability problem is* NP-*complete.*

Proof The lower bound follows from the complexity of the same problem for classical propositional logic. The upper bound is a consequence of the NP-complexity of the satisfiability problem for weight formulas from [2, Theorem 2.9].[7] ■

3.6 A Heuristic Approach to the $LPP_{2,\mathrm{Meas}}$-Satisfiability Problem PSAT

Since the $LPP_{2,\mathrm{Meas}}$-satisfiability problem PSAT is NP-complete, it is natural to try to solve its instances using heuristics. In this section we describe such an approach which is based on genetic algorithms [8–10].

Genetic algorithms (GA) use populations of individuals. Each individual (also called chromosome) is seen as a possible solution in the search space for the particular problem. Thus, a GA can be seen as a searching procedure for the global optima of the corresponding problem. Individuals are represented by genetic code over a finite alphabet. An evaluation function assigning fitness values to individuals has to be defined. Fitness values indicate quality of the corresponding individuals, while average fitness of entire populations may be good measures of obtained quality of the procedures. GA's consist of applications of the genetic operators to populations that must ensure that average fitness values are continually improved from each generation to subsequent. Basic genetic operators are selection, crossover, and mutation but some additional operators such as inversion, local search, etc., may be used.

Selection mechanism favors highly fitted individuals (as well as parts of genetic code of individuals, i.e., genes) to have better chances for reproduction into next generations. On the other hand, chances for reproduction for less fitted members are reduced, and they are gradually wiped out from populations. Crossover operator partitions a population into a set of pairs of individuals named parents. For each pair a recombination of their genetic material is performed with some probability. In that

---

[7] Statements about complexity of the satisfiability problem for weight formulas from [2]. $|A|$ and $\|A\|$ denote the length of A (the number of symbols required to write A), and the length of the longest coefficient appearing in A, when written in binary, respectively. The size of a rational number a/b, where a and b are relatively prime, is defined to be the sum of lengths of a and b, when written in binary.

Theorem 2.6 Suppose A is a weight formula that is satisfied in some measurable probability structure. Then A is satisfied in a structure (S, H, μ, v) with at most $|A|$ states where every set of states is measurable, and where the probability assigned to each state is a rational number with size $O(|A|\|A\| + |A|\log(|A|))$.

Lemma 2.7 If a system of r linear equalities and/or inequalities with integer coefficients each of length at most l has a nonnegative solution, then it has a nonnegative solution with at most r entries positive, and where the size of each member of the solution is $O(rl + r\log(r))$.

Lemma 2.8 Let A be a weight formula. Let $M = (S, H, \mu, v)$ and $M_0 = (S, H, \mu, v')$ be probability structures with the same underlying probability space (S, H, μ). Assume that $v(w, p) = v'(w, p)$ for every state w and every primitive proposition p that appears in A. Then $M \models A$ iff $M_0 \models A$.

Theorem 2.9 The problem of deciding whether a weight formula is satisfiable in a measurable probability structure is NPcomplete.

```
InputData();
PopulationInit();
while ( not FinishedGA() ) {
        for ( i = 0 ; i < N_pop ; i++) p_i = ObjectiveFunction();
        HeuristicImprovement();
        ComputeFitnesses();
        Selection();
        Crossover();
        Mutation();
}
OutputResults();
```

Fig. 3.2 A general description of GA's

way nondeterministic exchange of genetic material in populations is obtained. Multiple usage of selection and crossover operators may produce that the variety of genetic materials is lost. It means that some areas of search spaces become not reachable. This usually causes the convergence in local optimums far from the global optimal values. Mutation operator can help to avoid this shortcoming. Parts of individuals (genes) can be changed with some small probability to increase diversity of genetic material. An initial population is usually generated by random, although sometimes it may be fully or partially produced by an initial heuristic. A general description of GA's is given in Fig. 3.2, where N_{pop} and p_i denote the number of individuals and their objective values, respectively. The objective value of an individual corresponds to the value which the individual owns in the case of the considered problem. The for-loop is repeated until a finishing criterion (the global optima is found, the maximal number of iterations is reached, ...) is satisfied. Since the procedure is not complete, if the maximal number of iterations is reached, we do not know whether the considered problem is solvable. HeuristicImprovement() can be optionally included to improve efficiency of GA and/or to help the procedure to escape from local optima.

In this section, we slightly expand syntax of probabilistic formulas [2]. Namely, as we will mention below in Sect. 5.8, sometimes is suitable to consider Boolean combinations of basic weight formulas of the form:

$$a_1 w(\alpha_1) + \cdots + a_n w(\alpha_n) \geq c$$

where a_i's and c are rational numbers, and α_i's are classical propositional formulas containing primitive propositions from ϕ. The intended meaning of $w(\alpha)$ is "the probability of α." Note that $w(\alpha) \geq s$ can be written as $P_{\geq s}\alpha$ in our notation.

A weight literal is an expression of the form $\sum_i a_i w(\alpha_i) \geq c$ or $\sum_i a_i w(\alpha_i) < c$. The logic that allows such kind of formulas is still NP-complete—which can be proved as above, i.e., by reducing the $LPP_{2,\mathrm{Meas}}$-satisfiability problem to linear programming problem—so by using this logic we just add some expressiveness to our language.

Since For$_P$-formulas can be equivalently translated into their disjunctive normal forms, and a disjunction is satisfiable if at least one disjunct is satisfiable, in the sequel we will only consider formulas of the following form:

$$\bigwedge_{j=1}^{k} a_1^j w(\mathrm{CDNF}(\alpha_1^j)) + \cdots + a_{n_j}^j w(\mathrm{CDNF}(\alpha_{n_j}^j))\ \rho_j\ c^j$$

where $\rho_j \in \{\geq, <\}$, a_i^j's and c^j are rational numbers, and CDNF(α) denotes the complete disjunctive normal form of α. We say that such a formula is in the weight conjunctive form (wfc-form). Also, we will use $at \in \mathrm{CDNF}(\alpha)$ to denote that the atom at appears in CDNF(α).

Example 3.4 Let us consider the formula

$$A = w(p \to q) + w(p) \geq 1.7 \wedge w(q) \geq 0.6.$$

The set of atoms of A is

$$At(A) = \{p \wedge q, p \wedge \neg q, \neg p \wedge q, \neg p \wedge \neg q\}\ .$$

The classical formulas from A are $p \to q, p$, and q, while the sets of atoms satisfying them, or appearing in the corresponding complete disjunctive normal forms are

α	$p \to q$	p	q
$a \in \mathrm{CDNF}(\alpha)$	$p \wedge q, \neg p \wedge q, \neg p \wedge \neg q$	$p \wedge q, p \wedge \neg q$	$p \wedge q, \neg p \wedge q$

Now, the formula A is satisfiable since the same holds for the linear system

$$\begin{aligned}
&\mu(p \wedge q) + \mu(p \wedge \neg q) + \mu(\neg p \wedge q) + \mu(\neg p \wedge \neg q) = 1\\
&\mu(p \wedge q) \geq 0\\
&\mu(p \wedge \neg q) \geq 0\\
&\mu(\neg p \wedge q) \geq 0\\
&\mu(\neg p \wedge \neg q) \geq 0\\
&\mu(p \wedge \neg q) + \mu(\neg p \wedge q) + \mu(\neg p \wedge \neg q) + 2\mu(p \wedge q) \geq 1.7\\
&\mu(p \wedge q) + \mu(\neg p \wedge q) \geq 0.6.
\end{aligned}$$

For example the following assignment μ

a	$p \wedge q$	$p \wedge \neg q$	$\neg p \wedge q$	$\neg p \wedge \neg q$
$\mu(a)$	0.8	0.2	0	0

satisfies the formula. ■

The input for the $LPP_{2,\mathrm{Meas}}$-satisfiability checker based on genetic algorithms is a weight formula A in the wfc-form with L weight literals. Without loss of generality, we demand that classical formulas appearing in weight terms are in disjunctive normal form.

Let $\phi(A) = \{p_1, \ldots, p_N\}$ denote the set of all primitive propositions from A, and $|\phi(A)| = N$.

An individual M consists of L pairs of the form (atom, probability) that describe a probabilistic model. The first coordinate is given as a bit string of length N, where 1 at the position i denotes $\neg p_i$, while 0 denotes p_i. Probabilities are represented by floating point numbers.

For an individual

$$M = ((at_1, \mu(a_1)), \ldots, (at_N, \mu(at_N)))$$

the linear system is equivalent to

$$\bigvee_{i=1}^{L} \left(\sum_{j=1}^{L} a_{ij} \mu(at_j) \right) \rho_i \; c_i.$$

Note that it is possible that some $a_{ij} = 0$, though $[a_{ij}]$ matrix is usually not sparse.

The individuals are evaluated using function $d(M)$, which measures a degree of unsatisfiability of an individual M. Function $d(M)$ is defined as the distance between left and right hand side values of the weight literals not satisfied in the model described by M

$$d(M) = \sqrt{\sum_{M \not\models t_i \; \rho_i \; c_i} [a_1^i \sum_{at \in \mathrm{CDNF}(\alpha_1^i)} \mu(at) + \cdots + a_{n_i}^i \sum_{at \in \mathrm{CDNF}(\alpha_{n_i}^i)} \mu(at) - c_i]^2}.$$

If $d(M) = 0$, all the inequalities in the linear system are satisfied, hence the individual M is a solution.

Some features of GA have been set for all tests

- the population consists of 10 individuals,
- one set of tests has been performed with a population of 20 individuals,
- selection is performed using the rank-based roulette operator (with the rank from 2.5 for the best individual to 1.6 for the worst individual - the step is 0.1),
- The crossover operator is one-point, with the probability 0.85,
- the elitist strategy with one elite individual is used in the generation replacement scheme, and
- multiple occurrences of an individual are removed from the population.

Two problem-specific *two-parts* mutation operator are used:

- The first operator (*TP1*) features two different probabilities of mutation for the two parts *(atoms, probabilities)* of an individual; after mutation, the real numbers in *probabilities* part of an individual have to be scaled since their sum must equal 1.

- The second operator (*TP2*) is a combination of ordinary mutation on *atoms* part, and a special mutation on *probabilities* part of an individual. Instead of performing mutation on two bits in the representation of *probabilities* part, two members p_{i1}, p_{i2} of *probabilities* part are chosen randomly and then replaced with random p'_{i_1}, p'_{i_2}, such that $p_{i_1} + p_{i_2} = p'_{i_1} + p'_{i_2}$ and $0 \leq p'_{i_1}, p'_{i_2} \leq 1$. The sum of probabilities does not change and no scaling is needed.

We have experimented with the following choices in the local search procedure:

- LS1 (LS denotes "local search"): For an individual M all the weight literals are divided into two sets: the first set (B) contains all satisfied literals, while the second one (W) contains all the remaining literals. The literal

$$t_B \ \rho_B \ c_B \in B$$

(called the best one) with the biggest difference $|\mu(t_B) - c_B|$ between the left and the right side, and the literal

$$t_W \ \rho_W \ c_W \in W$$

(the worst one) with the biggest difference $|\mu(t_W) - c_W|$ are found. Two sets of atoms are determined: the first set

$$B_{\mathrm{At}(f)}$$

contains all the atoms from M satisfying at least one classical formula α_i^B from $t_B = a_1^B w(\alpha_1^B) + \cdots + a_{k_B}^B w(\alpha_{k_B}^B)$, while the second one

$$W_{\mathrm{At}(f)}$$

contains all the atoms from M satisfying at least one classical formula α_i^W from $t_W = a_1^W w(\alpha_1^W) + \cdots + a_{k_W}^W w(\alpha_{k_W}^W)$. The probabilities of a randomly selected atom from

$$B_{\mathrm{At}(f)} \setminus W_{\mathrm{At}(f)}$$

and a randomly selected atom from

$$W_{\mathrm{At}(f)} \setminus B_{\mathrm{At}(f)}$$

are changed so that $t_B \ \rho_B \ c_B$ remains satisfied, while the distance $|\mu(t_W) - c_W|$ is decreased or $t_W \ \rho_W \ c_W$ is satisfied.
- LS2: For na individual M, the *worst* weight literal

$$t_W \ \rho_W \ c_W$$

from W (the set of unsatisfied literals) with the biggest difference $|\mu(t_W) - c_W|$ is found. The literal can be represented as

$$\sum_{j=1}^{L} a_{Wj}\mu(at_j)\ \rho_W\ c_W.$$

We try to change the vector of probabilities $[\mu(at_j)]$, so that the linear equation

$$\sum_{j=1}^{L} a_{Wj}\mu(at_j) = c_W$$

is satisfied. The equation $\sum_{j=1}^{L} a_{Wj}\mu(at_j) = c_W$ represents a hyperplane in $\mathbb{R}^n$ while $[a_{Wj}]$ denotes a vector normal to the hyperplane. The projection of $[\mu(at_j)]$ to the hyperplane—which satisfies the equation—is

$$[\mu'(at_j)] = [\mu(at_j)] + k_W[a_{Wj}].$$

The calculation of k and the projection vector is simple and straightforward ($k = \frac{c_W - a_W \circ [\mu(at_j)]}{|a_W|^2} = \frac{c_W - \sum_{j=1}^{L}\mu(at_j)a_{Wj}}{\sum_{j=1}^{L} a_{Wj}{}^2}$). We set the new vector of probabilities to be

$$[\mu''(at_j)] = \frac{[\max\{\mu'(at_j), 0\}]}{\sum_{k=1}^{L}\max\{\mu'(at_k), 0\}}$$

(negative coordinates are replaced with 0, and the vector is scaled so that the sum of its coordinates $\sum_{j=1}^{L}\mu''(at_j)$ equals 1).

- LS3 is similar to LS2, with the difference being made when choosing the weight literal $t_W\ \rho_W\ c_W$ from W (the set of unsatisfied literals). The chosen literal is the one with the smallest difference $|\mu(t_W) - c_W|$; it is the *best bad literal*.
- LS4 is similar to LS2 and LS3. Instead of calculating the projection $[\mu'(at_j)] = [\mu(at_j)] + k_W[a_{Wj}]$ for one chosen weight literal $t_W\ \rho_W\ c_W$ from W, we calculate $k_{W_i}[a_{W_i j}]$ for each literal $t_{W_i}\ \rho_{W_i}\ c_{W_i}$ from W (the set of unsatisfied literals) and calculate the *intermediate* vector $[\mu'(at_j)]$, by adding the linear combination to the original vector

$$[\mu'(at_j)] = [\mu(at_j)] + \sum_{W_i} k_{W_i}[a_{W_i j}].$$

  The new vector of probabilities $[\mu''(at_j)]$ is then calculated in same fashion as in LS2.

In our methodology, introduced in [8], the performance of the system is evaluated on a set of PSAT-instances, i.e., on a set of randomly generated formulas in the wfc-form (with classical formulas in disjunctive normal form). The advantage of this approach is that a formula can be randomly generated according to the following parameters:

- N—the number of propositional letters,
- L—the number of weight literals,

- S—the maximal number of summands in weight terms, and
- D—the maximal number of disjuncts in DNF's of classical formulas.

The considered set of test problems contains 27 satisfiable formulas. Three PSAT-instances were generated for each of nine pairs of (N, L), where $N \in \{50, 100, 200\}$, and $L \in \{N, 2N, 5N\}$. For every instance $S = D = 5$.

Having the above parameters, L atoms and their probabilities (with the constraint that the sum of probabilities must be equal to 1) are chosen. Next, a formula f containing L basic weight formulas is generated. It contains primitive propositions from the set $\{p_1, \ldots, p_N\}$ only. Every weight literal contains at most S summands in its weight term. Every classical formula is in disjunctive normal form with at most D disjuncts, while every disjunct is a conjunction of at most N literals. For every weight term t coefficients are chosen, and the value of t is computed. Next, the sum $sp(t)$ of positive coefficients and the sum $sn(t)$ of negative coefficients are computed. Finally, the right side value of the weight literals between $sp(t)$ and $sn(t)$, and the relation sign are chosen such that f is satisfiable.

We prefer to test more problem instances of different sizes (even very large scale instances) rather than making more trials on a smaller set of instances (of smaller or average size). Since the tests are of large sizes, the necessity to perform them in a reasonable time imposed to set the maximal number of generations to be: 10000 for $N = 50$, 7000 for $N = 100$ and 5000 for $N = 200$.

As an illustration of the corresponding results we give Table 3.1 which contains the average running time of successful tests as measured on our test computer (a Pentium P4 2.4 GHz, 512 MB-based Linux station). The table shows running times only for selected tests.

Columns 2 and 3 show times for tests without LS's, with different population size (10 individuals vs 20 individuals). Increased population size does result in smaller number of iterations needed to find the solution, but the computational cost for each iteration is increased and the overall computational cost is greater than with smaller population size.

In columns 4–7 and 8–11 we can compare the efficiency of various LS's. It is clear that LS2 and LS3 are more efficient than LS1 and LS4 when used for large problem instances, however it is not clear which of them is the most efficient. The running times in columns 8–11 (LS's applied in each third generation) are on average smaller than times in columns 4–7 (LS's applied in each generation). However, this does not mean that the principle of reducing application of LS's to each third generation is always more efficient.

Finally, columns 12–14 show execution times for tests using combination of LS's. Combined usage of LS's is not justified in terms of time efficiency, but it is justified in terms of increased success rate. Higher mutation rate in this setup leads to better time efficiency and higher success rate, except for a few less complex problem instances.

Table 3.1 Average time (rounded to seconds) used by the test computer to execute *successful tests for some selected parameters*. (Note: Value 0 means that the average time was less than half second

L, N, inst. no.	Table 1		Table 2				Table 3				Table 5		
	TP2 (12, 4)		TP2 (12, 4)				TP2 (12, 4)				TP2 (12, 4)	TP2 (24, 8)	TP2 (48, 16)
	10 ind.	20 ind.	10 individuals				10 individuals				10 individuals		
	No LS		LS's applied in each generation				LS's applied in each third generation				Combination of LS's applied in each generation		
			LS1	LS2	LS3	LS4	LS1	LS2	LS3	LS4			
50, 50, 1	0	1	0	0	0	0	0	0	0	0	0	0	0
50, 50, 2	0	1	0	0	0	0	0	0	0	0	0	0	0
50, 50, 3	0	1	0	1	1	0	0	0	0	0	0	0	0
50, 100, 1	1	1	1	0	0	2	1	0	0	1	1	0	1
50, 100, 2	1	2	1	1	2	2	1	1	1	2	2	2	3
50, 100, 3	3	3	1	2	7	10	1	2	3	4	1	3	3
50, 250, 1	16	20	28	16	16	39	22	14	11	21	40	35	42
50, 250, 2	51	56	24	38	34	97	26	35	30	50	68	70	132
50, 250, 3	18	20	18	9	17	25	10	8	13	14	15	16	19
100, 100, 1	0	1	0	0	0	0	0	0	0	0	0	0	0
100, 100, 2	0	1	0	0	0	0	0	0	0	0	0	0	0
100, 100, 3	0	1	0	0	0	1	1	0	0	0	1	1	1
100, 200, 1	8	12	10	3	8	9	6	3	8	7	5	5	7
100, 200, 2	2	3	1	3	2	4	1	3	1	4	4	1	2
100, 200, 3	1	3	4	1	2	26	2	1	1	2	2	2	2
100, 500, 1	187	236	170	130	149	384	94	145	244	228	269	294	271
100, 500, 2	295	309	242	241	298	333	169	306	151	228	236	260	480
100, 500, 3	484	575	326	509	416	775	296	390	355	461	1019	777	671

(continued)

Table 3.1 (continued)

L, N, inst. no.	Table 1		Table 2				Table 3				Table 5		
	TP2 (12, 4)		TP2 (12, 4)				TP2 (12, 4)				TP2 (12, 4)	TP2 (24, 8)	TP2 (48, 16)
	10 ind.	20 ind.	10 individuals				10 individuals				10 individuals		
	No LS		LS's applied in each generation				LS's applied in each third generation				Combination of LS's applied in each generation		
			LS1	LS2	LS3	LS4	LS1	LS2	LS3	LS4			
200, 200, 1	58	91	71	108	56	134	34	78	66	3471	146	270	202
200, 200, 2	5	6	11	7	7	14	11	7	10	9	13	11	9
200, 200, 3	2	3	4	1	2	2	4	1	1	4	4	2	3
200, 400, 1	12	11	4	6	5	25	6	7	5	14	8	11	7
200, 400, 2	238	286	N/A	195	163	484	N/A	171	161	296	479	686	1128
200, 400, 3	205	230	N/A	174	205	247	N/A	153	201	208	419	334	374
200, 1000, 1	1593	2173	3064	888	1347	2972	2307	811	1271	1865	2363	2087	2032
200, 1000, 2	N/A	N/A	N/A	N/A	N/A	N/A	N/A	N/A	N/A	N/A	N/A	19582	19977
200, 1000, 3	1489	1861	3298	1364	792	3548	2456	1135	1080	2023	2818	2770	2778

3.6.1 *Other Heuristics for PSAT and Similar Problems*

Another heuristic for PSAT based on Variable neighborhood search (VNS) metaheuristic was presented in [4]. The corresponding solution space consists of 0–1 variables, while the associated probabilities are found by a fast approximate variable neighborhood descent procedure combined with the Nelder-Mead nonlinear optimization method.

The logic $LPCP_2^{[0,1]_{\mathbb{Q}(\varepsilon)},\approx}$, described in Sect. 5.7.1, is suitable for representing and reasoning with uncertain knowledge and for modeling default reasoning. The $LPCP_{2,\mathrm{Meas},\mathrm{Neat}}^{[0,1]_{\mathbb{Q}(\varepsilon)},\approx}$-satisfiability problem is related to PSAT, while the main differences are

- CPSAT-ε involves conditional probability operators, and
- probabilities of formulas in CPSAT-ε may take infinitesimal values.

A method for solving CPSAT-ε based on the bee colony optimization metaheuristic was proposed in [14].

References

1. Bhaskara Rao, K.P.S., Bhaskara Rao, M.: Theory of Charges. Academic Press, New York (1983)
2. Fagin, R., Halpern, J., Megiddo, N.: A logic for reasoning about probabilities. Inf. Comput. **87**(1–2), 78–128 (1990)
3. Ikodinović, N., Ognjanović, Z., Rašković, M., Perović, A.: Hierarchies of probabilistic logics. Int. J. Approx. Reason. **55**(9), 1830–1842 (2014)
4. Jovanović, D., Ognjanović, N.M.Z.: Variable neighborhood search for the probabilistic satisfiability problem. In: Doerner, K.F., Gendreau, M., Greistorfer, P., Gutjahr, W., Hartl, R.F., Reimann, M. (eds.) Metaheuristics: Progress in Complex Systems Optimization, Springer Series: Operations Research/Computer Science Interfaces Series, vol. 39, pp. 173–188. Springer, New York (2007)
5. Kokkinis, I., Maksimović, P., Ognjanović, Z., Studer, T.: First steps towards probabilistic justification logic. Logic J. IGPL **23**(4), 662–687 (2015)
6. Ognjanović, Z., Rašković, M.: Some probability logics with new types of probability operators. J. Logic Comput. **9**(2), 181–195 (1999)
7. Ognjanović, Z., Rašković, M.: Some first-order probability logics. Theor. Comput. Sci. **247**(1–2), 191–212 (2000)
8. Ognjanović, Z., Kratica J., Milovanović, M.: A genetic algorithm for satisfiability problem in a probabilistic logic: a first report. Lecture Notes in Computer Science (LNCS/LNAI), vol. 2143, pp. 805– 816. Springer, London (2001)
9. Ognjanović, Z., Midić, U., Kratica, J.: Artificial Intelligence and Soft Computing ICAISC 2004, Zakopane, Poland, 7–11 June, 2004. In: Rutkowski, L., Siekmann, J., Tadeusiewicz, R., Zadeh, L.A. (eds.) A Genetic Algorithm for Probabilistic SAT Problem. Lecture notes in artificial intelligence (LNCS/LNAI), vol. 3070, pp. 462–467. Springer, Heidelberg (2004)
10. Ognjanović, Z., Midić, U., Mladenović, N.: Proceedings of the second international workshop on hybrid metaheuristics HM2005, Barcelona, Spain, 29–31 August, 2005. In: Blesa, M.J. (ed.) A Hybrid Genetic and Variable Neighborhood Descent for Probabilistic SAT Problem. Lecture Notes in Computer Science, pp. 42–53. Springer, Heidelberg (2005)

11. Ognjanović, Z., Rašković, M., Marković, Z.: Probability logics. In: Ognjanović, Z. (ed.) Zbornik Radova, Subseries Logic in Computer Science 12 (20), pp. 35–111. Matematički institut, Beograd (2009)
12. Rašković, M., Ognjanović, Z.: Some propositional probabilistic logics. In: Kovačević-Vujičić, V., Mijajlović, Ž. (eds.) Proceedings of the Kurepa's symposium 1996, Belgrade, pp. 83–90. Scientific review 19-20 (1996)
13. Rašković, M., Ognjanović, Z.: A first order probability logic, LP_Q. Publications de L'Institute Matematique, ns. **65**(79), 1–7 (1999). http://elib.mi.sanu.ac.rs/files/journals/publ/85/n079p001.pdf
14. Stojanović, T., Davidović, T., Ognjanović, Z.: Bee Colony Optimization for the satisfiability problem in probabilistic logic. Appl. Soft Comput. **31**, 339–347 (2015)

Chapter 4
Probability Logics with Iterations of Probability Operators

Abstract The first order probability logic $LFOP_1$ is introduced. The logic allows formulas about higher-order probabilities. To give semantics to the logic we introduced first order probability models with constant domains and rigid terms. An infinitary axiom system for $LFOP_1$ is presented. The logic $LFOP_1$ inherits the main properties of LPP_2, so, in spite of all differences, we use the techniques from Chap. 3 to prove strong completeness. The same technique can be also applied to other kinds of probability logics: the first order $LFOP_2$, and propositional LPP_1, etc. In $LFOP_1$-formulas probabilistic operators and the classical quantifiers can be mixed and nested, so the logic is related to modal logics and we discuss how (translations of) some properties of modal logics (e.g., Barcan formula) behave in the probabilistic settings. Since $LFOP_1$ extends classical first order logic, it is undecidable, but we show that the monadic fragment of $LFOP_1$ without iterations of probability operators is decidable. The same hold for the propositional LPP_1. Finally, the logic LPP_1^{LTL} suitable for a combination of probability and temporal reasoning is presented. This chapter covers some results from Ikodinović, Proceedings of the 8th European conference symbolic and quantitative approaches to reasoning with uncertainty ECSQARU 2005, vol 3571, pp 726–736, 2005, [10], Marković et al. Int J Approx Reason, 49(1): 52–66, 2008, [11], Milošević, Ognjanović, Logic J Interes Group Pure Appl Logics, 20(1): 235–253, 2012, [12], Milošević, Ognjanović, Publications de L'InstituteMatematique, ns, 93(107): 19–27, 2013, [8], Ognjanović, J Logic Comput 16(2): 257–285, 2006, [15], Ognjanović, Publications de L'Institute Matematique (Beograd), ns, 82(96): 141–154, 2007, [16], Ognjanović, Rašković, Publications de L'Institute Matematique (Beograd), ns, 60(74): 1–4, 1996, [13], Ognjanović, Rašković, Theor Comput Sci, 247(1–2): 191–212, 2000, [14], Ognjanović et al. Fifth international conference on scalable uncertainty management SUM-2011, vol 6929, pp 219–232, 2011, [18], Ognjanović et al. Zbornik radova. Subseries logic in computer science, vol 12(20), pp 35-111, [17], Perović et al. Proceedings of the 5th international symposium foundations of information and knowledge systems FoIKS 2008, vol 4932, pp 239–252, 2008, [21], Rašković, Ognjanović , Proceedings of the Kurepa's symposium 1996, pp 83–90, 1996, [19].

Z. Ognjanović et al., *Probability Logics*,
DOI 10.1007/978-3-319-47012-2_4

4.1 Introduction

In [1], Abadi and Halpern provided a comprehensive study of decidability of first order probability logics and showed that in the general case the set of first order valid formulas is not recursively enumerable, which means that no complete *finitary* axiomatization is possible. However, it is possible to extend our *infinitary* axiomatization from Chap. 3 to a strongly complete axiom system for first order probability logic. Abadi and Halpern considered two kinds of first order logics:

- with probabilities on domains, i.e., Keisler-like logics, and
- with probabilities on possible worlds, which is close to our approach.

The formal language of the latter logic is more expressive than the language used in our systems. We show elsewhere [19] how to extend our syntax to the same level of expressiveness, while still being able to obtain the results about strong completeness. So, for ease of exposition we will present the results in this Chapter in our standard notations, but note that the same can be done in the extended language of [1].

4.2 Syntax and Semantics of $LFOP_1$

This section is devoted to a probabilistic extension of first order classical logic. In this case, interleaving of the probabilistic operators and the classical quantifiers is essential, especially when we compare first order probability logics to first order modal logics. Thus, contrary to Chap. 3, we will start here with a logic in which iterations of quantifiers and probability operators are allowed.

4.2.1 Syntax

The symbols of the first order language L of $LFOP_1$ are:

- for each $k \in \mathbb{N}$, k-ary relation symbols $P_0^k, P_1^k, \ldots, Q_0^k, Q_1^k, \ldots, R_0^k, R_1^k, \ldots,$
- for each $k \in \mathbb{N}$, k-ary function symbols $F_0^k, F_1^k, \ldots, G_0^k, G_1^k, \ldots, H_0^k, H_1^k, \ldots,$
- the connectives $\wedge$, and $\neg$, and the universal quantifier $\forall$,
- a list of unary probability operators $P_{\geq s}\alpha$, for every $s \in [0, 1]_{\mathbb{Q}}$,
- a list of individual variables $x, y, z, \ldots$, and
- auxiliary symbols commas and parentheses.

We will occasionally omit superscripts and subscripts from the function and predicate symbols, and write simply f, g, h, and P, Q, R, respectively. The function symbols of the arity 0 are called (individual) constant symbols. The constants will be denoted by $a, b, c, \ldots$

Terms are defined as in first order classical logic and denoted by $t_1, t_2, \ldots$. The set $\text{For}_{LFOP_1}(L)$ of all formulas of the language L is the smallest set:

- containing all atomic formulas, i.e., if P is a relation symbol of arity k, and $t_1, \ldots, t_k$ are terms, than $P(t_1, \ldots, t_k) \in \mathrm{For}_{LFOP_1}(L)$, and
- closed under the formation rules: if $\alpha, \beta \in \mathrm{For}_{LFOP_1}(L)$, then $\neg\alpha$, $P_{\geq s}\alpha$, $\alpha \wedge \beta$ and $(\forall x)\alpha \in \mathrm{For}_{LFOP_1}(L)$.

We will use α, β, γ, indexed if necessary, to denote formulas, while $(\exists x)\alpha$ abbreviates the formula $\neg(\forall x)\neg\alpha$.

Example 4.1 An example of a formula is:

$$P_{\geq s}(\forall x)P_1^1(x) \rightarrow P_3^2(y, F_0^0) \wedge P_{\geq r}P_{\geq t}P_1^1(F_1^0).$$

■

Since formulas with iterations of probability operators are allowed, it is possible to formalize reasoning about higher-order probabilities, e.g., "The probability is s that the event has the probability r."

In a formula of the form $(\forall x)\alpha$, α is said to be in the scope of that quantifier. An occurrence of a variable x in a formula α is bound if it occurs in a part of α which is of the form $(\forall x)\beta$, otherwise, the occurrence is called free. If α is a formula and t is a term, then t is said to be free for x in α if no free occurrences of x lie in the scope of any quantifier $(\forall y)$, where y is a variable in t. $\alpha(x_1, \ldots, x_m)$ indicates that free variables of the formula α form a subset of $\{x_1, \ldots, x_m\}$, while $\alpha(t/x)$ denotes the result of substituting in α the term t for all free occurrences of x. We can also use the shorter form $\alpha(t)$ to denote the same substitution. A formula α is a sentence if no variable is free in α.

4.2.2 Semantics

Similarly as in Sect. 3.1.2, the semantics for $\mathrm{For}_{LFOP_1}(L)$ will be based on the possible-world approach, with an important difference that worlds of models are now classical first order models. Also, note that, since formulas with iterated probability operators are allowed, probability measures are associated to worlds of a model, and not to a model, as in Definition 3.2.

Definition 4.1 An *$LFOP_1$-model* is a structure $\mathbf{M} = \langle W, D, I, Prob\rangle$ where:

- W is a non empty set of objects called worlds,
- D associates a non empty domain $D(w)$ with every world $w \in W$,
- I associates an interpretation $I(w)$ with every world $w \in W$ such that:
  - $I(w)(F_i^k)$ is a function from $D(w)^k$ to $D(w)$, for all i, and k,
  - $I(w)(P_i^k)$ is a relation over $D(w)^k$, for all i, and k.
- $Prob$ is a probability assignment which assigns to every $w \in W$ a probability space, such that $Prob(w) = \langle W(w), H(w), \mu(w)\rangle$, where:

– $W(w)$ is a non empty subset of W,
– $H(w)$ is an algebra of subsets of $W(w)$ and
– $\mu(w) : H(w) \rightarrow [0, 1]$ is a finitely additive probability measure. ■

The next definitions reflect the mentioned fact that worlds in $LFOP_1$-models are classical first order models.

Definition 4.2 Let $\mathbf{M} = \langle W, D, I, Prob \rangle$ be an $LFOP_1$-model. A *variable valuation* v assigns some element of the corresponding domain to every world w and every variable x, i.e., $v(w)(x) \in D(w)$. If $w \in W$, $d \in D(w)$, and v is a valuation, then $v_w[d/x]$ is a valuation like v except that $v_w[d/x](w)(x) = d$. ■

Definition 4.3 For an $LFOP_1$-model $\mathbf{M} = \langle W, D, I, Prob \rangle$ and a valuation v, the *value of a term t* (denoted by $I(w)(t)_v$) is:

- if t is a variable x, then $I(w)(x)_v = v(w)(x)$, and
- if $t = F_i^m(t_1, \ldots, t_m)$, then $I(w)(t)_v = I(w)(F_i^m)(I(w)(t_1)_v, \ldots, I(w)(t_m)_v)$. ■

Definition 4.4 The truth value of a formula α in a world $w \in W$ for a given $LFOP_1$-model $\mathbf{M} = \langle W, D, I, Prob \rangle$, and a valuation v (denoted by $I(w)(\alpha)_v$) is:

- if $\alpha = P_i^m(t_1, \ldots, t_m)$, then $I(w)(\alpha)_v = true$, if $\langle I(w)(t_1)_v, \ldots, I(w)(t_m)_v \rangle \in I(w)(P_i^m)$, otherwise $I(w)(\alpha)_v = false$,
- if $\alpha = \neg\beta$, then $I(w)(\alpha)_v = true$, if $I(w)(\beta)_v = false$, otherwise $I(w)(\alpha)_v = false$,
- if $\alpha = P_{\geq s}\beta$, then $I(w)(\alpha)_v = true$, if $\mu(w)\{u \in W(w) : I(u)(\beta)_v = true\} \geq s$, otherwise $I(w)(\alpha)_v = false$,
- if $\alpha = \beta \wedge \gamma$, then $I(w)(\alpha)_v = true$, if $I(w)(\beta)_v = true$, and $I(w)(\gamma)_v = true$, otherwise $I(w)(\alpha)_v = false$, and
- if $\alpha = (\forall x)\beta$, then $I(w)(\alpha)_v = true$, if for every $d \in D$, $I(w)(\beta)_{v_w[d/x]} = true$, otherwise $I(w)(\alpha)_v = false$. ■

Another consequence of the definition of $\mathrm{For}_{LFOP_1}(L)$-formulas is that satisfiability is considered in worlds of models, and not in models, as was introduced in Definition 3.5.

Definition 4.5 A formula is *satisfied in a world* w from an $LFOP_1$-model $\mathbf{M} = \langle W, D, I, Prob \rangle$ (denoted by $(\mathbf{M}, w) \models \alpha$, or simply $w \models \alpha$ if $\mathbf{M}$ is clear from the context) if for every valuation v, $I(w)(\alpha)_v = true$. If $d \in D(w)$, we will use $(\mathbf{M}, w) \models \alpha(d)$ to denote that for every valuation v, $I(w)(\alpha(x))_{v_w[d/x]} = true$.

A formula is *valid in a $LFOP_1$-model* $\mathbf{M} = \langle W, D, I, Prob \rangle$ (denoted by $\mathbf{M} \models \alpha$), if it is satisfied in every world w from W.

A formula α is *valid* if for every $LFOP_1$-model $\mathbf{M}$, $\mathbf{M} \models \alpha$.

A sentence α is *satisfiable* if there is a world w in an $LFOP_1$-model $\mathbf{M}$ such that $(\mathbf{M}, w) \models \alpha$. A set T of sentences is satisfiable if there is a world w in an $LFOP_1$-model $\mathbf{M}$ such that for every $\alpha \in T$, $(\mathbf{M}, w) \models \alpha$ (also denoted $w \models T$). ■

In this chapter we will consider a class of all $LFOP_1$-models that satisfy:

- all the worlds from a model have the same domain, i.e., for all $v, w \in W$, $D(v) = D(w)$,
- for every sentence α, and every world w from a model $\mathbf{M}$ the set of all worlds from $W(w)$ that satisfy α, i.e., that is definable by α, $[\alpha]_{\mathbf{M},w} = \{u \in W(w) : I(u)(\alpha)_v = true\}$, is measurable, and
- the terms are rigid, i.e., for every model their meanings are the same in all worlds.

We use $LFOP_{1,\mathrm{Meas}}$ to denote the class of all constant domain measurable models with rigid terms. Some of the introduced notions will be illustrated in the next example. In Sect. 4.5, some consequences of the rejection of the assumption about rigidness of terms will be addressed.

Example 4.2 Let us consider an $LFOP_{1,\mathrm{Meas}}$-model $\mathbf{M} = \langle W, D, I, Prob\rangle$, $w \in W$, and suppose that $(\mathbf{M}, w) \models P_{\geq s}P_1^1(x)$.

By the Definitions 4.4 and 4.5, this holds iff for every valuation v:

- $I(w)(P_{\geq s}P_1^1(x))_v = \top$, iff
- $\mu(w)\{u \in W(w) : I(u)(P_1^1(x))_v = true\} \geq s$, iff
- $\mu(w)\{u \in W(w) : v(w)(x) \in I(u)(P_1^1)$.

The last condition means that the element of $D(w)$ assigned to x by $v(w)$, should belong to the interpretation of P_1^1 in u. It means that the formula $P_{\geq s}P_1^1(x)$ connects elements of two different domains. Since for $LFOP_{1,\mathrm{Meas}}$ terms are rigid and domains are equal, this will not cause any problems. However, as we discuss in Sect. 4.5, the situation is much more complicated without these constrains. ■

Note that $(\mathbf{M}, w) \models P_{\geq s}P_1^1(x)$ iff $(\mathbf{M}, w) \models (\forall x)P_{\geq s}P_1^1(x)$, but, as we will show in Example 4.3, $P_{\geq s}P_1^1(x)$ and $P_{\geq s}(\forall x)P_1^1(x)$ have quite different meanings.

4.3 Axiom System Ax_{LFOP_1}

The axiom system Ax_{LFOP_1} is a combination of a classical first order axiomatization and the probability system introduced in Chap. 3. It involves the following axiom schemata:

1. all $\mathrm{For}_{LFOP_1}(L)$-substitutional instances of the axioms of classical propositional logic
2. $(\forall x)(\alpha \to \beta) \to (\alpha \to (\forall x)\beta)$, where x is not free in α
3. $(\forall x)\alpha(x) \to \alpha(t/x)$, where $\alpha(t/x)$ is obtained by substituting all free occurrences of x in $\alpha(x)$ by the term t which is free for x in $\alpha(x)$
4. $P_{\geq 0}\alpha$
5. $P_{\leq r}\alpha \to P_{<s}\alpha$, $s > r$
6. $P_{<s}\alpha \to P_{\leq s}\alpha$
7. $(P_{\geq r}\alpha \wedge P_{\geq s}\beta \wedge P_{\geq 1}(\neg\alpha \vee \neg\beta)) \to P_{\geq \min(1,r+s)}(\alpha \vee \beta)$
8. $(P_{\leq r}\alpha \wedge P_{<s}\beta) \to P_{<r+s}(\alpha \vee \beta)$, $r + s \leq 1$

and inference rules:

1. From α and $\alpha \to \beta$ infer β.
2. From α infer $(\forall x)\alpha$
3. From α infer $P_{\geq 1}\alpha$.
4. From $\beta \to P_{\geq s - \frac{1}{k}}\alpha$, for every integer $k \geq \frac{1}{s}$, infer $\beta \to P_{\geq s}\alpha$.

We use the notions of proofs, theorems, and deducibility introduced in Chap. 3, with an important exception related to applications of Rule 3:

Definition 4.6 A formula α is *deducible from a set T of formulas* ($T \vdash \alpha$) if there is a sequence $\alpha_0, \alpha_1, \ldots, \alpha_{\lambda+1}$ (λ is a finite or countable ordinal) of $\mathrm{For}_{LFOP_1}(L)$-formulas, such that:

- $\alpha_{\lambda+1} = \alpha$, and
- every α_i, $i \leq \lambda + 1$, is an axiom-instance, or $\alpha_i \in T$, or α_i is derived by an inference rule applied on some previous members of the sequence, with the proviso that *Rule 3 can be applied on the theorems only.* ■

Additionally, since the set of formulas $\mathrm{For}_{LFOP_1}(L)$ is considered, we have:

Definition 4.7 A set T of $\mathrm{For}_{LFOP_1}(L)$-formulas is *consistent* if $T \nvdash \bot$, otherwise T is *inconsistent*.

A consistent set T of formulas is *maximal consistent* if for every $\alpha \in \mathrm{For}_{LFOP_1}(L)$, either $\alpha \in T$ or $\neg\alpha \in T$.

A set T of $\mathrm{For}_{LFOP_1}(L)$-formulas is *saturated* if it is maximal consistent and satisfies:

- if $\neg(\forall x)\alpha(x) \in T$, then for some term t, $\neg\alpha(t) \in T$. ■

4.4 Soundness and Completeness

The proofs of soundness and completeness for $LFOP_{1,\mathrm{Meas}}$ extend the corresponding proofs from Chap. 3, and we will only emphasize the main differences.

Theorem 4.1 (Soundness theorem) *The axiom system Ax_{LFOP_1} is sound with respect to the $LFOP_{1,\mathrm{Meas}}$ class of models.*

Proof Let $\mathbf{M} = \langle W, D, I, Prob \rangle$ be an $LFOP_{1,\mathrm{Meas}}$-model, and $w \in W$ such that

- $(\mathbf{M}, w) \models (\forall x)\alpha(x)$.

It means that $I(w)((\forall x)\alpha(x))_v = true$ for every valuation v. For each v, among all valuations there must be one (denoted v') such that $v'(w)(x) = d = I(w)(t)_v$ and $I(w)(\alpha(x))_{v'} = true$. Since $I(w)(\alpha(x))_{v'} = I(w)(\alpha(t/x))_v$, we have $I(w)(\alpha(t/x))_v = true$ for every valuation. Thus, every instance of Axiom 3 is valid. ■

Note that the assumptions about constant domains and rigidness of terms are crucial. If it is not the case, and $\alpha(t/x)$ is of the form $P_{\geq s}\beta(t/x)$, the term t refers to objects in other worlds (different from w). Example 4.4 illustrates that it can have a consequence that $I(w)(\alpha(t/x))_v = false$.

Deduction Theorem 3.3 and the Lemmas 3.1, 3.2, and 3.3 can be proved as above, while for Lindenbaum's theorem we need modifications, according to the Henkin construction of saturated extensions of consistent sets.

Theorem 4.2 (Lindenbaum's theorem) *Let T be a consistent set of* $\mathrm{For}_{LFOP_1}(L)$-*sentences, and C be a countably infinite set of new constant symbols ($C \cap L = \emptyset$). Then T can be extended to a saturated set $\overline{\mathcal{T}}$ in the language $\overline{L} = L \cup C$.*

Proof The main novelty in the proof concerns the saturation property. The completion of T is accomplished using the following method. Let $\alpha_0, \alpha_1, \ldots$be an enumeration of all $\mathrm{For}_{LFOP_1}(\overline{L})$-sentences. The sequence of sets T_i, $i = 0, 1, 2, \ldots$is defined such that:

1. $T_0 = T$
2. for every $i \geq 0$,
   a. if $T_i \cup \{A_i\}$ is consistent, then $T_{i+1} = T_i \cup \{A_i\}$, otherwise
      i. if α_i is of the form $\beta \to P_{\geq s}\gamma$, then $T_{i+1} = T_i \cup \{\neg\alpha_i, \beta \to \neg P_{\geq s-\frac{1}{n}}\gamma\}$, for some positive $n \in \mathbb{N}$, so that T_{i+1} is consistent, otherwise
      ii. $T_{i+1} = T_i \cup \{\neg\alpha_i\}$.
   b. If T_{i+1} is obtained by adding a formula of the form $\neg(\forall x)\beta(x)$ to T_i, then for some $c \in C$, $\neg\beta(c)$ is also added to T_{i+1}, so that T_{i+1} is consistent,
3. $\overline{\mathcal{T}} = \cup_{i=0}^{\infty} T_i$.

Let us consider the only new step 2b. Suppose that for some $i > 0$, a formula of the form $\neg(\forall x)\beta(x)$ is consistently added (in the steps 2a, or 2(a)ii) to T_i. If there is a constant symbol $c \in C$ such that $\neg\beta(c) \in T_i$, then obviously $T_i \cup \{\neg(\forall x)\beta(x), \neg\beta(c)\}$ is consistent. Suppose that there is no such c. Since $T_i \cup \{\neg(\forall x)\beta(x)\}$ is obtained by adding only finitely many formulas to T, and T does not contain constants from C, there is at least one constant $c \in C$ which does not appear in $T_i \cup \{\neg(\forall x)\beta(x)\}$. If $T_i \cup \{\neg(\forall x)\beta(x), \neg\beta(c)\} \vdash \bot$, then by Deduction theorem

- $T_i, \neg(\forall x)\beta(x) \vdash \beta(c)$.

Since c does not appear in $T_i \cup \{\neg(\forall x)\beta(x)\}$, we have

- $T_i, \neg(\forall x)\beta(x) \vdash (\forall x)\beta(x)$, and
- $T_i \vdash (\forall x)\beta(x)$.

But, by the hypothesis $T_i \cup \{(\forall x)\beta(x)\} \vdash \bot$, which means that T_i is not consistent, a contradiction. Thus, the step 2b produces consistent sets. The rest of the proof is the same as in Theorem 3.4. ■

Let the tuple $\mathbf{M}_T = \langle W, D, I, Prob\rangle$ be defined in the following way:

- W is the set of all saturated sets in the language $\overline{L} = L \cup C$,
- D is the set of all variable-free terms in $\overline{L}$,
- for every $w \in W$, $I(w)$ is an interpretation such that:
  - for every function symbol F_i^m, $I(w)(F_i^m)$ is a function from D^m to D such that for all variable-free terms $t_1, \dots, t_m$ in $\overline{L}$, $F_i^m : \langle t_1, \dots, t_m \rangle \to F_i^m(t_1, \dots, t_m)$, and
  - for every relation symbol P_i^m, $I(w)(P_i^m) = \{\langle t_1, \dots, t_m \rangle$ for all variable-free terms $t_1, \dots, t_m \in \overline{L} : P_i^m(t_1, \dots, t_m) \in w\}$.
- for every $w \in W$, $Prob(w) = \langle W(w), H(w), \mu(w) \rangle$ such that:
  - $W(w) = W$,
  - $H(w)$ is the class of sets $[\alpha] = \{w \in W : \alpha \in w\}$, for every sentence α, and
  - for every set $[\alpha] \in H(w)$, $\mu(w)([\alpha]) = \sup_s \{P_{\geq s}\alpha \in w\}$.

We can prove (as in Theorem 3.5) that for every $w \in W$, $Prob(w)$ is a probability assignment, i.e., that:

- $H(w)$ is an algebra of subsets of $W(w)$, and
- $\mu(w)$ is a finitely additive probability defined on $H(w)$.

Note that above in the algebras $H(w)$'s, sets $[\alpha]$ are defined using $\alpha \in w$, and not $(\mathbf{M}_T, w) \models \alpha$ as it is required for $LFOP_{1,\mathrm{Meas}}$-models. So, we have first to show

Theorem 4.3 $\mathbf{M}_T$ *is an* $LFOP_{1,\mathrm{Meas}}$*-model.*

Proof Actually, we have to proof prove that for every formula α, and every $w \in W$, $(\mathbf{M}_T, w) \models \alpha$ iff $\alpha \in w$, which can be done as in Theorem 3.6. The only new case concerns $\alpha = (\forall x)\beta$.

If $\alpha \in w$, then, because of Axiom 3, $\beta(t) \in w$ for every $t \in D$. By the induction hypothesis $(\mathbf{M}_T, w) \models \beta(t)$ for every $t \in D$, and $(\mathbf{M}_T, w) \models (\forall x)\beta$.

On the other hand, let $\alpha \notin w$. Since w is saturated, there is some $t \in D$ such that $(\mathbf{M}_T, w) \models \neg\beta(t)$. It follows that $(\mathbf{M}_T, w) \not\models (\forall x)\beta$. ■

From the Theorems 4.2 and 4.3 we obtain:

Theorem 4.4 (Strong completeness theorem for $LFOP_{1,\mathrm{Meas}}$) *A set* T *of sentences is* Ax_{LFOP_1}*-consistent iff it is* $LFOP_{1,\mathrm{Meas}}$*-satisfiable.*

Proof By Theorem 4.2, T can be extended to a saturated set, while by Theorem 4.3 the corresponding $\mathbf{M}_T$ is an $LFOP_{1,\mathrm{Meas}}$-model such that for some world w from $\mathbf{M}_T$, $w \models T$. Hence, T is $LFOP_{1,\mathrm{Meas}}$-satisfiable. ■

4.4.1 Semantical Consequences

In the modal-like context of the probability logic $LFOP_1$, determining the semantical counterpart of the relation of syntactical consequence can be ambiguous (Melvin Fitting (1942) analyzed this dichotomy for modal logics in [3]):

- $T \models \alpha$, if in every world w in which all formulas from the set T are satisfied, $w \models \alpha$, or
- $T \models \alpha$, if for every model $\mathbf{M}$ in which all formulas from the set T are valid, $\mathbf{M} \models T$.

The relation of syntactical consequence introduced in Definition 4.6 corresponds to the latter option, in which, according to Fitting, members of the set of formulas T can be seen as logical truths. Now, following Theorem 3.7, we can prove

Theorem 4.5 *Let T be a set of formulas, and α a formula. Then:*

$$T \models \alpha \quad \textit{iff} \quad T \vdash \alpha.$$

Proof The only specific case concerns the application of the necessitation rule:

$\vdash \alpha$
$T \vdash P_{\geq 1}\alpha$ (by Rule 3).

Since α is a Ax_{LFOP_1}-theorem, by Theorem 4.4, for every world w from any $LFOP_{1,\text{Meas}}$-model $\mathbf{M} = \langle W, D, I, Prob \rangle$, $w \models \alpha$, and $\mu(w)([\alpha]) = 1$. Thus, $\mathbf{M} \models P_{\geq 1}\alpha$. It follows that for every $LFOP_{1,\text{Meas}}$-model $\mathbf{M}$, if $\mathbf{M} \models T$, then $\mathbf{M} \models P_{\geq 1}\alpha$. ■

4.4.2 Completeness for Other Classes of Measurable First-Order and Propositional Models

The classes of models

- $LPFOP_{1,\text{Meas},\text{All}}$,
- $LFOP_{1,\text{Meas},\sigma}$, and
- $LFOP_{1,\text{Meas},\text{Neat}}$

can be defined similarly as in Sect. 3.1.2. $LFOP_1$ can be restricted to the propositional logic LPP_1 with iterations of probability operators so that we can consider the classes of models $LPP_{1,\text{Meas}}$, $LPP_{1,\text{Meas},\text{All}}$, $LPP_{1,\text{Meas},\sigma}$, and $LPP_{1,\text{Meas},\text{Neat}}$. For all those classes of models the corresponding strong completeness theorems can be proved in the same way as in the Sects. 3.4.4 and 4.4. The same holds for $LFOP_2$, the first order probability logic without iterations of probability operators (that generalized the propositional logic LPP_2), and the corresponding classes of models $LFOP_{2,\text{Meas},\text{All}}$, $LFOP_{2,\text{Meas},\sigma}$, and $LFOP_{2,\text{Meas},\text{Neat}}$.

The language of $LFOP_1$ can be further extended to involve conditional probability operators of the forms $CP_{\geq s}(\alpha, \beta)$ and $CP_{\leq s}(\alpha, \beta)$ with the intended meaning "the conditional probability of α given β is at least s", and "at most s", respectively. Complete axiom systems for the first order logics $LFOCP_2$ without iterations of conditional probability operators, and $LFOCPE_1$ with iterations and with the symbol of equality, are given in [11, 12], while the propositional counterparts are analyzed in [10] (with nonstandard-valued probabilities), [8] (coherent conditional probabilities), and [14] (higher order conditional probabilities).

4.5 Modal Logics Versus Probability Logics

Semantics of modal and probability formulas is given using models with possible worlds, i.e., modal and probability Kripke models, respectively. The main difference between the definitions of those models concerns accessibility relation between worlds which in the modal case is described in binary terms: two worlds either are connected, or they are not connected. On the other hand, probability Kripke models give more refined quantitative characterization in the sense that worlds are connected with some probability. Beside that distinction the two approaches coincide which gives us opportunity to use ideas and results about modal logics in the probability logics framework. In Sect. 4.7 we present a logic involving both modal and probability operators and analyze their relationship.

Example 4.3 Let us consider the well known Barcan formula of the first order modal logic:

$$\mathbf{BF} \qquad (\forall x)\Box\alpha(x) \rightarrow \Box(\forall x)\alpha(x).$$

It is proved that **BF** holds in the class of all first-order fixed domain modal models, and that it is independent from the other first order modal axioms [6, 7]. However, the behavior of the probabilistic analogon of that formula:

$$BF(s) \qquad (\forall x)P_{\geq s}\alpha(x) \rightarrow P_{\geq s}(\forall x)\alpha(x)$$

is quite different.

First, if $s = 0$, $\mathbf{BF}(0)$ is valid, because $P_{\geq 0}(\forall x)\alpha(x)$ always holds since probability functions are nonnegative.

Next, suppose that $0 < s < 1$. Let us consider the $LFOP_{1,\mathrm{Meas}}$-model $\mathbf{M} = \langle W, D, I, Prob\rangle$ such that:

- $W = \{w_1, w_2, w_3, w_4\}$,
- $D = \{d_1, d_2\}$,
- $(\mathbf{M}, w_1) \not\models P(d_1)$, $(\mathbf{M}, w_1) \not\models P(d_2)$,
  $(\mathbf{M}, w_2) \models P(d_1)$, $(\mathbf{M}, w_2) \not\models P(d_2)$,
  $(\mathbf{M}, w_3) \models P(d_1)$, $(\mathbf{M}, w_3) \models P(d_2)$, and
  $(\mathbf{M}, w_4) \not\models P(d_1)$, $(\mathbf{M}, w_4) \models P(d_2)$,
- $\mu(w_1)(\{w_1\}) = 1 - (s + \frac{1}{n})$, $\mu(w_1)(\{w_2\}) = \frac{1}{n}$, $\mu(w_1)(\{w_3\}) = s - \frac{1}{n}$, and $\mu(w_1)(\{w_4\}) = \frac{1}{n}$.

Then:

- $(\mathbf{M}, w_1) \models (\forall x)P_{\geq s}P(x)$, since $\mu(w_1)(\{w : w \models P(d_1)\}) = \mu(w_1)(\{w_2, w_3\}) = s$, and $\mu(w_1)(\{w : w \models P(d_2)\}) = \mu(w_1)(\{w_3, w_4\}) = s$, and
- $(\mathbf{M}, w_1) \not\models (\forall x)P(x)$, $(\mathbf{M}, w_2) \not\models (\forall x)P(x)$, and $(\mathbf{M}, w_4) \not\models (\forall x)P(x)$, whilst $(\mathbf{M}, w_3) \models (\forall x)P(x)$.

Since

- $\mu(w_1)(\{w_3\}) = s - \frac{1}{n}$, and
- $(\mathbf{M}, w_1) \not\models P_{\geq s}(\forall x)P(x)$

it follows that

$$(\mathbf{M}, w_1) \not\models \mathbf{BF}(s).$$

Finally, let $s = 1$ and $\mathbf{M}' = \langle W, D, I, Prob\rangle$ be the following infinitary $LFOP_{1,\mathrm{Meas}}$-model with a constant countable domain:

- $W = \{w_0, w_1, w_2, \ldots, w_n, \ldots\}$,
- $D = \{d_0, d_1, d_2, \ldots, d_n, \ldots\}$,
- for every i, $w_i \models P(d_j)$ iff $i \neq j$,
- $W(w_0) = W$, while the algebra $H(w_0)$ contains all singletons $\{w_i\} = [\neg P(d_i)]_{w_0}$, all finite and all co-finite subsets of W, and
- $\mu(w_0)$ is a finitely additive probability such that $\mu(w_0)(H') = 0$ for every finite $H' \in H(w_0)$, and $\mu(w_0)(H") = 1$ for every co-finite $H" \in H(w_0)$.

Obviously, for every $w \in W$, $w \not\models (\forall x)P(x)$, and $\mu(w_0)(\{u : u \models (\forall x)P(x)\}) = 0$. For every $d \in D$, the set $\{w : w \models P(d)\}$ is co-finite, and $\mu(w_0)(\{w : w \models P(d)\} = 1$. Thus, $w_0 \models (\forall x)P_{\geq s}P(x)$, and $w_0 \not\models \mathbf{BF}(1)$.

Here, the fact that all finite sets have the zero measure (from w_0) makes the key difference in comparison to modal logics. Namely:

- in the modal framework the left hand side of **BF** says that for every element of the domain d, the formula $\alpha(d)$ holds necessarily, i.e., in every accessible world, while
- in the model $\mathbf{M}'$ the same $\alpha(d)$ holds with the probability 1, but still there is a world with the zero probability in which $\alpha(d)$ does not hold.

So, although probability and modal logics are closely related, modal necessity (denoted by $\Box$) is a stronger notion than probability necessity (probability one, $P_{\geq 1}$).

On the other hand, the inverse of **BF** is a theorem in first order modal logics, and the same holds for the inverse of $\mathbf{BF}(s)$:

$$\mathbf{CBF}(s) \qquad P_{\geq s}(\forall x)\alpha(x) \rightarrow (\forall x)P_{\geq s}\alpha(x)$$

which is a theorem of Ax_{LFOP_1}:

$\vdash (\forall x)\alpha(x) \rightarrow \alpha(x)$, by Axiom 3
$\vdash P_{\geq 1}[(\forall x)\alpha(x) \rightarrow \alpha(x)]$, by Rule 3
$\vdash [P_{\geq s}(\forall x)\alpha(x)] \rightarrow P_{\geq s}\alpha(x)$, by Lemma 3.1.1 and Rule 1.
$\vdash [P_{\geq s}(\forall x)\alpha(x)] \rightarrow (\forall x)P_{\geq s}\alpha(x)$, by Rule 2, Axiom 2 and Rule 1. ■

The assumptions about constant domains and rigidness of terms allow us to give semantics of probabilistic formulas which is similar to the objectual interpretation for first order modal logics [4]. In this case, as we illustrated in Sect. 4.4, it is possible to obtain axiomatization for $LFOP_1$ by extending the standard first order axiom system with axioms and rules about probability operators. Furthermore, the same method can be used if the first order probability language contains the relation symbol $=$ interpreted as equality [12].

The assumption about constant domains can be relaxed in the modal framework such that models with monotone domains are considered. Domains in a model are monotone if:

- if u is accessible from w, then $W(w) \subset W(u)$.

In the considered first order probability models there is no accessibility relation between worlds, but monotonicity can be formulated in the following way:

- if u belongs to at least one member of $H(w)$, then $W(w) \subset W(u)$.

Under such assumption it is still possible to use the same technique to axiomatize valid formulas of the first order modal models with rigid terms. On the other hand, if the assumption about rigidness of terms is discarded, classical Axiom 3 will be no more valid (the example is given in [5]).

Example 4.4 Let us consider the class of probability models without the condition about rigidness of terms. Let **M** be the following model:

- $W = \{w_1, w_2\}$,
- $D(w_1) = D(w_2) = \{d_1, d_2\}$,
- $w_1 \models P(d_1)$, $w_1 \not\models P(d_2)$, $w_2 \not\models P(d_1)$, and $w_2 \models P(d_2)$, and
- $\mu(w_1)(\{w_1\}) = \mu(w_1)(\{w_2\}) = \frac{1}{2}$.

Let c be a constant symbol and $I(w_1)(c) = d_2$ i $I(w_2)(c) = d_1$. Then:

- $\mu(w_1)(\{w : w \models P(d_1)\}) = \mu(w_1)(\{w_1\}) = \frac{1}{2}$,
- $\mu(w_1)(\{w : w \models P(d_2)\}) = \mu(w_1)(\{w_2\}) = \frac{1}{2}$,
- $w_1 \not\models P(c)$,
- $w_2 \not\models P(c)$, and
- $w_1 \not\models P_{\geq \frac{1}{2}} P(c)$.

It follows that:

- $w_1 \models (\forall x) P_{\geq \frac{1}{2}} \alpha(x)$, and
- $w_1 \not\models (\forall x) P_{\geq \frac{1}{2}} P(x) \rightarrow P_{\geq \frac{1}{2}} P(c)$.

So, the classical first order axiom $(\forall x)\alpha \rightarrow \alpha(t/x)$, where the term t is free for x, is not valid. ■

Finally, we note that our proof of completeness in Sect. 4.4 differs from the usual completeness proofs for modal logics [6, 7] since in our approach we rely on Deduction theorem.

4.6 (Un)decidability

4.6.1 The First-Order Case

Since the considered first order probability logics $LFOP_1$ and $LFOP_2$ contain classical first order logic, they are undecidable. Even stronger, we already noted that [1] proved that the set of valid $LFOP_{1,\mathrm{Meas}}$-formulas is not recursively enumerable. Furthermore, using a procedure due to Saul Kripke [6, 9], it can be shown that the monadic fragment[1] of $LFOP_1$ is undecidable, too. Kripke considered a translation of classical first order formulas that contain only one binary relation symbol P^2 to monadic modal formulas. The translation replaces every expression of the form $P^2(t_1, t_2)$ in a classical formula by $\Diamond(P_1^1(t_1) \wedge P_2^1(t_2))$, such that a classical first order formula is valid if and only if its translation is a valid modal formula. As the similar translation can be constructed for probability logics, the undecidability of the monadic fragment of $LFOP_1$ follows from undecidability of the fragment of the classical first order logic with a single binary relation symbol.

On the other hand, the monadic fragment of $LFOP_2$ is decidable. Using the same procedure as in Sect. 3.5, it can be proved that every monadic probability sentence $A \in \mathrm{For}_{P,LFOP_2}$ can be equivalently transformed to a disjunctive normal form

$$\mathrm{DNF}(A) = \bigvee_{i=1}^{m} \bigwedge_{j=1}^{k_i} \pm P_{\geq s_{i,j}} \alpha^{i,j}$$

where $\pm P_{\geq s_{i,j}}$ is either $P_{\geq s_{i,j}}$ or $\neg P_{\geq s_{i,j}}$, and $\alpha^{i,j}$'s are classical first order sentences. Let us consider an arbitrary disjunct $\mathscr{D}^i = \bigwedge_{j=1}^{k_i} \pm P_{\geq s_{i,j}} \alpha^{i,j}$ and an $LFOP_{2,\mathrm{Meas}}$ model[2] $\mathbf{M} = \langle W, D, I, H, \mu\rangle$ such that $\mathbf{M} \models \mathscr{D}^i$. We can suppose, without the loss of generality, that all quantifiers in $\mathscr{D}^i$ refer to different variables. Let m_i be the number of different variables in $\mathscr{D}^i$. In every world from W holds exactly one conjunction of the form $\wedge_{j=1}^{k_i} \pm \alpha^{i,j}$.

It is well known that a classical monadic formula with m variables is satisfiable if and only if it is satisfiable in a model of cardinality 2^m. For every world $w \in W$ we consider a classical first order model $w' = \langle D(w'), I(w')\rangle$ such that $D(w')$ contains 2^m elements and $I(w') \models \wedge_{j=1}^{k_i} \pm\alpha^{i,j}$ if and only if $w \models \wedge_{j=1}^{k_i} \pm\alpha^{i,j}$. For the model $\mathbf{M}$ we can construct a model $\mathbf{M}' = \langle W', D', I', H', \mu'\rangle$ whose worlds are the considered classical first order models with the domains of the cardinality 2^m. For every $H_1 \in H$, we can consider $H_1' \in H'$ such that $H_1' = \{w' : w \in H_1\}$, while $\mu'(H_1') = \mu(H_1)$. It follows that $\mathbf{M}' \models D^i$ if and only if $\mathbf{M} \models D^i$. Let $d_1, d_2, \ldots, d_{2^m}$ denote the elements in the domains of worlds from M'. Let $c_1, c_2, \ldots, c_{2^m}$ be a sequence of

[1]The formal language of the monadic fragment of first order logic contains only relation symbols of arity 1.

[2]Since there is no iterations of probability operators, instead of a function *Prob* in an $LFOP_{1,\mathrm{Meas}}$-models, it is enough to specify an algebra H and a probability measure μ, as in Chap. 3.

2^m new constant symbols. The definition of $\mathbf{M}'$ is extended so that for every w', $I(w')(c_i) = d_i$. For every $\alpha^{i,j}$ we consider a formula $\#\alpha^{i,j}$ without quantifiers which contains variable-free formulas only. Namely, we replace every universal quantifier by a conjunction, and every existential quantifier by a disjunction of variable-free formulas. For example,

$$\#((\forall x)P_1^1(x) \wedge (\exists y)P_2^1(y)) = \wedge_{k=1}^{2^m} P_1^1(c_k) \wedge (\vee_{k=1}^{2^m} P_2^1(c_k)).$$

Since domains of worlds from $\mathbf{M}'$ are finite, for every $w' \in W'$, $w' \models \alpha^{i,j}$ if and only if $w' \models \#\alpha^{i,j}$. Thus,

$$\mathbf{M} \models D^i \quad \text{iff} \quad \mathbf{M}' \models \bigwedge_{j=1}^{k_i} \pm P_{\geq s_{i,j}} \#\alpha^{i,j} \qquad (= \#D^i).$$

Finally, variable-free formulas can be treated as propositional formulas, and in this way the problem PSAT for the monadic fragment of $\mathrm{For}_{P,LFOP_2}$ is reduced to the decidable problem of satisfiability of the For_{LPP_2}-formulas.

Decidability of some fragments of the first order logics $LFOCP_2$ and $LFOCPE_1$ with conditional probability operators, is discussed in [11, 12].

4.6.2 The Propositional Case

To prove decidability of the propositional LPP_1-logic we follow the approach for LPP_2 from Sect. 3.5. However, the situation is a little bit more complicated here since in LPP_1-logic iterations of probability operators are allowed. So, decidability for LPP_1 will be proved in two steps:

- first, we show that an LPP_1-formula is satisfiable iff it is satisfiable in a model with a finite[3] (and bounded by a function of the length of the formula) number of worlds, and
- second, we have to reduce the satisfiability problem in those finite models to decidable linear programming problem.

Note that in the propositional case we do not need domains as in Definition 4.1, so the models from the class $LPP_{1,\mathrm{Meas}}$ are of the form $\langle W, I, Prob\rangle$, where I assigns propositional valuations to worlds from W.

[3] Note that, while in the modal framework this is enough to prove decidability, since for every $k \in \mathbb{N}$ there are only finitely many modal models with k worlds, this is not the case for probability logics, i.e., since probability models involve probabilities, for every finite set of k worlds, there are uncountable many probability measures defined on them, and uncountably many models with k worlds.

Theorem 4.6 *If a formula α is $LPP_{1,\mathrm{Meas}}$-satisfiable, then it is satisfied in an $LPP_{1,\mathrm{Meas}}$-model with a finite number of worlds. The number of worlds in that model is at most $2^{\mathrm{len}(\alpha)}$.*

Proof We will use the method of filtration [7]. Assume that $w \models \alpha$, where w is a world from an $LPP_{1,\mathrm{Meas}}$-model $\mathbf{M} = \langle W, I, Prob\rangle$. Let $\mathrm{Subf}(\alpha)$ denote the set of all subformulas of α, and $k = |\mathrm{Subf}(\alpha)|$. Let $\approx\subset W \times W$ denote the equivalence relation such that $w \approx u$ iff for every $\beta \in \mathrm{Subf}(\alpha)$, $w \models \beta$ iff $u \models \beta$. The quotient set $W_{/\approx}$ is finite with the cardinality at most $2^{|\mathrm{Subf}(\alpha)|}$. From every class C_i we choose an element and denote it w_i. We consider the model $\mathbf{M}^* = \langle W^*, Prob^*, I^*\rangle$, where:

- $W^* = \{w_i\}$,
- $Prob^*$ is defined as follows:
  - $W^*(w_i) = \{w_j \in W^* : (\exists u \in C_{w_j})u \in W(w_i)\}$
  - $H^*(w_i)$ is the powerset $\mathbb{P}(W^*(w_i))$,
  - $\mu^*(w_i)(\{w_j\}) = \mu(w_i)(C_{w_j} \cap W(w_i))$, and for any $D \in H^*(w_i)$, $\mu^*(w_i)(D) = \sum_{w_j \in D} \mu^*(w_i)(\{w_j\})$,
- $I^*(w_i)(p) = I(w_i)(p)$, for every primitive proposition p.

It is straightforward to show that $\mathbf{M}^*$ is an $LPP_{1,\mathrm{Meas}}$-model. For example, for every w_i, $\mu^*(w_i)$ is a finitely additive probability measure, since

$$\mu^*(w_i)(W^*(w_i)) = \sum_{w_j \in W^*(w_i)} \mu^*(w_i)(\{w_j\}) = \sum_{C_{w_j} \in W_{/\approx}} \mu^*(w_i)(C_{w_j} \cap W(w_i)) = 1.$$

Now, we can prove that for every $\beta \in \mathrm{Subf}(\alpha)$, $(\mathbf{M}, w) \models \beta$ iff β is satisfied in the world w_i which represents C_w in $\mathbf{M}^*$. If β is a primitive proposition, $(\mathbf{M}, w) \models \beta$ iff for $w_i \in C_w$, $(\mathbf{M}, w_i) \models \beta$ iff $(\mathbf{M}^*, w_i) \models \beta$. The cases related to $\wedge$ and $\neg$ can be proved as usual. Finally, let $\beta = P_{\geq s}\gamma$. Then, $(\mathbf{M}, w) \models P_{\geq s}\gamma$ iff for $w_i \in C_w$, $(\mathbf{M}, w_i) \models P_{\geq s}\gamma$ iff

$$\begin{aligned} s &\leq \mu(w_i)([\gamma]_{\mathbf{M},w_i}) \\ &= \sum_{C_u:\mathbf{M},(\mathbf{M},u)\models\gamma} \mu(w_i)(C_u \cap W(w_i)) \\ &= \sum_{u:(\mathbf{M}^*,u)\models\gamma} \mu^*(w_i)(\{u\}) \\ &= \mu^*(w_i)([\gamma]_{\mathbf{M}^*,w_i}) \end{aligned}$$

iff $\mathbf{M}^*, w_i \models P_{\geq s}\gamma$.

Finally, since the number of different classes in $W_{/\approx}$ is at most $2^{|\mathrm{Subf}(\alpha)|} \leq 2^{\mathrm{len}(\alpha)}$, the same holds for the number of worlds in $\mathbf{M}^*$. ■

As we noted above, there is uncountable many finite $LPP_{1,\mathrm{Meas}}$-models with no more than $2^{\mathrm{len}(\alpha)}$ worlds, so Theorem 4.6 does not directly imply decidability of the PSAT problem for $LPP_{1,\mathrm{Meas}}$. However, we can again use a reduction to linear programming to solve PSAT in a finite number of steps, but the role which atoms have in Chap. 3, will be played here by conjunctions of subformulas of the considered formula.

Theorem 4.7 *The PSAT problem for LPP_1-formulas is decidable.*

Proof Let

$$\mathrm{Subf}(\alpha) = \{\beta_1, \beta_2, \ldots, \beta_k\}$$

denote the set of all subformulas of α, and let $\pm\beta_i$ denote β_i or $\neg\beta_i$. In every world w in every model $\mathbf{M}$ exactly one of the formulas of the form

$$\delta_w = \pm\beta_1 \wedge \cdots \wedge \pm\beta_k$$

is satisfied.

For every $l \leq 2^k$ we will consider l formulas of the above form such that

- The chosen formulas are not necessarily different, but they are propositionally consistent, i.e., we consider only formulas δ_w such that there is no different conjuncts β and $\neg\beta$ appearing in δ_w.
- At least one then must contain the examined formula α.

Using probabilistic constraints (i.e., formulas of the form $P_{\geq s}\beta$, $\neg P_{\geq s}\beta$) from the formulas δ_w we shall examine whether there is an $LPP_{1,\mathrm{Meas}}$-model $\mathbf{M}$ with l which contains a world w satisfying α. We do not try to determine probabilities precisely. Rather, we just check whether there are probabilities such that probabilistic constraints are satisfied in the corresponding world. To do that, for every world w_i, $i < l$, we consider a system of linear equalities and inequalities of the form (we write $\beta \in \delta_w$ to denote that β occurs positively in the top conjunction of δ_w, i.e., if δ_w can be seen as $\wedge_i \delta_i$, then for some i, $\beta = \delta_i$):

$$\sum_{j=1}^{l} \mu(w_i)(\{w_j\}) = 1$$

$$\mu(w_i)(\{w_j\}) \geq 0 \ , \ \text{for every world } w_j$$

$$\sum_{w_j : \beta \in \delta_{w_j}} \mu(w_i)(\{w_j\}) \geq s \ , \ \text{for every } P_{\geq s}\beta \in \delta_{w_i}$$

$$\sum_{w_j : \beta \in \delta_{w_j}} \mu(w_i)(\{w_j\}) < s \ , \ \text{for every } \neg P_{\geq s}\beta \in \delta_{w_i}$$

The first two rows correspond to the general constraints: the probability of the set of all worlds must be 1, while the probability of every measurable set of worlds must be nonnegative. The last two rows correspond to the probabilistic constraints, because

$$\sum_{w_j:\beta\in\delta_{w_j}} \mu(w_i)(\{w_j\}) = \mu(w_i)([\beta]_{w_i}).$$

Such a system is solvable iff there is a probability $\mu(w_i)$ satisfying all probabilistic constraints that appear in δ_{w_i}. Note that there are finitely many such systems that can be solved in a finite number of steps.

If the above test is positively solved there is an $LPP_{1,\mathrm{Meas}}$-model in which for every world w_i, $w_i \models \delta_{w_i}$. Since α belongs to at least one of the formulas δ_{w_i}, we have that α is satisfiable. If the test fails, and there is another possibility of choosing l and/or the set of l formulas δ_w, we continue with the procedure, otherwise we conclude that α is not satisfiable.

It is easy to see that the procedure terminates in a finite number of steps. Thus, the problem PSAT for the class $LPP_{1,\mathrm{Meas}}$ is decidable. Since $\models \alpha$ iff $\neg\alpha$ is not satisfiable, the $LPP_{1,\mathrm{Meas}}$-validity problem is also decidable. ■

4.7 A Discrete Linear-Time Probabilistic Logic

In this section we provide a look at a logic in which the probability and the modal (i.e., temporal) operators are mixed, so that probabilistic reasoning is enriched with temporal features [13]. The temporal part of the logics is a standard discrete linear-time logic LTL [2, 20], where the flow of time is isomorphic to natural numbers, i.e., each moment of time has a unique possible future, while the corresponding language contains the "next" operator ($\bigcirc$) and the reflexive strong "until" operator (U), (the operators "sometime" F and "always" G are definable: $\mathrm{F}\alpha = \top \mathrm{U} \alpha$ and $\mathrm{G}\alpha = \neg\mathrm{F}\neg\alpha$). Similarly as above, nesting of the probabilistic and temporal operators is important and we will start from the logic LPP_1. A first-order branching time counterpart is presented in [17].

In our logic, denoted LPP_1^{LTL}, the probabilistic operators quantify events along a single time line. It allows us to express sentences such as "(according to the current set of information) the probability that, sometime in the future, α is true is at least s". And, as the knowledge can evolve during the time, the probability of α might change too. As we noted in Sect. 4.5, the operators "sometime" and "always" can be seen as two extreme cases of probabilistic quantification of the future time instants definable by formulas. We may try to motivate the proposed semantics in the following way.

Example 4.5 A suitable representation of all possible outcomes of an infinite sequence of probabilistic experiments (let us say that experiments A and B are permanently repeated resulting in a or $\neg a$, and b or $\neg b$, respectively) could be an infinite tree, where every branch corresponds to a possible realization of the sequence of the experiments, and every time instant is described in the form $\pm a$, $\pm b$ depending on obtaining (or not obtaining) a and b in the corresponding experiment. We might be interested in probabilistic properties that hold for all branches. In that case we can reason about an arbitrary branch and need ability to express probabilities of events along it, for example that the probability of the event a is at least s, or some more complicated conditions, like that in every time instant, if the probability of a is less than r, then b must hold forever. ■

The set $\mathrm{For}_{LPP_1^{\mathrm{LTL}}}$ of formulas is defined inductively as the smallest set containing primitive propositions and closed under formation rules:

- if α and β are formulas, then
  - $\neg\alpha, \alpha \wedge \beta$,
  - $P_{\geq s}\alpha$, and for every $s \in [0, 1]_{\mathbb{Q}}$,
  - $\bigcirc\alpha$ (α holds in the next moment), and $\alpha \mathrm{U} \beta$ (α holds until β becomes true)

  are formulas.

We will use the following notational definition:

- $\bigcirc^0\alpha = \alpha$, and $\bigcirc^{i+1}\alpha = \bigcirc\bigcirc^i\alpha$ for $i \geq 0$,
- $\mathrm{F}\alpha = \top\mathrm{U}\alpha$, i.e., α is true or will be true in a future moment,
- $\mathrm{G}\alpha = \neg\mathrm{F}\neg\alpha$, i.e., α is true or will be always true, and
- if $T = \{\alpha_1, \alpha_2, \ldots\}$ is a set of formulas, then $\bigcirc T$ denotes $\{\bigcirc\alpha_1, \bigcirc\alpha_2, \ldots\}$.

Example 4.6 An example of a formula is

$$(\bigcirc P_{\geq r}p \wedge \mathrm{F}P_{<s}(p \to q)) \to \mathrm{G}P_{=t}q$$

which can be read as "if the probability of p in the next moment is at least r and sometime in the future q follows from p with the probability less than s, then the probability of q will always be equal to t." ■

4.7.1 Semantics

The semantics for LPP_1^{LTL} is a Kripke-style one using sequences of natural numbers as frames. Let ϕ denote the set of primitive propositions.

Definition 4.8 An LPP_1^{LTL}*-model* is a structure $\mathbf{M} = \langle W, Prob, v\rangle$ where:

- $W = \{w_0, w_1, \ldots\}$ is a sequence of *time instants*,
- $Prob$ is a probability assignment which assigns to every $w \in W$ a probability space, such that $Prob(w) = \langle W(w), H(w), \mu(w)\rangle$, where:
  - $W(w) = \{w_j : j \geq i\}$,
  - $H(w)$ is an algebra of subsets of $W(w)$ and
  - $\mu(w) : H(w) \to [0, 1]$ is a finitely additive probability measure.
- $v : W \times \phi \to \{true, false\}$. ■

Definition 4.9 Let $\mathbf{M} = \langle W, Prob, v\rangle$ be a LPP_1^{LTL}-model, $i \in \mathbb{N}$ and α be a formula. The *satisfiability relation* $\models$ is inductively defined as follows:

- if $p \in \phi$ is a primitive proposition, $w_i \models p$ if $v(w_i)(p) = true$,
- $w_i \models \neg\alpha$ if $w_i \not\models \alpha$,
- $w_i \models P_{\geq s}\alpha$ if $\mu(w_i)(\{w_{i+j}, j \geq 0 : w_{i+j} \models \alpha\}) \geq s$,
- $w_i \models \bigcirc\alpha$ if $w_{i+1} \models \alpha$,
- $w_i \models \alpha \wedge \beta$ if $w_i \models \alpha$ and $w_i \models \beta$.
- $w_i \models \alpha \mathrm{U} \beta$ if there is an integer $j \geq 0$ such that $w_{i+j} \models \beta$, and for every $k \in \mathbb{N}$ such that $0 \leq k < j$, $w_{i+k} \models \alpha$. ■

We concern a reflexive, strong version of the until operator, i.e., if $\alpha \mathrm{U} \beta$ holds in a time instant, β must eventually hold. In the above definition the future includes the present, so that:

- $w_i \models \mathrm{F}\alpha$ if there is $j \geq 0$ such that $w_{i+j} \models \alpha$, and
- $w_i \models \mathrm{G}\alpha$ if for every $j \geq 0$, $w_{i+j} \models \alpha$.

Also, the present time instant is included when the probability of formulas are considered. All the presented results then can be proved with essentially no change if we use the temporal and probabilistic operators referring to the strict future that does not concern the present.

Again, we will consider *measurable* models only, i.e., the class $LPP^{\mathrm{LTL}}_{1,\mathrm{Meas}}$ of all LPP^{LTL}_1-models such that for every $w_i \in W$ the set $H(w_i)) = \{[\alpha]_{w_i} : \alpha \in For_{LPP^{\mathrm{LTL}}_1}\}$, where $[\alpha]_{w_i} = \{w_{i+j} : j \geq 0, w_{i+j} \models \alpha\}$.

The notions of *satisfiable* and *valid* formulas and *satisfiable sets* of formulas are defined as in Sect. 4.2.2.

Note that, similarly to the probabilistic logics, compactness does not hold for LTL. For example, we can consider the set of temporal formulas of the form

$$\{\neg \mathrm{G}\alpha\} \cup \{\bigcirc^i \alpha : i \in \mathbb{N}\}$$

which is unsatisfiable but finitely satisfiable, similarly to the set considered in Sect. 3.3. As a consequence, in the next section we will introduce another infinitary rule which eliminates possible inconsistent sets of temporal formulas (Rule 3).

4.7.2 Axiomatization

An axiomatization $Ax_{LPP^{\mathrm{LTL}}_1}$ that characterizes the set of all $LPP^{\mathrm{LTL}}_{1,\mathrm{Meas}}$-valid formulas extends the system Ax_{LPP_2} (having in mind that instances of the axiom schemas and rules must obey the syntactical rules for LPP^{LTL}_1) with the following axiom schemas:

8. $\bigcirc(\alpha \to \beta) \to (\bigcirc\alpha \to \bigcirc\beta)$
9. $\neg \bigcirc \alpha \leftrightarrow \bigcirc\neg\alpha$
10. $\alpha \mathrm{U} \beta \leftrightarrow \beta \vee (\alpha \wedge \bigcirc(\alpha \mathrm{U} \beta))$
11. $\alpha \mathrm{U} \beta \to F\beta$
12. $\mathrm{G}\alpha \to P_{\geq 1}\alpha$

while the inference rules should be rewritten in the following form:

1. from α and $\alpha \rightarrow \beta$ infer β
2. from α infer $\bigcirc\alpha$
3. from $\gamma \rightarrow \neg((\bigwedge_{k=0}^{i-1} \bigcirc^k\alpha) \wedge \bigcirc^i\beta)$ for all $i \geq 0$, infer $\gamma \rightarrow \neg(\alpha \mathrm{U} \beta)$
4. from $\beta \rightarrow \bigcirc^m P_{\geq s-\frac{1}{k}}\alpha$, for any $m \geq 0$, and for every $k \geq \frac{1}{s}$, infer $\beta \rightarrow \bigcirc^m P_{\geq s}\alpha$.

The main novelty in $Ax_{LPP_1^{\mathrm{LTL}}}$ concerns axioms about temporal reasoning (the axioms 8 and 9 are the usual axioms for the next operator $\bigcirc$, as well as the axioms 10 and 11 for the until operator) and mixing of probabilistic and temporal reasoning (Axiom 12). There are two infinitary inference rules: 3 and 4. The former one characterizes the until operator. The temporal part of $Ax_{LPP_1^{\mathrm{LTL}}}$ offers possibility to prove extended completeness which cannot be proved using finitary means.

In this framework we can use the Definition 4.6 of deductions and consistency.

Modifications of $Ax_{LPP_1^{\mathrm{LTL}}}$ according to ideas presented in the previous sections could produce the corresponding axiomatic systems for a first order logic for reasoning about discrete linear time and probability. Also, we can specify additional relationships between the flow of time and the probability measures by adding new axioms.

Example 4.7 The formula $\neg\alpha \rightarrow (P_{\geq s}\alpha \rightarrow \bigcirc P_{\geq s}\alpha)$, considered as an additional axiom scheme, characterizes models with the property that if a formula does not hold in a time instant, then in the next time instant its probability will be not decreased. ■

4.7.2.1 Completeness and Decidability

The proof of extended completeness again follows the ideas given in the previous sections, so we only outline the main new details.

Theorem 4.8 (Strong completeness theorem for $LPP_{1,\mathrm{Meas}}^{\mathrm{LTL}}$) *A set T of formulas is $Ax_{LPP_1^{\mathrm{LTL}}}$-consistent iff it is $LPP_{1,\mathrm{Meas}}^{\mathrm{LTL}}$-satisfiable.*

Proof We start with Deduction theorem and consider the temporal part only. For example, assume that $T, \delta \vdash \gamma \rightarrow \neg(\alpha \mathrm{U} \beta)$ is obtained by Rule 3. Then:

1. $T, \delta \vdash \gamma \rightarrow \neg((\bigwedge_{k=0}^{i-1} \bigcirc^k\alpha) \wedge \bigcirc^i\beta)$, for $i \geq 0$,
2. $T \vdash \delta \rightarrow (\gamma \rightarrow \neg((\bigwedge_{k=0}^{i-1} \bigcirc^k\alpha) \wedge \bigcirc^i\beta))$, for $i \geq 0$, by the induction hypothesis,
3. $T \vdash (\delta \wedge \gamma) \rightarrow \neg((\bigwedge_{k=0}^{i-1} \bigcirc^k\alpha) \wedge \bigcirc^i\beta)$, for $i \geq 0$,
4. $T \vdash (\delta \wedge \gamma) \rightarrow \neg(\alpha \mathrm{U} \beta)$, by Rule 3,
5. $T \vdash \delta \rightarrow (\gamma \rightarrow \neg(\alpha \mathrm{U} \beta))$.

The axioms and rules imply some auxiliary statements (T denotes a consistent set of formulas):

1. the following inference rule is derivable: From $\beta \to \bigcirc^i\alpha$ for all $i \geq 0$, infer $\beta \to G\alpha$,
2. $\vdash G\alpha \leftrightarrow \alpha \wedge \bigcirc G\alpha$,
3. $\vdash G \bigcirc \alpha \leftrightarrow \bigcirc G\alpha$,
4. $\vdash (\bigcirc\alpha \to \bigcirc\beta) \to \bigcirc(\alpha \to \beta)$,
5. $\vdash \bigcirc(\alpha \wedge \beta) \leftrightarrow (\bigcirc\alpha \wedge \bigcirc\beta)$,
6. $\vdash \bigcirc(\alpha \vee \beta) \leftrightarrow (\bigcirc\alpha \vee \bigcirc\beta)$,
7. $G\alpha \vdash \bigcirc^i\alpha$ for every $i \geq 0$,
8. if $\vdash \alpha$, then $\vdash G\alpha$,
9. if $T \vdash \alpha$, where T is a set of formulas, then $\bigcirc T \vdash \bigcirc\alpha$.
10. for $j \geq 0$, $\bigcirc^j\beta, \bigcirc^0\alpha, \ldots, \bigcirc^{j-1}\alpha \vdash \alpha U\beta$,
11. For any formula α, either $T \cup \{\alpha\}$ is consistent or $T \cup \{\neg\alpha\}$ is consistent.
12. If $\gamma \to \neg(\alpha U\beta) \in T$, then there is $j_0 \geq 0$ such that $T \cup \{\gamma \to \neg((\bigwedge_{k=0}^{j_0-1} \bigcirc^k\alpha) \wedge \bigcirc^{j_0}\beta)\}$ is consistent.
13. If $\neg(\alpha \to \bigcirc^m P_{\geq s}\beta) \in T$, then there is $j_0 > \frac{1}{s}$ such that $T \cup \{\alpha \to \neg\bigcirc^m P_{\geq s-\frac{1}{j_0}}\beta\}$ is consistent.

For example, the statement (10) follows in the following way. Assume $\vdash \alpha$. By application of Rule 2, we get $\vdash \bigcirc^k\alpha$, for every $k \in \omega$. We obtain $\vdash G\alpha$ by the derivable rule (1). From Axiom 12 and by application of Modus Ponens, we have $\vdash P_{\geq 1}\alpha$.

Then we can show that every consistent set T of formulas can be extended to a maximal consistent set. Let $\alpha_0, \alpha_1, \ldots$ be an enumeration of all formulas. A maximal consistent extension $\mathscr{T}$ of T can be obtained as follows:

1. $T_0 = T$.
2. For every $i \geq 0$ if $T_i \cup \{\alpha_i\}$ is consistent, then $T_{i+1} = T_i \cup \{\alpha_i\}$. Otherwise, if α_i is of the form $\gamma \to \neg(\alpha U\beta)$, then $T_{i+1} = T_i \cup \{\neg\alpha_i, \gamma \to \neg((\bigwedge_{k=0}^{j_0-1} \bigcirc^k\alpha) \wedge \bigcirc^{j_0}\beta)\}$ for some $j_0 \geq 0$ such that T_{i+1} is consistent. Otherwise, α_i is of the form $\gamma \to \bigcirc^m P_{\geq s}\beta$, then $T_{i+1} = T_i \cup \{\neg\alpha_i, \gamma \to \neg \bigcirc^m P_{\geq s-\frac{1}{j_0}}\beta\}$ for some $j_0 > 0$ such that T_{i+1} is consistent. Otherwise, $T_{i+1} = T_i \cup \{\neg\alpha_i\}$.
3. $\mathscr{T} = \bigcup_{i=0}^{\infty} T_i$.

For a maximal consistent extension $\mathscr{T}$ of a consistent set T of formulas we define the canonical model $\mathbf{M}_T = \langle W, Prob, v\rangle$ such that:

- $W = w_0, w_1, \ldots, w_0 = \mathscr{T}$, and for $i > 0$, $w_i = \{\alpha : \bigcirc\alpha \in w_{i-1}\}$,
- for $i \geq 0$, $Prob(w_i) = \langle W(w_i), H(w_i), \mu(w_i)\rangle$ is defined as follows:
  - $W(w_i) = \{w_{i+j} : j \geq 0\}$,
  - $H(w_i) = \{\{w_{i+j} : j \geq 0, \alpha \in w_{i+j}\}\}$,
  - for $\mu(w_i)(\{w_{i+j} : j \geq 0, \alpha \in w_{i+j}\}) = \sup_s\{P_{\geq s}\alpha \in w_i\}$,
- for every primitive proposition $p \in \phi$, and every $w_i \in W$, $v(w_i)(p) = \top$ iff $p \in w_i$.

First of all, we can prove that for every $i \geq 0$, w_i is a maximal consistent set. By hypothesis, w_0 is maximal and consistent. Suppose that w_{i+1} is not maximal. There is a formula α such that $\{\alpha, \neg\alpha\} \cap w_{i+1} = \emptyset$. Consequently, $\{\bigcirc\alpha, \bigcirc\neg\alpha\} \cap w_i = \emptyset$. We obtain that $\{\bigcirc\alpha, \neg\bigcirc\alpha\} \cap w_i = \emptyset$ which is in contradiction with the maximality of w_i. Suppose that w_{i+1} is not consistent, i.e., that $w_{i+1} \vdash \alpha \wedge \neg\alpha$. Then, $w_i \vdash \bigcirc(\alpha \wedge \neg\alpha)$, and $w_i \vdash \bigcirc\alpha \wedge \neg \bigcirc \alpha$ which is in contradiction with consistency of w_i.

Then, similarly as in the previous sections, we can show that $\mathbf{M}_T$ is an $LPP^{\mathrm{LTL}}_{1,\mathrm{Meas}}$-model such that for all w_i and α, $\alpha \in w_i$ iff $w_i \models \alpha$. For example, if $\alpha = \bigcirc\beta$, we have $w_i \models \alpha$ iff $w_{i+1} \models \beta$ iff $\beta \in w_{i+1}$ iff $\alpha \in w_i$ (by the construction of w_{i+1}). ■

For the previously presented logics as the first step in the proofs of their decidability we have used some kind of the filtration technique which helps as to show that every formula is satisfiable iff it is satisfiable in a finite model. The problem is that the filtration cannot be used here since the $LPP^{\mathrm{LTL}}_{1,\mathrm{Meas}}$-models are (by their definition) infinite. However, we can show (following the ideas presented in [20]) that a formula is satisfiable if and only if it is satisfiable in a model such that the sequence of time instants of the model has a finite initial sequence of time instants followed by another finite sequence of time instants, which permanently repeats and in that way forms the rest of the whole time-line. The lengths of both sequences are bounded by functions of the size of the considered formula. The full proof of decidability and complexity of the $LPP^{\mathrm{LTL}}_{1,\mathrm{Meas}}$-satisfiability problem can be found in [13]. As it is rather long, we give only the corresponding main statements:

Theorem 4.9 *Every $LPP^{\mathrm{LTL}}_{1,\mathrm{Meas}}$-satisfiable formula α is satisfiable in a model with the starting sequence of time instants, followed by the sequence of time instants which permanently repeats. The length of the former sequence is $\leq 2^{2|\alpha|} + 1$, and the length of the later sequence is $\leq (2^{|\alpha|} + 1) \times 2^{|\alpha|}$, where $|\alpha|$ denotes the length of α.*

Theorem 4.10 (Decidability and complexity for LPP^{LTL}_1) *The LPP^{LTL}_1 is decidable. The $LPP^{\mathrm{LTL}}_{1,\mathrm{Meas}}$-satisfiability problem is* PSPACE*-hard and in nondeterministic exponential time.*

References

1. Abadi, M., Halpern, J.Y.: Decidability and expressiveness for first-order logics of probability. Inf. Comput. **112**, 1–36 (1994)
2. Burgess, J.: Basic tense logic. In: Gabbay, D., Guenthner, F. (eds.) Handbook of Philosophical Logic, vol. II, pp. 89–133. Reidel Publishing Company, Dordrecht (1984)
3. Garson, J.W.: Basic modal logic. In: Gabbay, D., Hogger, C., Robinson, J. (eds.) Handbook of Logic in Artificial Intelligence and Logic Programming, vol. 1, pp. 368–448. Oxford University Press Inc., New York (1993)
4. Garson, J.W.: Quantification in modal logic. In: Gabbay, D., Guenthner, F. (eds.) Handbook of Philosophical Logic, vol. II, pp. 249–307. D. Reidel Publishing Company, Dordrecht (1984)
5. Halpern, Joseph Y.: An analysis of first-order logics of probability. Artif. Intell. **46**, 311–350 (1990)
6. Hughes, G.E., Cresswell, M.J.: Modal Logic. Methuen, London (1968)
7. Hughes, G.E., Cresswell, M.J.: A Companion to Modal logic. Methuen, London (1984)
8. Ikodinović, N., Ognjanović, Z.: A logic with coherent conditional probabilities. In: Godo, L. (ed.), Proceedings of the 8th European Conference Symbolic and Quantitative Approaches

to Reasoning with Uncertainty ECSQARU 2005, Barcelona, Spain, 6–8 July 2005. Lecture Notes in Computer Science (Subseries: Lecture Notes in Artificial Intelligence), vol. 3571, pp. 726–736. Springer (2005)

9. Kripke, S.: The undecidability of monadic modal quantification theory. Zeitschrift für Matematische Logik **8**, 113–116 (1962)
10. Marković, Z., Rašković, M., Ognjanović, Z.: A logic with approximate conditional probabilities that can model default reasoning. Int. J. Approx. Reason. **49**(1), 52–66 (2008)
11. Milošević, M., Ognjanović, Z.: A first-order conditional probability logic. Logic J. Interes. Group Pure Appl. Logics **20**(1), 235–253 (2012)
12. Milošević, M., Ognjanović, Z.: A first-order conditional probability logic with iterations. Publications de L'Institute Matematique, n.s. **93**(107), 19–27 (2013). http://elib.mi.sanu.ac.rs/files/journals/publ/113/n107p019.pdf
13. Ognjanović, Z.: Discrete linear-time probabilistic logics: completeness, decidability and complexity. J. Logic Comput. **16**(2), 257–285 (2006)
14. Ognjanović, Z., Ikodinović, N.: A logic with higher order conditional probabilities. Publications de L'Institute Matematique (Beograd), ns. **82**(96), 141–154 (2007).http://elib.mi.sanu.ac.rs/files/journals/publ/102/n096p141.pdf
15. Ognjanović, Z., Rašković, M.: A logic with higher order probabilities. Publications de L'Institute Matematique (Beograd), ns. **60**(74), 1–4 (1996). http://elib.mi.sanu.ac.rs/files/journals/publ/80/n074p001.pdf
16. Ognjanović, Z., Rašković, M.: Some first-order probability logics. Theor. Comput. Sci. **247**(1–2), 191–212 (2000)
17. Ognjanović, Z., Doder, D., Marković, Z.: A branching time logic with two types of probability operators. In: Benferhat, S., Grant, J. (eds.) Fifth International Conference on Scalable Uncertainty Management SUM-2011, 10–12 October 2011, Dayton, Ohio, USA. Lectuer Notes in Computer Science, vol. 6929, pp. 219–232. Springer (2011)
18. Ognjanović, Z., Rašković, M., Marković, Z.: Probability logics. In: Ognjanović, Z. (ed.) Zbornik radova. Subseries Logic in Computer Science, vol. 12(20), pp. 35–111. Matematički institut, Beograd (2009). http://elib.mi.sanu.ac.rs/files/journals/zr/20/n020p035.pdf
19. Perović, A., Ognjanović, Z., Rašković, M., Marković, Z.: A probabilistic logic with polynomial weight formulas. In: Proceedings of the 5th International Symposium Foundations of Information and Knowledge Systems FoIKS 2008, Pisa, Italy, 11–15 February 2008. Lecture Notes in Computer Science, vol. 4932, pp. 239–252 (2008)
20. Prasad Sistla, A., Clarke, E.M.: The complexity of propositional linear temporal logic. J. ACM **32**(3), 733–749 (1985)
21. Rašković, M., Ognjanović, Z.: Some propositional probabilistic logics. In: Kovačević-Vujičić, V., Mijajlović, Ž. (eds.) Proceedings of the Kurepa's Symposium 1996, Belgrade, pp. 83–90. Scientific review 19-20 (1996)

Chapter 5
Extensions of the Probability Logics LPP_2 and $LFOP_1$

Aleksandar Perović, Dragan Doder, Nebojša Ikodinović and Angelina Ilić Stepić

Abstract We describe various extensions of logics introduced in the Chaps. 3 and 4 that concern introduction of new types of probability operators, and various ranges of probability functions (finite ranges, non-Archimedean ranges, and unordered ranges). We outline the general features of the corresponding completeness-proof techniques. We present finitary probability logics for reasoning about probability measures with fixed finite ranges, and an infinitary logic with probability functions with arbitrary (not fixed) finite ranges. We introduce logics with the additional probability operators of the form Q_F. The intended meaning of $Q_F\alpha$ is that the probability of α is in F. A characterization of the hierarchy of logics with Q_F-operators is provided. We give strongly complete axiomatization for a logic with the qualitative probability operator $\preceq$. A probability extension of the intuitionistic logic is presented. Logics that correspond to Kolmogorov's and de Finetti's notions of conditional probabilities, and a logic with $[0, 1]_{\mathbb{Q}(\varepsilon)}$-valued probability functions with binary operators for conditional and approximate probabilities are presented. We describe strongly complete propositional axiomatizations for logics with linear and polynomial weight formulas. We consider axiomatization of probability functions with unordered ranges, and illustrate that using p-adic valued probabilities. This Chapter covers some results from Doder et al., Publications de L'Institut Mathematique (N.S.), 87(101), 85–96 (2010), [2], Doder and Ognjanović, Probabilistic logics with independence and probabilistic support, (2015), [3], Dordević et al. Arch. Math. Logic, 43, 557–563 (2004), [4], Ghilezan et al. Proceedings of the 22nd international conference on types for proofs and programs, TYPES (2016), [6], Ikodinović, Some Probability and Topological Logics (2005), [7], Ikodinović and Ognjanović, Proceedings of the 8th European Conference Symbolic and Quantitative Approaches to Reasoning with Uncertainty, ECSQARU (2005), [8], Ikodinović, J Multiple Valued Logic Soft Comput., 20(5–6), 527–555 (2013), [9], Ikodinović, Int. J. Approx. Reason. 55(9), 1830–1842, 2014), [13] (Ilić-Stepić, Math. Logic Q. 58(4–5), 63–280 (2012), [10], Ilić-Stepić, Int. J. Approx. Reason., 55(9), 1843–1865 (2014), [14], Ilić-Stepić,

The original version of this chapter was revised: The coauthors' names were added: The erratum to this chapter is available at: 10.1007/978-3-319-47012-2_8

Z. Ognjanović et al., *Probability Logics*,
DOI 10.1007/978-3-319-47012-2_5

Ognjanović, Publications de l'Institut Mathematique, N.s. tome, 95(109), 73–86 (2014), [11], Ilić-Stepić and Ognjanović, Studia Logica, 103, 145–174 (2015), [12], Kokkinis et al. Logic J. IGPL, 23(4), 662–687 (2015), [16], Kokkinis et al. roceedings of Logical Foundations of Computer Science International Symposium, LFCS, (2016), [17], Marković et al. Math. Logic Q. 49, 415–424 (2003), [19], Marković et al. Publications de L'Institute Matematique (N.S.), 73(87), 31–38 (2003), [20], Marković et al. IPMU 2004, 443–450 (2004), [21], Milošević and Ognjanović, Logic J. Interest Gr. Pure Appl. Logics, 20(1), 235–25 (2012), [22], Milošević and Ognjanović, Publications de L'Institute Matematique, N.S., 93(107), 19–27 (2013), [23], Ognjanović and Rašković, J. Logic Comput., 9(2), 181–195 (1999), [25], Ognjanović, Publications de L'Institute Matematique Ns., 78(92), 35–49 (2005), [26], Ognjanović and Ikodinović, Publications de L'Institute Matematique (Beograd), ns., 82(96), 141–154 (2007), [24], Ognjanović et al. Logic J. IGPL, 16(2), 105–120 (2008), [27], Perović, Some applications of the formal method in set theory, model theory, probabilistic logics and fuzzy logics, (2008), [28], Perović et al. 5th International Symposium on Foundations of Information and Knowledge Systems, FoIKS 2008, Proceedings, (2008), [29], Perović, et al. 11th European Conference on Logics in Artificial Intelligence, JELIA 2008, Proceedings, (2008), [30], Perović, et al. Fuzzy Sets Syst., 169, 65–90 (2011), [31], Rašković, J. Symb. Logic 51(3), 586–590 (1986), [32], Rašković et al. Int. J. Approx. Reason., 49(1), 52–66 (2008), [33], Savić et al. Proceedings of the 9th International Symposium on Imprecise Probability: Theories and Applications, ISIPTA, (2015), [34], Tomović, Proceedings of the 13th European Conference Symbolic and Quantitative Approaches to Reasoning with Uncertainty ECSQARU (2015), [35].

5.1 Generalization of the Completeness-Proof Technique

In order to avoid the repetition of essentially the same arguments, in this section we shall present a general outline of the completeness-proof technique for probability logics with countable inference rules and denumerable sets of formulas. Here we assume that the underlying formal system allows the basic properties of the inference relation such as weakening and deduction theorem.

Let *IR* be an arbitrary sound infinitary inference rule of the general form

$$\frac{\{\varphi_n\ :\ n \in \mathbb{N}\}}{\varphi}\ (IR).$$

The first technical step in the completeness proof is to show that if a consistent set of formulas T is incompatible with a particular conclusion of *IR*, then there is at least one premise φ_n such that $T \cup \{\neg\varphi_n\}$ is consistent. The proof is essentially the same as the one presented in Lemma 3.2.

Lemma 5.1 *Suppose that T is a consistent set of formulas, φ is a formula that can be obtained from $\{\varphi_n\ :\ n \in \mathbb{N}\}$, and that $T \cup \{\varphi\}$ is inconsistent. Then there is $n \in \mathbb{N}$ such that $T \cup \{\neg\varphi_n\}$ is consistent.* ■

The second technical step in the completeness proof (the proof of the corresponding Lindenbaum theorem) is to apply the Lemma 5.1 in all iterations where the current extension T_n is incompatible with the current formula φ, provided that φ is obtainable by some infinitary inference rule *IR*. Using an adaptation of the proof of Theorem 3.4 we obtain the following theorem:

Theorem 5.1 *Every consistent set of formulas T can be extended to a maximal consistent set of formulas.*

5.2 Logic $LPP_2^{\mathrm{Fr}(n)}$

The logic $LPP_2^{\mathrm{Fr}(n)}$ provides arguably the simplest method of resolving the non-compactness phenomenon of the logic LPP_2. Namely, for the given positive integer n, an LPP_2 consistent set of formulas

$$T_{\mathrm{Fr}(n)} = \left\{ \bigvee_{i=0}^{n} P_{=i/n}\alpha \ : \ \alpha \in For_C \right\}$$

ensures that all formulas have the probability in the set

$$\mathrm{Fr}(n) = \left\{ 0, \frac{1}{n}, \frac{2}{n}, \ldots, \frac{n-1}{n}, 1 \right\}$$

and makes any use of the Archimedean rule in the proofs from $T_{\mathrm{Fr}(n)}$ redundant. More precisely, the following theorem holds:

Theorem 5.2 *Suppose that T is a consistent set of LPP_2-formulas and that $T \cup T_{\mathrm{Fr}(n)} \vdash_{LPP_2} \varphi$. Then, there is a finite LPP_2-proof of φ from $T \cup T_{\mathrm{Fr}(n)}$.*

Proof The standard proof technique is induction on the length of the inference. The only slightly nontrivial case is the application of the Archimedean rule. So suppose that

$$\{\theta \to P_{>r}\alpha \ : \ r < s, \ r \in [0, 1_{\mathbb{Q}}]\} \cup \{\theta \to P_{\geq s}\alpha\}$$

is an LPP_2 proof of $\theta \to P_{\geq s}\alpha$ from $T \cup T_{\mathrm{Fr}(n)}$. There is a unique $k \in \{0, 1, 2, \ldots, n\}$ such that $\frac{k}{n} \leq s \leq \frac{k+1}{n}$. By the completeness theorem for LPP_2,

$$T_{\mathrm{Fr}(n)} \vdash (\theta \to P_{\geq s}\alpha) \leftrightarrow (\theta \to P_{\geq \frac{k+1}{n}}\alpha)$$

and

$$T_{\mathrm{Fr}(n)} \vdash (\theta \to P_{>r}\alpha) \leftrightarrow (\theta \to P_{\geq \frac{k+1}{n}}\alpha)$$

for all $r \geq \frac{k}{n}$, so the above infinite proof can be reduced to the pair of formulas

$$\{\theta \to P_{>\frac{k}{n}}\alpha, \theta \to P_{\geq\frac{k+1}{n}}\alpha\}.$$ ■

In the axiomatization of $LPP_2^{\mathrm{Fr}(n)}$ logics (varying n produces different logics) we shall use the fact that any $\mathrm{Fr}(n)$ is a discrete ordering, so successor and predecessor functions are defined on it (the only exceptions are the endpoints 0 and 1). If $s \in \mathrm{Fr}(n)$, then its immediate successor will be denoted by s^+ (here 1 is excluded), while its immediate predecessor will be denoted by s^- (here 0 is excluded).

The axiomatic system $Ax_{LPP_2^{\mathrm{Fr}(n)}}$ differs from the system Ax_{LPP_2} in the following way:

- It has one additional axiom $P_{>s}\alpha \to P_{\geq s^+}\alpha$, where $s \in \mathrm{Fr}(n) \setminus \{1\}$;
- The Archimedean rule is excluded.

Note that $Ax_{LPP_2^{\mathrm{Fr}(n)}}$ is a finitary axiomatic system. The next lemma shows that $Ax_{LPP_2^{\mathrm{Fr}(n)}}$ has essentially the same properties as the above introduced LPP_2 set of formulas $T_{\mathrm{Fr}(n)}$.

Lemma 5.2 *Let α be a formula. Then*

1. $\vdash P_{<r}\alpha \to P_{\leq r^-}\alpha$,
2. $\vdash P_{>r}\alpha \leftrightarrow P_{\geq r^+}\alpha$,
3. $\vdash P_{\leq r^-}\alpha \leftrightarrow P_{<r}\alpha$,
4. $\vdash \bigvee_{s\in\mathrm{Fr}(n)} P_{=s}\alpha$,
5. $\vdash \underline{\bigvee}_{s\in\mathrm{Fr}(n)} P_{=s}\alpha$, *where* $\underline{\bigvee}$ *denotes the exclusive disjunction.*

Proof It is essentially the same argument as in the proof of Theorem 5.2. All five statements are LPP_2 consequences of the additional axiom schemata $P_{>s}\alpha \to P_{\geq s^+}\alpha$ treated as a set of LPP_2 formulas. By Theorem 5.2, all applications of the Archimedean rule can be replaced by finite sets of formulas, so any LPP_2 proof from $P_{>s}\alpha \to P_{\geq s^+}\alpha$ can be transformed into an $LPP_2^{\mathrm{Fr}(n)}$ proof. ■

The completeness proofs for the classes $LPP^{\mathrm{Fr}(n)}_{2,\mathrm{Meas}}$, $LPP^{\mathrm{Fr}(n)}_{2,\mathrm{Meas,All}}$, $LPP^{\mathrm{Fr}(n)}_{2,\mathrm{Meas},\sigma}$, and $LPP^{\mathrm{Fr}(n)}_{2,\mathrm{Meas,Neat}}$ are similar to the corresponding proofs in the case of LPP_2 logic. Theorem 5.3 announces a property that does not hold for the infinitary systems considered before. Another difference between logics from this and the previous sections is illustrated in Example 5.1.

Theorem 5.3 (Compactness theorem for $LPP_2^{\mathrm{Fr}(n)}$) *Let L be any class of models considered in this section and T be a set of formulas. If every finite subset of T is L-satisfiable, then T is L-satisfiable.*

Proof If T is not L-satisfiable, then it is not $Ax_{LPP_2^{\mathrm{FR}(n)}}$-consistent. It follows that $T \vdash \bot$. Since the axiomatic system $Ax_{LPP_2^{\mathrm{FR}(n)}}$ is finitary one, there must be a finite set $T' \subset T$ such that $T' \vdash \bot$. It is a contradiction because every finite subset of T is both L-satisfiable and $Ax_{LPP_2^{\mathrm{FR}(n)}}$-consistent. ■

Example 5.1 For every positive integer n and Range defined as above, it is easy to construct an $LPP_{2,\mathrm{Meas}}$-model $\mathbf{M} = \langle W, H, \mu, v \rangle$ which does not satisfy Axiom $P_{>s}\alpha \to P_{\geq s^+}\alpha$. For example, let $n = 3$, and $p \in \varphi$

- $W = \{w_1, w_2\}$,
- H is the power set of W
- $\mu(w_1) = 1/2$, $\mu(w_2) = 1/2$ and
- $v(w_1, p) = true$, $v(w_2, p) = false$.

Since $\mu([p]_{\mathbf{M}}) = 1/2$, obviously $\mathbf{M} \models P_{>1/3}p$, and $\mathbf{M} \not\models P_{\geq 2/3}p$, so the instance

$$P_{>1/3}p \to P_{\geq 2/3}p$$

of the considered Axiom does not hold in $\mathbf{M}$. ■

Finally, decidability of the satisfiability problem for the classes of models considered in this section can be proved similarly as in the LPP_2 case. The only difference is in the fact that the measures of atoms must be in the set $\mathrm{Fr}(n)$. Since that set is always finite, there are only finitely many possibilities for such distributions, and decidability easily follows.

5.3 Logic $LPP_2^{A,\omega_1,\mathrm{Fin}}$

The present section offers a complete axiomatization of probability functions with arbitrary finite ranges, which substantially generalize the case of a fixed finite range treated in Sect. 5.2.

Arguably, the most interesting challenge in the development process was to provide an adequate formal representation of the finiteness of range.

Though the solution is not unique, the following characterization from [1] provides a nice foundation for implementation of techniques from infinitary logics and admissible set theory (for more information on infinitary logic and admissible set theory, see the corresponding sections in the introductory chapter.)

Theorem 5.4 *Let $\langle W, H, \mu \rangle$ be a probability space. Then, the following statements are equivalent:*

1. *μ has a finite range;*
2. *There is a real number c from the open unit interval $(0, 1)$ such that*

$$\mu(X) > 0 \Rightarrow \mu(X) > c$$

for all $X \in H$. ■

The complete axiomatization of the logic $LPP_2^{A,\omega_1,\mathrm{Fin}}$ presented here was published in [4]. It is a somewhat extensive modification of our early work on probability logics with probability quantifiers published in [32].

Throughout this section $\mathbb{A}$ will denote a large enough countable admissible set that contains all objects relevant for our purpose. As it is usual in infinitary logics, finitary and infinitary formulas are treated as sets. Consequently, admissible sets are closed for Boolean combinations of formulas and countable conjunctions and disjunctions of formulas.

The basic syntactical objects of $LPP_2^{A,\omega_1,\mathrm{Fin}}$ are the countable set *Var* of propositional variables and the countable set of probabilistic operators $\{P_{\geq s} : s \in [0,1] \cap \mathbb{A}\}$. The inductive definition of the set of all $LPP_2^{A,\omega_1,\mathrm{Fin}}$-formulas is given below

- propositional variables are $LPP_2^{A,\omega_1,\mathrm{Fin}}$-formulas;
- the set of $LPP_2^{A,\omega_1,\mathrm{Fin}}$-formulas is closed for Boolean combinations;
- if $G \in \mathbb{A}$ is a countable set of $LPP_2^{A,\omega_1,\mathrm{Fin}}$-formulas, then both $\bigwedge G$ and $\bigvee G$ are $LPP_2^{A,\omega_1,\mathrm{Fin}}$-formulas.

A standard method for formalization of De Morgan laws for infinitary logics requires introduction of the novel syntactical object $\varphi\neg$. The corresponding inductive definition goes as follows:

- $p\neg =_{\mathrm{def}} \neg p$ for any propositional variable p;
- $(\varphi \wedge \psi)\neg =_{\mathrm{def}} (\varphi\neg) \vee (\psi\neg)$;
- $(\varphi \vee \psi)\neg =_{\mathrm{def}} (\varphi\neg) \wedge (\psi\neg)$;
- $(\bigwedge G)\neg =_{\mathrm{def}} \bigvee_{\varphi \in G}(\varphi\neg)$;
- $(\bigvee G)\neg =_{\mathrm{def}} \bigwedge_{\varphi \in G}(\varphi\neg)$.

Here we consider a particular subclass of the class $LPP_{2,\mathrm{Meas}}$ of all measurable probabilistic models. We denote it $LPP_{2,\mathrm{Meas}}^{A,\omega_1,\mathrm{Fin}}$,and it contains all measurable models whose measures have finite ranges. The satisfiability relation $\models$ generalizes the corresponding relation from Definition 3.4. The new cases are related to infinitary formulas

- if G is a finite or countable set of For_P-formulas, $\mathbf{M} \models \bigwedge G$ iff for every $B \in G$, $\mathbf{M} \models B$, and
- if G is a finite or countable set of For_P-formulas, $\mathbf{M} \models \bigvee G$ iff there is some $b \in G$ so that $\mathbf{M} \models B$.

The axiomatic system $Ax_{LPP_2^{A,\omega_1,\mathrm{Fin}}}$ contains all the axioms and rules from the system Ax_{LPP_2}, and also the following new axioms:

7. $(\neg\varphi) \leftrightarrow (\varphi\neg)$
8. $(\bigwedge_{B \in G} B) \rightarrow C$, $C \in G$, $G \in A$, G is a set of probability formulas
9. $\bigvee_{c>0} \bigwedge_{\alpha \in G}(P_{>0}\alpha \rightarrow P_{>c}\alpha)$, $G \in A$, G is a set of classical propositional formulas

and the rule

4. From $B \to C$, for all $C \in G$, infer $B \to \bigwedge_{C \in G} C$, G is a set of probability formulas

introduced in [15]. In the completeness proof a result from [1] and the weak–strong model construction from [32] will be used.

Theorem 5.5 *An $LPP_2^{A,\omega_1,\mathrm{Fin}}$-formula φ is consistent iff it is satisfiable in a weak model in which every $LPP_2^{A,\omega_1,\mathrm{Fin}}$-theorem is true.*

Proof The simpler direction follows from the soundness of the axiomatic system. For the other direction, we use argument presented in Theorem 5.1 to construct a maximal consistent set T that contains φ. Then, we can follow the completeness prof for $LPP_{2,\mathrm{Meas}}$, and construct the canonical model $\mathbf{M}_\varphi$. The axioms guarantee that $\mathbf{M}_\varphi$ is a weak model in which every $LPP_2^{A,\omega_1,\mathrm{Fin}}$-theorem is true, and that $\mathbf{M}_\varphi \models \varphi$ iff $\varphi \in T$. ■

Note that, although in a weak model (since Axiom 9 holds) for every For_C-formula α the following condition is satisfied:

$$\text{if } \mathbf{M} \models P_{>0}\alpha \text{ then } \mathbf{M} \models P_{>c}\alpha. \tag{5.1}$$

It may be the case that there is no single $c > 0$ such that the condition (5.1) holds for all formulas. Thus, we will now construct the corresponding strong model, i.e., a weak model $\mathbf{M}$ which satisfies that there is a $c > 0$ such that for every For_C-formula α the condition (5.1) holds. By Theorem 3.2.10 from [1] (see Theorem 5.4), measures from a strong model have finite ranges, and the model belongs to the $LPP_{2,\mathrm{Meas}}^{A,\omega_1,\mathrm{Fin}}$-class.

Theorem 5.6 *An $LPP_2^{A,\omega_1,\mathrm{Fin}}$-formula φ is consistent iff it is satisfiable in a strong model in which every $LPP_2^{A,\omega_1,\mathrm{Fin}}$-theorem is true.*

Proof Again, the simpler direction follows from the soundness of the axiomatic system. To prove another part of the statement we consider a language L_A containing

- the following three kinds of variables
  - variables for sets $(X, Y, Z, \ldots)$,
  - variables for elements $(x, y, z, \ldots)$,
  - variables for reals from $[0, 1]$ $(r, s, \ldots)$, and
  - variables for positive reals greater than 1 $(u, v, \ldots)$
- the predicates: $\leq$ for reals, $V(u, u)$, $E(x, X)$ and $\mu(X, r)$,
- a set constant symbol W_α for every $LPP_2^{A,\omega_1,\mathrm{Fin}}$-$For_C$-formula α,
- a constant symbol r' for every real number $r \in [0, 1] \cap A$, and
- two function symbols for additions and multiplications for reals.

The intended meaning of $E(x, X)$ is $x \in X$, $V(u, u)$ means that a formula φ with the Gödel-number u (denoted $gb(\varphi) = u$) holds in the model, while $\mu(X, r)$ can be understood as "r is the measure of X". We use $\mu(X) \geq r$ to denote $(\exists s)(s \geq r \wedge \mu(X, s))$, and $V(\varphi)$ to denote $V(gb(\varphi), gb(\varphi))$.

We define a theory T of $L_{\omega_1\omega} \cap \mathbb{A}$ which contains the following formulas:

1. $(\forall X)(\forall Y)((\forall x)(E(x, X) \leftrightarrow E(x, Y)) \leftrightarrow X = Y)$
2. $(\forall x)(E(x, W_{\alpha\wedge\beta}) \leftrightarrow (E(x, W_\alpha) \wedge E(x, W_\beta)))$ for every $\alpha \wedge \beta \in For_C$
3. $(\forall x)(E(x, W_{\neg\alpha}) \leftrightarrow \neg E(x, W_\alpha))$, for every $\alpha \in For_C$
4. $(\forall x)(E(x, W_{p\vee\neg p})$
5. $V(\alpha) \leftrightarrow W_\alpha = W_{p\vee\neg p}$, for every $\alpha \in For_C$
6. $V(P_{\geq s}\alpha) \leftrightarrow \mu(W_\alpha) > s$, for every $\alpha \in For_C$
7. $V(\bigwedge_{B\in G} B) \leftrightarrow \bigwedge_{B\in G} V(B)$, for every set of probability formulas $G \in A$
8. $V(\neg B) \leftrightarrow \neg V(B)$, for every $LPP_2^{A,\omega_1,\mathrm{Fin}}$-$For_P$-formula B
9. $(\forall X)(\exists_1 r)\mu(X, r)$
10. $(\forall X)(\forall Y)((\mu(X, r) \wedge \mu(Y, s) \wedge \neg(\exists y)(E(y, X) \wedge E(y, Y))) \rightarrow$
    $\rightarrow (\exists Z)((\forall y)((E(y, X) \vee E(y, Y)) \leftrightarrow E(y, Z)) \wedge \mu(Z, r + s)))$
11. $(\forall X)((\forall y)E(y, X) \rightarrow \mu(X, 1))$
12. $(\exists r > 0)(\forall X)(\mu(X) > 0 \rightarrow \mu(X) > r)$
13. Axioms for Archimedian fields for real numbers
14. $(\forall x)E(x, W_\psi)$ where ψ is an axiom of $LPP_2^{A,\omega_1,\mathrm{Fin}}$
15. $(\exists x)E(x, W_\varphi)$ where φ is the formula from the formulation of the statement.

Let a standard model for $L_\mathbb{A}$ be

$$\langle W, H, F, V, E, \mu, +, \cdot, \leq, W_\alpha, r\rangle_{\alpha\in For_C, r\in F},$$

where $H \subset \mathbb{P}(W)$, $F = F' \cap [0, 1]$, $F' \subset \mathbb{R}$ a field, $V \subset \mathbb{R} \times \mathbb{R}$, $E \subset W \times H$, $\mu : H \rightarrow F, +, \cdot : F^2 \rightarrow F, \leq \subset F^2$, and $W_\alpha \in H$.

Let $\mathbf{M} = \langle W, H, \mu, v\rangle$ be a weak model for $LPP_2^{A,\omega_1,\mathrm{Fin}}$. If we define $W_\alpha = \bigcup_{w\in W}[\alpha]_w$, and $H = \{W_\alpha : \alpha \in For_C\}$, it easy to show that $\mathbf{M}$ can be transformed to a standard model. On the other hand, if ψ is a consistent $LPP_2^{A,\omega_1,\mathrm{Fin}}$-formula, then there is a weak model in which it is satisfied, and consequently there is a standard model in which $V(\psi)$ holds.

Let $T_0 \subset T$, $T_0 \in \mathbb{A}$. Since Axiom 9 holds in the weak model M it follows that every T_0 has a model. Hence, by the Barwise compactness theorem, T has a model $\mathbf{M}' = \langle W, H, F, V, E, \mu, +, \cdot, \leq, W_\alpha, r\rangle_{\alpha\in For_C, r\in F}$. Finally, we use $\mathbf{M}'$ to obtain the strong model $\mathbf{M}'' = \langle W, H, \mu, v\rangle$ of ϕ in the following way:

- for every $w \in W$, $v(w, p) = true$ iff $w \in W_p$ for every primitive proposition p,
- $H = \{W_\alpha : \alpha \in For_C\}$,
- $\mu(X) = r$ iff $\mu(X, r)$ holds in $\mathbf{M}'$.

Indeed, in order to verify that $\mathbf{M}''$ has required properties, it remains to check whether $\mathbf{M}'' \models \varphi$, which is an immediate consequence of the fact that (15) holds in $\mathbf{M}'$. ■

Completeness also holds for Σ_1 definable sets of formulas (i.e., recursively enumerable sets), but it is possible to show that it cannot be generalized to arbitrary sets of formulas.

5.4 Probability Operators of the Form Q_F

In this section, we will analyze an extensions of LPP_2. We will use $LPP_{2,P,Q,O}$ to denote a probability logic which depends on a recursive family O of recursive subsets of $[0, 1]$ in a manner which will be explained below, while P and Q in the index means that two kinds of probabilistic operators will be used. More precisely, the language of $LPP_{2,P,Q,O}$ extends the LPP_2-language with a list of unary probabilistic operators of the form Q_F, where $F \in O$. For example, the set $For_{LPP_{2,P,Q,O}}$ of formulas contains $Q_F\alpha \rightarrow \neg P_{\geq s}\beta$. Note that every particular choice of the family O of sets produces a different probability language, a different set of probability formulas and a distinct $LPP_{2,P,Q,O}$-logic.

To give semantics to formulas, we use the class $LPP_{2,\mathrm{Meas}}$ of measurable LPP_2-models, and the corresponding satisfiability relation (from Definition 3.4) with additional requirement that

- $\mathbf{M} \models Q_F\alpha$ iff $\mu([\alpha]) \in F$, for every $F \in O$

which covers the case of the new operators. Note that $\neg Q_F\alpha$ is not equivalent to $Q_{[0,1]\setminus F}\alpha$ because $[0, 1] \setminus F \notin O$, and the later is not a well formed formula.

5.4.1 Complete Axiomatization

Let us consider a fixed recursive family O of recursive subsets of S and the corresponding $LPP_{2,P,Q,O}$-logic. The axiomatic system $Ax_{LPP_{2,P,Q,O}}$ extends the system Ax_{LPP_2} with the following axiom:

7. $P_{=s}\alpha \rightarrow Q_F\alpha$, where $F \in O$ and $s \in F$

and the inference rule

4. Q_F-**rule.** From $P_{=s}\alpha \rightarrow \varphi$, for all $s \in F$, infer $Q_F\alpha \rightarrow \varphi$.

As an illustration we give a list of useful theorems of $Ax_{LPP_{2,P,Q,O}}$

Theorem 5.7 *If all the mentioned formulas belong to the set* $For_{LPP_{2,P,Q,O}}$, *the following holds in the corresponding* $LPP_{2,P,Q,O}$*-logic:*

1. $\vdash Q_F\alpha \rightarrow Q_G\alpha$, *for* $F \subset G$
2. $\vdash (Q_F\alpha \wedge Q_G\alpha) \leftrightarrow Q_{F\cap G}\alpha$
3. $\vdash (Q_F\alpha \vee Q_G\alpha) \leftrightarrow Q_{F\cup G}\alpha$
4. $\vdash (Q_F\alpha \wedge P_{\geq s}\alpha) \leftrightarrow Q_{[s,1]\cap F}\alpha$, *and similar for* $P_{\geq s}\alpha$, $P_{\leq s}\alpha$, $P_{\leq s}\alpha$
5. $\vdash Q_F\alpha \leftrightarrow Q_{1-F}\neg\alpha$, *where* $1 - F = \{1 - f : f \in F\}$
6. $\vdash (Q_F\alpha \wedge \neg Q_G\alpha) \leftrightarrow Q_{F\setminus G}\alpha$

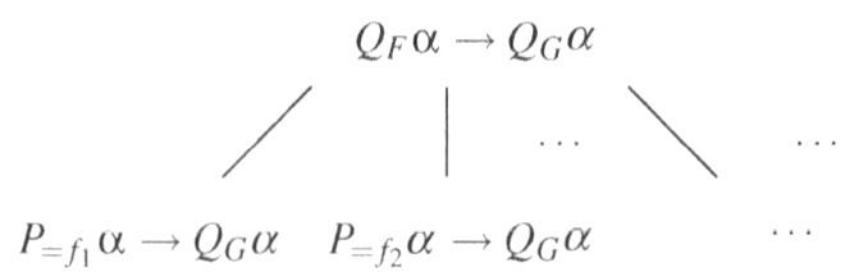

Fig. 5.1 Tree-like representation of the proof for $\vdash Q_F\alpha \to Q_G\alpha$, $F = \{f_1, f_2, \ldots\} \subset G$

Proof Let us consider the case (1). If $F, G \in O$:

- $\vdash P_{=s}\alpha \to Q_G\alpha$ for every $s \in F \subset G$, by Axiom 7
- $\vdash Q_F\alpha \to Q_G\alpha$, by Rule 4.

This proof is illustrated in Fig. 5.1. The other statements follow similarly. ■

The strong completeness of $Ax_{LPP_{2,P,Q,O}}$ can be shown following the completeness-proof technique outlined in Sect. 5.1. In particular, $Ax_{LPP_{2,P,Q,O}}$ has two infinitary inference rules so the step 4 in the proof of Theorem 5.1 has two instances: one for the Archimedean rule, and the other for the Q_F-rule 4.

5.4.2 Decidability

In Sect. 3.5, we proved decidability of the LPP_2 logic which can be seen as an $LPP_{2,P,Q,O}$-logic with the empty family O. The proof involves a reduction of a formula to a system of linear (in)equalities. A look on this method indicates that the similar procedure might be applied for an arbitrary $LPP_{2,P,Q,O}$-logic. However, since there are also the operators of the form Q_F, instead of the system (3.13), we have to consider linear systems of the following form:

$$
\begin{array}{rl}
\sum_{i=1}^{2^n} y_i & = 1 \\
y_i & \geq 0, \text{ for } i = 1, \ldots, 2^n \\
\sum_{a_t \in X^1(p_1,\ldots,p_n) \in D} y_t & \begin{cases} \geq s_1 & \text{if } X^1 = P_{\geq s_1} \\ < s_1 & \text{if } X^1 = P_{<s_1} \\ \in F_1 & \text{if } X^1 = Q_{F_1} \\ \notin F_1 & \text{if } X^1 = \neg Q_{F_1} \end{cases} \\
\ldots & \\
\sum_{a_t \in X^k(p_1,\ldots,p_n) \in D} y_t & \begin{cases} \geq s_k & \text{if } X^k = P_{\geq s_k} \\ < s_k & \text{if } X^k = P_{<s_k} \\ \in F_k & \text{if } X^1 = Q_{F_k} \\ \notin F_k & \text{if } X^1 = \neg Q_{F_k} \end{cases}
\end{array}
\tag{5.2}
$$

An obvious statement holds

Theorem 5.8 *An $LPP_{2,P,Q,O}$-logic is decidable iff for every probabilistic formula $A \in For_{LPP_{2,P,Q,O}} \setminus For_C$ there is at least one disjunct from $DNF(A)$ such that the corresponding system (5.2) is solvable.*

The requirement from Theorem 5.8 is very strong. For example, consider the system

$$\begin{aligned} y_1 + y_2 &= 1 \\ y_i &\geq 0, \text{ for } i = 1, 2 \\ y_1 &\geq s \\ y_1 &\in F \end{aligned}$$

obtained from the formula $P_{\geq s}p \wedge Q_F p$. The system is solvable only if $F \cap [s, 1] \neq \emptyset$ is decidable, and this depends on the set F. If F is a codomain of a suitable rational-valued function, the system can be solved, but, in the general case, decidability of the set F does not imply that either the system is solvable or that the $LPP_{2,P,Q,O}$-logic is decidable. However, there are recursive families O such that the corresponding probabilistic logics are decidable. A trivial example of this kind is any recursive $O \subseteq [S]^{<\omega}$, where $[S]^{<\omega}$ is the family of all finite subsets of S. A nontrivial example of a decidable logics concerns the logic which is characterized by the family O such that each $F \in O$ is definable (with rational parameters) in the language of ordered groups.

5.4.3 The Lower and the Upper Hierarchy

Using the semantics of $P_{\geq s}$ and Q_F-operators, it obviously holds that for a set $F \in O$,

$$Q_F\alpha \Leftrightarrow \bigvee_{f_i \in F} P_{=f_i}\alpha.$$

However, if the set F is not finite, the right-hand side of this equality is an infinitary disjunction that is clearly not an $LPP_{2,P,Q,O}$-formula. Similarly goes for the formula $P_{\geq s}\alpha \leftrightarrow Q_{[s,1]}\alpha$, where s is a rational number from $[0, 1)$, the formula $Q_{[s,1]}\alpha \notin For_{LPP_{2,P,Q,O}}$. More formally:

Definition 5.1 Let $\varphi, \Psi \in For_{LPP_{2,P,Q,O}}$. Let $\mathrm{Mod}(\varphi) = \{\mathbf{M} \in LPP_{2,\mathrm{Meas}} : M \models \varphi\}$. φ is definable from Ψ if $\mathrm{Mod}(\varphi) = \mathrm{Mod}(\Psi)$. ■

The above discussion suggests that generally neither the $P_{\geq .}$-operators are definable from the $Q_.$-operators (i.e., some formulas on the language $\{\neg, \wedge, P_{\geq .}\}$ are not definable from the formulas on the language $\{\neg, \wedge, Q_.\}$), nor are the $Q_.$-operators definable from the $P_{\geq .}$-operators. The next theorems formalize these conclusions.

Theorem 5.9 *Let O be a recursive family of recursive rational subsets of* $[0, 1]$, *$F \in O$ an infinite set, and $LPP_{2,P,Q,O}$ the corresponding logic. For an arbitrary primitive proposition p, there is no probabilistic formula A on the sublanguage $\{\neg, \wedge, P_{\geq .}\}$ such that $Q_F p$ is definable from A.*

Proof Suppose that there is a formula A on the language $\{\neg, \wedge, P_{\geq .}\}$ such that

$$\mathrm{Mod}(Q_F p) = \{\langle W, H, \mu, v\rangle : \mu([p]) \in F\} = \mathrm{Mod}(A).$$

Recall that A is satisfiable iff at least a system from the set of all linear systems that correspond to $DNF(A)$ is satisfiable. Let a_t's be the atoms of A and y_t's be the corresponding measures. The solutions of any of those systems must satisfy $\sum_{a_t \in DNF(p)} y_t \in F$. But, the solutions of the systems are of the following form: $y_t \in (r, s)$, $y_t \in [r, s)$, $y_t \in (r, s]$, and $y_t \in [r, s]$. Such sets of solutions cannot produce the infinite, but denumerable set F. Hence, $Q_F p$ is not definable over A. ■

Theorem 5.10 *Let O be a recursive family of recursive rational subsets of* $[0, 1]$, *$LPP_{2,P,Q,O}$ the corresponding logic, and $s \in S \setminus \{1\}$. For an arbitrary primitive proposition $p \in \varphi$, there is no probabilistic formula A on the sublanguage $\{\neg, \wedge, Q_.\}$ such that $P_{\geq s}p$ is definable from A.*

Proof Suppose that there is a formula A on the language $\{\neg, \wedge, Q_.\}$ such that

$$\mathrm{Mod}(P_{\geq s}p) = \mathrm{Mod}(A).$$

The models of A are exactly those that satisfy that $\mu[p] \geq s$. But, similarly as above, the set of values for $\mu[p]$ produced by $\mathrm{Mod}(A)$ can be either denumerable, or its complement is denumerable. Hence, $P_{\geq s}p$ cannot be definable over A. ■

Example 5.2 Formulas with the new probabilistic operators are suitable for reasoning about discrete sample spaces. For example, consider an experiment which consists of tossing a fair coin an arbitrary, but finite number of times. Then, $Q_F\alpha$ holds in this model, where α means that only heads (i.e., no tails) is observed in the experiment, and F denotes the set $\{\frac{1}{2}, \frac{1}{2^2}, \frac{1}{2^3}, \ldots\}$. Since Q_F is not definable over the probability language $\{\neg, \wedge, P_{\geq .}\}$, this sentence cannot be described in the probability logics used so far. ■

5.4.4 Representability

The central part in the classification of $LPP_{2,P,Q,O}$ logics is the characterization of their mutual expressiveness. We say that the logic L_2 is more expressive than the logic L_1 iff for each L_1-formula φ exists an L_2-formula ψ such that $\mathrm{Mod}(\varphi) = \mathrm{Mod}(\psi)$.

The main result of this subsection is the proof of the characterization theorem (Theorem 5.11). We start with some notational remarks that will be used later on.

Note that the following semantical equivalences hold:

- $Q_{X \cap Y}\alpha \Leftrightarrow Q_X\alpha \wedge Q_Y\alpha$;
- $Q_{X \cup Y}\alpha \Leftrightarrow Q_X\alpha \vee Q_Y\alpha$;

- $Q_{X \setminus Y}\alpha \Leftrightarrow Q_X\alpha \wedge \neg Q_Y\alpha$;
- $Q_{1-X}\alpha \Leftrightarrow Q_X\neg\alpha$. Here the set $1 - X$ is defined by

$$1 - X = \{1 - a \,:\, a \in X\}.$$

The introduced operation $1-$ will be called the quasi-complement. It is easy to verify the following properties for every formula α:

1. $Q_{1-(F\cap G)}\alpha \Leftrightarrow Q_{(1-F)\cap(1-G)}\alpha$;
2. $Q_{1-(F\cup G)}\alpha \Leftrightarrow Q_{(1-F)\cup(1-G)}\alpha$;
3. $Q_{1-(F\setminus G)}\alpha \Leftrightarrow Q_{(1-F)\setminus(1-G)}\alpha$;
4. $Q_{1-(1-F)}\alpha \Leftrightarrow Q_F\alpha$.

For example, suppose that $M = \langle W, H, v, \mu\rangle$ is an arbitrary model of $Q_{1-(F\cap G)}\alpha$. Then $\mu[\alpha] \in 1 - (F \cap G)$, so there is $x \in \mu[\alpha] \in F \cap G$ such that $\mu[\alpha] = 1 - x$. Then, $1 - x \in 1 - F$, and $1 - x \in 1 - G$. Thus, $1 - x \in (1 - F) \cap (1 - G)$, and $M \models Q_{(1-F)\cap(1-G)}\alpha$. The converse direction follows similarly.

Moreover, we define operators $Q_{[r,s]}$, $Q_{(r,s]}$, $Q_{[r,s)}$, $Q_{(r,s)}$ and $Q_{[0,1]\setminus F}$ as follows:

- $Q_{[r,s]}\alpha \Leftrightarrow P_{\geq r}\alpha \wedge P_{\leq s}\alpha$;
- $Q_{(r,s]}\alpha \Leftrightarrow P_{>r}\alpha \wedge P_{\leq s}\alpha$;
- $Q_{[r,s)}\alpha \Leftrightarrow P_{\geq r}\alpha \wedge P_{<s}\alpha$;
- $Q_{(r,s)}\alpha \Leftrightarrow P_{>r}\alpha \wedge P_{<s}\alpha$;
- $Q_{[0,1]\setminus F}\alpha \Leftrightarrow \neg Q_F\alpha$.

Definition 5.2 Let O be a recursive family of recursive subsets of $[0, 1]_{\mathbb{Q}}$. By $\overline{O}$ we will denote the minimal superset of the family

$$O \cup \{[r, s] \,:\, r, s \in [0, 1]_{\mathbb{Q}} \text{ and } r < s\}$$

that is closed under the operations $\cup$, $\cap$, $\setminus$ and $1-$. ■

Definition 5.3 Let O_1 and O_2 be recursive families of recursive subsets of $[0, 1]_{\mathbb{Q}}$. We say that O_1 is representable in O_2 iff $O_1 \subseteq \overline{O_2}$. ■

As an example, consider a positive integer $k > 0$, the sets

- $F_1 = \{\frac{1}{2^i} : i = k, k+1, \ldots\} \cup \{\frac{3^i-1}{3^i} : i = k, k+1, \ldots\}$,
- $F_2 = \{\frac{1}{2^i} : i = 1, 2, \ldots\}$,
- $F_3 = \{\frac{1}{3^i} : i = 1, 2, \ldots\}$,

and the family $O_2 = \{F_2, F_3\}$. By Definition 5.3, F_1 is representable in O_2 because $F_1 = (F_2 \cap [0, \frac{1}{2^k}]) \cup ((1 - F_3) \cap [\frac{3^k-1}{3^k}, 1])$. On the other hand, the set

$$F_4 = \left\{\frac{1}{2^{2i}} : i = 1, 2, \ldots\right\}$$

is not representable (i.e., LPP_{2,P,Q,O_2}-definable) in O_2.

Definition 5.4 Let O_1 and O_2 be recursive families of recursive subsets of $[0,1]_{\mathbb{Q}}$, and L_1 and L_2 be the corresponding $LPP_{2,P,Q,O}$-logics. The logic L_2 is more expressive than the logic L_1 ($L_1 \leq L_2$) iff for every formula $\varphi \in For(P,Q,O_1)$ there is a formula $\psi \in For(P,Q,O_2)$ such that $\mathrm{Mod}(\varphi) = \mathrm{Mod}(\psi)$. ■

Theorem 5.11 *Let $L_1 \leq L_2$. Then, $\overline{O}_1 \subseteq \overline{O}_2$.*

Proof Clearly, it is enough to prove that $O_1 \subseteq \overline{O}_2$ follows from $L_1 \leq L_2$. Moreover, any finite subset of $[0,1]_{\mathbb{Q}}$ is also in $\overline{O}_2$, so it remains to prove that for any countably infinite $F = \{f_n : n \in \mathbb{N}\} \in O_1$ we have that $F \in \overline{O}_2$. Since $L_1 \leq L_2$, there is an L_2 formula φ such that $\mathrm{Mod}(Q_F p) = \mathrm{Mod}(\varphi)$ (p is a propositional letter).

First we will show that p must be an element of $Var(\varphi)$. Indeed, if $p \notin Var(\varphi) = \{p_1, \ldots, p_n\}$, then for any $a \in [0,1]$ there is a model $N = \langle W_N, H_N, \mu_N, v_N\rangle$ of φ such that $\mu_N([p]) = a$. In order to see that we shall start from the arbitrary model $M = \langle W, H, \mu, v\rangle$ of φ. Let $\alpha_1, \ldots, \alpha_{2^n}$ be all atoms over $Var(\varphi)$. Since $\sum_{i=1}^{2^n} \mu([\alpha_i]) = 1$, there are real numbers $a_1, \ldots, a_{2^n} \in [0,1]$ so that $a_i \leq \mu([\alpha_i])$ for all i and $\sum_{i=1}^{2^n} a_i = a$. Let $W_N = W$, $H_N = H$, $v_N = v$ and let μ_N be any finitely additive probability measure on H such that $\mu_N([p \wedge \alpha_i]) = a_i$ and $\mu_N([\neg p \wedge \alpha_i]) = \mu([\alpha_i]) - a_i$ for all i. Note that for all i

$$\mu_N([\alpha_i]) = \mu_N([p \wedge \alpha_i]) + \mu_N([\neg p \wedge \alpha_i]) = a_i + \mu([\alpha_i]) - a_i = \mu([\alpha_i]),$$

so $N \in \mathrm{Mod}(\varphi)$ and $\mu_N([p]) = a$. However, this contradicts the condition $\mathrm{Mod}(Q_F p) = \mathrm{Mod}(\varphi)$ since we can chose any $a \notin F$.

Hence, $p \in Var(\varphi)$. Without the loss of generality, we may assume that $p = p_1$. For any model M, in order to decide whether it satisfies φ or not, the only relevant information is the probabilities that M designates to atoms $\alpha_1, \ldots, \alpha_{2^n}$. The mapping $M \mapsto \langle \mu_M([\alpha_1]), \ldots, \mu_M([\alpha_{2^n}])\rangle$ allows us to see $\mathrm{Mod}(\varphi)$ as a subset of the intersection of the hyperplane $\Gamma : \sum_{i=1}^{2^n} x_i = 1$ and the hypercube $E : [0,1]^{2^n}$.

From now on, for any probabilistic formula ψ we shall identify $\mathrm{Mod}(\psi)$ with the corresponding subset of $\Gamma \cap E$.

Moreover, we shall assume that atoms are ordered lexicographically with respect to appearance of negation. For example, if $Var(\varphi) = \{p_1, p_2\}$, then $\alpha_1 = p_1 \wedge p_2$, $\alpha_2 = p_1 \wedge \neg p_2$, $\alpha_3 = \neg p_1 \wedge p_2$ and $\alpha_4 = \neg p_1 \wedge \neg p_2$.

Let β be any classical propositional formula such that $Var(\beta) \subseteq Var(\varphi)$. By $At(\beta)$ we shall denote the unique subset of $\{1, \ldots, 2^n\}$ such that

$$\beta \Leftrightarrow \bigvee_{i \in At(\beta)} \alpha_i.$$

By CDNF(φ) we shall denote the complete disjunctive normal form of the formula φ. CDNF(φ) has the form $\bigvee_{i=1}^{k} \bigwedge_{j=1}^{m_i} Q_{H_{i,j}} \beta_{i,j}$ which can be transformed to

$$\bigvee_{i=1}^{k} \bigwedge_{j=1}^{m_i} Q_{H_{i,j}} \left(\bigvee_{l \in At(\beta_{i,j})} \alpha_l \right),$$

where each $H_{i,j}$ is either an element of O_2, or it is an interval with rational endpoints, or $H_{i,j} = [0, 1] \setminus G$ for some $G \in O_2$.

Let $a \in [0, 1]$. Since $\mathrm{Mod}(Q_F p) = \mathrm{Mod}(\varphi)$, we have that $a \in F$ iff

$$\langle \underbrace{a, 0, \ldots, 0}_{2^{n-1}}, 1 - a, 0, \ldots, 0\rangle \in \mathrm{Mod}(\varphi).$$

Moreover, in each model of $\varphi \wedge \bigwedge_{i \neq 1, 2^{n-1}+1} P_{=0}\alpha_i$ the formula $Q_{H_{i,j}}\left(\bigvee_{l \in At(\beta_{i,j})} \alpha_l\right)$ is equivalent to

- $Q_{H_{i,j}}\alpha_1$, if $1 \in At(\beta_{i,j})$ and $2^{n-1} + 1 \notin At(\beta_{i,j})$;
- $Q_{1-H_{i,j}}\alpha_1$, if $1 \notin At(\beta_{i,j})$ and $2^{n-1} + 1 \in At(\beta_{i,j})$;
- $P_{=1}(p_2 \wedge \cdots \wedge p_n)$, if $1 \in At(\beta_{i,j})$ and $2^{n-1} + 1 \in At(\beta_{i,j})$;
- $P_{=0}\beta_{i,j}$, if $1 \notin At(\beta_{i,j})$ and $2^{n-1} + 1 \notin At(\beta_{i,j})$.

Note that $\bigwedge_{i \neq 1, 2^{n-1}+1} P_{=0}\alpha_i$ implies $P_{=0}\beta_{i,j}$ for $1, 2^{n-1}+1 \notin At(\beta_{i,j})$ and $P_{=1}(p_2 \wedge \cdots \wedge p_n)$. Hance, we have that

$$\begin{aligned} a \in F &\Leftrightarrow \langle a, 0, \ldots, 0, 1 - a, 0, \ldots, 0\rangle \in \mathrm{Mod}(\varphi) \\ &\Leftrightarrow \langle a, 0, \ldots, 0, 1 - a, 0, \ldots, 0\rangle \in \mathrm{Mod}\left(Q_H\alpha_1 \wedge \left(\bigwedge_{i \neq 1, 2^{n-1}+1} P_{=0}\alpha_i\right)\right). \end{aligned}$$

Here $H = \bigcup_{i=1}^{k} \bigcap_{j=1}^{m_i} H'_{i,j}$, where

$$H'_{i,j} = \begin{cases} H_{i,j} & , 1 \in At(\beta_{i,j}), 2^{n-1} + 1 \notin At(\beta_{i,j}) \\ 1 - H_{i,j} & , 1 \notin At(\beta_{i,j}), 2^{n-1} + 1 \in At(\beta_{i,j}) \\ [0,1] & , \text{otherwise} \end{cases}.$$

Hence, $F = H$, so $F \in \overline{O}_2$. ■

As a consequence, we have the following theorem:

Theorem 5.12 *Let O_1 and O_2 be recursive families of recursive rational subsets of $[0, 1]$, and L_1 and L_2 be the corresponding $LPP_{2,P,Q,O}$-logics. The family O_1 is representable in the family O_2 iff $L_1 \leq L_2$.* ■

5.4.5 *The Upper Hierarchy*

Theorem 5.12 correlates the relations of "being more expressive" between the $LPP_{2,P,Q,O}$-logics, and "being representable in" between the corresponding families of sets. In the sequel, we investigate the latter relation having in mind the former one. The relation "being more expressive" describes the hierarchy of expressiveness of the $LPP_{2,P,Q,O}$-logics.

Definition 5.5 Let O_1 and O_2 be recursive families of recursive subsets of $[0, 1]_{\mathbb{Q}}$. The binary relation $\sim$ is defined such that $O_1 \sim O_2$ iff $\overline{O_1} = \overline{O_2}$. ■

The relation $\sim$ is an equivalence relation on the set $\mathcal{O}$ of all recursive families of subsets of $[0, 1]_{\mathbb{Q}}$. We use $\mathcal{O}_{/\sim}$ to denote the corresponding quotient set. Each equivalence class $o \in \mathcal{O}_{/\sim}$ contains a unique maximal family O_o such that $O_o = \overline{O_o}$. For such an equivalence class o and the corresponding family O_o we say that O_o represents o. Let the set $\{\overline{O_o} : o \in \mathcal{O}_{/\sim}\}$ be denoted by $\mathcal{O}^*$. Clearly, $\mathcal{O}$ and $\mathcal{O}^*$ are countable.

Definition 5.6 Let O_1 and O_2 be different families from $\mathcal{O}^*$. Then $O_1 \leq O_2$ iff O_1 is representable in O_2. ■

We state some properties of the introduced hierarchy relation. Detailed proofs can be found in [10, 25].

1. Let O_1 and O_2 be different families from $\mathcal{O}^*$. Then $O_1 \leq O_2$ iff $\overline{O_1} \subseteq \overline{O_2}$;
2. The structure $(\mathcal{O}^*, \leq)$ is a lattice. The meet $(\cdot)$ and join $(+)$ operations can be defined as follows:
   - $O_1 \cdot O_2 = \overline{O_1 \cap O_2}$;
   - $O_1 + O_2 = \overline{O_1 \cup O_2}$;
3. The lattice $(\mathcal{O}^*, \leq)$ is non-modular;
4. $\overline{\emptyset}$ is the smallest element of $(\mathcal{O}^*, \leq)$;
5. A necessary and sufficient condition that an $O \in \mathcal{O}^*$ be an atom is that $O = \overline{\{F\}}$, where F is a recursive set with only one accumulation point. The lattice $(\mathcal{O}^*, \leq)$ is non-atomic;
6. There is no greatest element in $(\mathcal{O}^*, \leq)$. Consequently, the lattice $\mathcal{O}^*$ is σ-incomplete, i.e., there exists a countable increasing chain $L_0 < L_1 < L_2 < \cdots$ that has no upper bound in $\mathcal{O}^*$.

Thus, we can define a hierarchy of the $LPP_{2,P,Q,O}$-logics, so that a logic L_1 is less expressive than a logic L_2 ($L_1 \leq L_2$) iff the corresponding families O_1 and O_2 of subsets of $[0, 1]_{\mathbb{Q}}$ satisfy a similar requirement ($O_1 \leq O_2$). The hierarchy of the probability logics is isomorphic to $\langle \mathcal{O}^*, \leq \rangle$. For instance, the probability logic LPP_2 is the minimum of the hierarchy of the $LPP_{2,P,Q,O}$-logics and corresponds to the 0-element of $\langle \mathcal{O}^*, \leq \rangle$.

As we have seen, for all $LPP_{2,P,Q,O}$-logics L_1 and L_2, $L_1 \leq L_2$ iff $\overline{O_1} \subseteq \overline{O_2}$. The natural maximum of $\langle \mathcal{O}^*, \leq \rangle$ would be the minimal extension of all $LPP_{2,P,Q,O}$ logics. Such logic can be obtained as follows:

1. the set of $LPP_{2,P,Q,O}$-formulas is the smallest superset of the set

$$\{P_{\geq s}\alpha, Q_F\alpha \mid s \in [0, 1]_{\mathbb{Q}}, \alpha \in For_C, F \subseteq [0, 1]_{\mathbb{Q}} \text{ is recursive}\}$$

   that is closed for Boolean connectives;

2. axioms and inference rules are the same as for any $LPP_{2,P,Q,O}$ logic.

That logic will be denoted by $LPP_{2,P,Q,\mathrm{all}}$. Here "all" stands for the family of all recursive subsets of $[0, 1]_{\mathbb{Q}}$. Though the set of $LPP_{2,P,Q,\mathrm{all}}$-formulas is not recursive, from now on we will assume that $LPP_{2,P,Q,\mathrm{all}}$ is also an $LPP_{2,P,Q,O}$-logic. The strong completeness of $LPP_{2,P,Q,\mathrm{all}}$ can be straightforwardly derived from the corresponding argumentation for $LPP_{2,P,Q,O}$-logics that is presented here.

5.4.6 The Lower Hierarchy

Here we shall study the so-called lower hierarchy of $LPP^{\mathrm{Fr}(n)}_{2,\mathrm{Meas}}$ logics. It is defined in the same manner as the upper one (see Definition 5.4).

Definition 5.7 Let L_1 and L_2 be arbitrary $LPP^{\mathrm{Fr}(n)}_2$-logics. We say that the logic L_2 is more expressible than L_1 and write $L_1 \leq L_2$ iff for each L_1-formula φ exists an L_2 formula ψ such that $\mathrm{Mod}(\varphi) = \mathrm{Mod}(\psi)$ (i.e. φ and ψ have the same models). ■

It is easy to see that the introduced relation is reflexive and transitive. Furthermore, for any $LPP^{Fr(n)}_2$-formula φ, an LPP_2-formula ψ defined by

$$\psi =_{\mathrm{def}} \varphi \wedge \bigwedge_{\alpha \in For_C(\varphi)} \bigvee_{k=0}^{n} P_{=\frac{k}{n}} \alpha$$

have the same models as φ (here $For_C(\varphi)$ is the set of all classical propositional formulas appearing in φ), so we can naturally consider the upper hierarchy as an end-extension of the lower hierarchy.

We shall show that the characterization theorem for the upper hierarchy (Theorem 5.11) has the natural counterpart in the lower hierarchy.

Theorem 5.13 *Suppose that L_1 and L_2 are arbitrary $LPP^{\mathrm{Fr}(n)}_2$-logics. Then, $L_1 \leq L_2$ if and only if* $\mathrm{Fr}(n_1) \subseteq \mathrm{Fr}(n_2)$.

Proof Suppose that $\mathrm{Fr}(n_1) \subseteq \mathrm{Fr}(n_2)$ and let φ be an arbitrary L_1-formula. As above, we define an L_2-formula ψ by

$$\psi =_{\mathrm{def}} \varphi \wedge \bigwedge_{\alpha \in For_C(\varphi)} \bigvee_{k=0}^{n_1} P_{=\frac{k}{n_1}} \alpha.$$

Clearly, φ and ψ have the same models, so $L_1 \leq L_2$.

Conversely, let $\mathrm{Fr}(n_1) \nsubseteq \mathrm{Fr}(n_2)$. Then, we can chose $s \in \mathrm{Fr}(n_1) \setminus \mathrm{Fr}(n_2)$. Let p be an arbitrary propositional letter. Then, $P_{=s}p$ is satisfiable as L_1-formula, while by A8, $\vdash_{L_2} \neg P_{=s}p$. Hence, $L_1 \nleq L_2$. ■

Since $\mathrm{Fr}(1) \subseteq \mathrm{Fr}(n)$ for all positive integers n, the $LPP_2^{\mathrm{Fr}(1)}$ logic is the minimum of the lower hierarchy. Moreover, $\mathrm{Fr}(n)$ is a proper subset of $\mathrm{Fr}(2n)$ for all positive integers n, so the lower hierarchy has no maximal elements.

Note that logics L_1 and L_2 are incomparable if and only if the symmetric difference $\mathrm{Fr}(n_1) \bigtriangleup \mathrm{Fr}(n_1) \neq \emptyset$. Thus, the hierarchy contains incomparable elements (for instance, $\mathrm{Fr}(2) = \{0, \frac{1}{2}, 1\}$ and $\mathrm{Fr}(3) = \{0, \frac{1}{3}, \frac{2}{3}, 1\}$).

Another immediate consequence of Theorem 5.13 is that the lower hierarchy is a lattice. Namely, the greatest lower bound of L_1 and L_2 is determined by $\mathrm{Fr}(n_1) \cap \mathrm{Fr}(n_2) = \mathrm{Fr}(\mathrm{GCD}(n_1, n_2))$. The least upper bound of L_1 and L_2 is determined by $\mathrm{Fr}(n_1) \cup \mathrm{Fr}(n_2) = \mathrm{Fr}(\mathrm{LCM}(n_1, n_2))$. Notice that $L_1 \leq L_2$ iff $n_1 | n_2$ (n_1 divides n_2).

Theorem 5.14 *The lower hierarchy is atomic and non-modular.*

Proof Concerning non-modularity, it is well known that any lattice is non-modular iff the pentagon lattice N_5 can be embedded into it. In particular, logics

$$LPP_2^{\mathrm{Fr}(1)}, LPP_2^{\mathrm{Fr}(2)}, LPP_2^{\mathrm{Fr}(3)}, LPP_2^{\mathrm{Fr}(4)}, LPP_2^{\mathrm{Fr}(12)}$$

form the N_5 lattice so we have established non-modularity (see Fig. 5.2). Moreover, by Theorem 5.13, the logics L_1 and L_2 are incomparable iff $\mathrm{Fr}(n_1) \bigtriangleup \mathrm{Fr}(n_1) \neq \emptyset$. As a consequence, atoms of the lower hierarchy are determined by $\mathrm{Fr}(n)$, where n is a prime number. ■

As we have mentioned earlier, it is quite natural to merge the upper and the lower hierarchy into a single hierarchy of probability logics due to the same definition of $\leq$. Since each $LPP_2^{\mathrm{Fr}(n)}$-logic can be embedded into any $LPP_{2,P,Q,O}$-logic in the same manner as we have demonstrated for the LPP_2 logic, the upper hierarchy is an end-extension of the lower hierarchy (see Fig. 5.3).

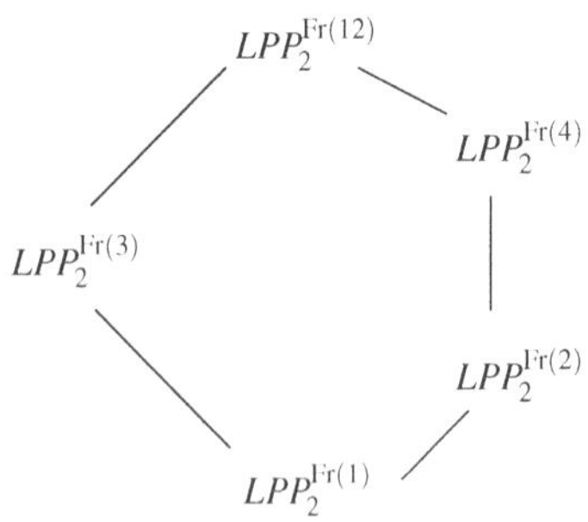

Fig. 5.2 N_5–lattice embedded in the lower hierarchy

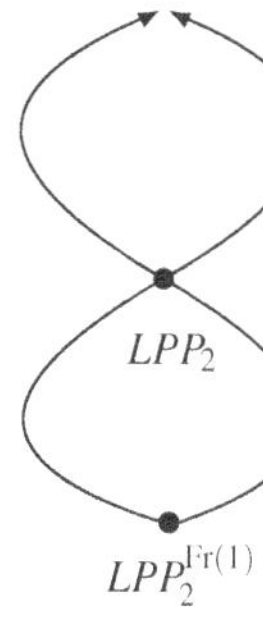

Fig. 5.3 The lower and the upper hierarchy

5.5 Qualitative Probabilities

Reasoning about qualitative probabilities is one of the most common cases of qualitative reasoning. Here we offer the first strongly complete formalization of the notion of qualitative probability within the framework of probabilistic logic [27, 28]. We obtain the language of the corresponding logic (denoted $LPP_{2,\preceq}$) by extending the LPP_2-language with an additional binary operator $\preceq$, such that for some For_C-formulas α and β, $\alpha \preceq \beta$ means "β is at least probable as α." Similarly as in Sect. 5.4, we use the class $LPP_{2,\mathrm{Meas}}$ of measurable LPP_2-models, and the corresponding satisfiability relation with another additional requirement that

- If $\alpha, \beta \in For_C$, $\mathbf{M} \models \alpha \preceq \beta$ iff $\mu([\alpha]) \leq \mu([\beta])$,

The axiom system $Ax_{LPP_{2,\preceq}}$ extends Ax_{LPP_2} with the following axioms:

7. $(P_{\leq s}\alpha \wedge P_{\geq s}\beta) \rightarrow \alpha \preceq \beta$
8. $(\alpha \preceq \beta \wedge P_{\geq s}\alpha) \rightarrow P_{\geq s}\beta$,

and the inference rule

4. From $A \rightarrow (P_{\geq s}\alpha \rightarrow P_{\geq s}\beta)$ for every $s \in [0, 1]_{\mathbb{Q}}$, infer $A \rightarrow \alpha \preceq \beta$.

The next theorem gives us some useful properties of the probability operator $\preceq$.

Theorem 5.15 *Suppose that T is a set of formulas and that $\alpha, \beta, \gamma \in For_C$. Then the following holds*

1. $T \vdash \alpha \preceq \beta$ *if and only if* $T \vdash P_{\geq s}(\alpha) \rightarrow P_{\geq s}(\beta)$ *for all* $s \in [0, 1]_{\mathbb{Q}}$;
2. $\vdash \alpha \preceq \beta \vee \beta \preceq \alpha$;
3. $\vdash (\alpha \preceq \beta \wedge \beta \preceq \gamma) \rightarrow \alpha \preceq \gamma$;
4. $\vdash \alpha \preceq \alpha$;
5. *If* $T \vdash P_{\geq 1}(\alpha \rightarrow \beta)$ *then* $T \vdash \alpha \preceq \beta$;
6. *If* $T \vdash \alpha \rightarrow \beta$ *then* $T \vdash \alpha \preceq \beta$. ■

The corresponding completeness and decidability results can be obtained in the similar way as it was the case for the LPP_2-logic.

Note that the qualitative probability operator is definable in the logic $LPP_2^{\mathrm{Fr}(n)}$

$$\alpha \preceq \beta \Leftrightarrow_{\mathrm{def}} \bigwedge_{s \in \mathrm{Fr}(n)} (P_{\geq s}\alpha \rightarrow P_{\geq s}\beta).$$

Thus, the notion of the qualitative probability is definable in $LPP_2^{\mathrm{Fr}(n)}$, and $LPP_2^{\mathrm{Fr}(n)}$ and $LPP_{2,\preceq}^{\mathrm{Fr}(n)}$ coincide.

5.6 An Intuitionistic Probability Logic

Intuitionistic logic arises quite naturally from a conception of mathematics as a human endeavor not pertaining to some outside reality. Since the statements of mathematics are not about something which exists out there, they cannot be true or false but only proved or disproved. This leaves another category of statements, those which are as yet undetermined. Thus intuitionistic logic may be viewed as the logic of the growth of human knowledge (as opposed to the classical logic which we may regard as the logic of the static Platonic universe of mathematical objects).

Thanks to this, intuitionistic logic has less consequences which would seem rather unintuitive in a real-life situation (e.g., $(p \rightarrow q) \vee (q \rightarrow p)$ and $(p \rightarrow (q \vee r)) \rightarrow ((p \rightarrow q) \vee (p \rightarrow r))$ are not intuitionistic theorems, i.e., there are models in which they are false).

In reality, there is the fact that the intuitionistic logic might be the least popular non-classical logic among the practitioners of artificial intelligence and computer science in general. However, for those comfortable with the ubiquitous S4-modal logic and uncomfortable with intuitionism, we should emphasize that these two logics are practically the same: their models are the same, while the Gödel translation enables us to interpret syntax. Furthermore, as we shall show in the Remark at the end of this section, intuitionistic logic arises naturally whenever we deal with possible worlds semantics. In any case, starting with intuitionistic logic, we naturally have, besides proved statements (probability is 1) and disproved statements (probability is 0), undetermined statements whose probability should range between 0 and 1.

This is more obvious if we consider a Kripke model in which we can assign a probability to a formula on the basis of the number of possible worlds in which it is true. In our approach the probabilistic operators have the classical treatment. As a justification, we may say that once we determine the probability of an uncertain proposition α, it should be either greater or equal to some $s \in [0, 1]$ or not, so it is not unreasonable to assume $P_{\geq s}\alpha \vee \neg P_{\geq s}\alpha$ (even if we reject $\alpha \vee \neg\alpha$).

We use LPP_2^I to denote the corresponding intuitionistic probability logic. At the propositional level, the language contains the connectives $\neg$, $\wedge$, $\vee$ and $\rightarrow$, while on the probabilistic level we have two lists of unary probabilistic operators $(P_{\geq s})$, and $(P_{\leq s})$, $s \in [0, 1]_{\mathbb{Q}}$, and the connectives $\neg$ and $\wedge$. Note that, since we have the intuitionistic base

- at the propositional level, the propositional connectives are independent, and
- at the probabilistic level, the probabilistic operators $P_{\geq \cdot}$, and $P_{\leq \cdot}$ are independent, but $\vee$ and $\rightarrow$ can be defined from $\neg$ and $\wedge$.

Similarly as for the logic LPP_2, we do not allow iterations of probabilistic operators, and define the sets For_I of propositional formulas, For_P of probabilistic formulas, and $For_{LPP_2^I}$ of all formulas, as in Sect. 3.1.1.

5.6.1 Semantics

Corresponding to the structure of the set $For_{LPP_2^I}$, there are two levels in the definition of models. At the first level there is the notion of intuitionistic Kripke models [18], while probability comes in the picture at the second level.

Definition 5.8 An *intuitionistic Kripke* model for the language For_I is a structure $\langle W, \leq, v\rangle$ where

- $\langle W, \leq\rangle$ is a partially ordered set of possible worlds which is a tree, and
- v is a valuation function, i.e., v maps the set W into $\mathbb{P}(\phi)$, which satisfies the condition: for all $w, w' \in W$, $w \leq w'$ implies $v(w) \subseteq v(w')$. ■

The last requirement from Definition 5.8 allows that v does not determine the status of some primitive propositions from ϕ in some worlds. In each Kripke model we define the forcing relation $\Vdash \subset W \times For_I$ by the following:

Definition 5.9 Let $\langle W, \leq, v\rangle$ be an intuitionistic Kripke model. *The forcing relation* $\Vdash$ is defined by the following conditions for every $w \in W$, α, $\beta \in For_I$:

- if $\alpha \in \phi$, $w \Vdash \alpha$ iff $\alpha \in v(w)$,
- $w \Vdash \alpha \wedge \beta$ iff $w \Vdash \alpha$ and $w \Vdash \beta$,
- $w \Vdash \alpha \vee \beta$ iff $w \Vdash \alpha$ or $w \Vdash \beta$,
- $w \Vdash \alpha \rightarrow \beta$ iff for every $w' \in W$ if $w \leq w'$ then $w' \nVdash \alpha$ or $w' \Vdash \beta$, and
- $w \Vdash \neg\alpha$ iff for every $w' \in W$ if $w \leq w'$ then $w' \nVdash \alpha$. ■

We read $w \Vdash \alpha$ as "w *forces* α" or "α is true in the world w". *Validity in the intuitionistic Kripke model* $\langle W, \leq, v\rangle$ is defined by

$$\langle W, \leq, v\rangle \models \alpha \text{iff } (\forall w \in W) w \Vdash \alpha.$$

A formula α is *valid* ($\models \alpha$) if it is valid in every intuitionistic Kripke model.

Let $\mathbf{M}_I = \langle W, \leq, v\rangle$ be an intuitionistic Kripke model. Let $[\alpha]$ denote $\{w \in W : w \Vdash \alpha\}$ for every $\alpha \in For_I$. The family $H_I = \{[\alpha]_{\mathbf{M}_I} : \alpha \in For_I\}$ is a Heyting algebra with operations

- $[\alpha] \cup [\beta] = [\alpha \vee \beta]$,
- $[\alpha] \cap [\beta] = [\alpha \wedge \beta]$,

- $[\alpha] \Rightarrow [\beta] = [\alpha \rightarrow \beta]$, and
- $\sim [\alpha] = [\neg\alpha]$.

Thus, H_I is a lattice on W, but it may be not closed under complementation.

Definition 5.10 A *measurable probabilistic model* is a structure $\mathbf{M} = \langle W, \leq, v, H, \mu\rangle$ where

- $\mathbf{M}_I = \langle W, \leq, v\rangle$ is an intuitionistic Kripke model,
- H is an algebra on W containing $H_I = \{[\alpha] : \alpha \in For_I\}$,
- $\mu : H \rightarrow [0, 1]$ is a finitely additive probability. ■

Note that H contains all sets of the form $W \setminus [\alpha]$, even if for some $\alpha \in For_I$ it may be that $W \setminus [\alpha] \neq [\neg\alpha]$. The fact that $[\neg\alpha]$ does not have to contain the complement of $[\alpha]$ is the reason why we need both $P_{\geq s}$ and $P_{\leq s}$ operators since $P_{\leq s}\alpha$ will not imply $P_{\geq 1-s}\neg\alpha$.

We use $LPP^I_{2,\mathrm{Meas}}$ to denote the class of all measurable probabilistic models.

Definition 5.11 *The satisfiability relation* $\models$ is defined by the following conditions for every $LPP^I_{2,\mathrm{Meas}}$-model $\mathbf{M} = \langle W, \leq, v, H, \mu\rangle$:

- if $\alpha \in For_I$, $\mathbf{M} \models \alpha$ if $(\forall w \in W) w \Vdash \alpha$,
- $\mathbf{M} \models P_{\geq s}\alpha$ if $\mu([\alpha]) \geq s$,
- $\mathbf{M} \models P_{\leq s}\alpha$ if $\mu([\alpha]) \leq s$,
- if $A \in For_P$, $\mathbf{M} \models \neg A$ if $\mathbf{M} \models A$ does not hold, and
- if $A, B \in For_P$, $\mathbf{M} \models A \wedge B$ if $\mathbf{M} \models A$, and $\mathbf{M} \models B$. ■

Definition 5.12 A formula $\varphi \in For_{LPP^I_2}$ is *satisfiable* if there is a $LPP^I_{2,\mathrm{Meas}}$-model $\mathbf{M}$ such that $\mathbf{M} \models \varphi$; φ is *valid* if for every $LPP^I_{2,\mathrm{Meas}}$-model $\mathbf{M}$, $\mathbf{M} \models \varphi$; a set of formulas is *satisfiable* if there is an $LPP^I_{2,\mathrm{Meas}}$-model $\mathbf{M}$ such that for every formula φ from the set, $\mathbf{M} \models \varphi$. ■

5.6.2 Axiomatization, Completeness, Decidability

An axiomatization that characterizes the set of all $LPP^I_{2,\mathrm{Meas}}$-valid formulas can be obtain by combining

- any propositional intuitionistic axiomatization for For_I,
- any classical propositional axiomatization for For_P and
- probabilistic axioms and rules from Sect. 3.2

with the proviso that in this framework the probabilistic operators $P_{\geq .}$, and $P_{\leq .}$ are independent. Thus, Axiom 3 from the system Ax_{LPP_2} should be rewritten in the form

$$P_{\geq 1-r}\neg\alpha \rightarrow \neg P_{\geq s}\alpha \text{ for } s > r.$$

Strong completeness and decidability theorems can be obtained in the similar way as it was shown earlier. Details can be found in [19–21].

We will show here that even if we start with classical logic, possible worlds semantics naturally produces intuitionistic logic. It turns out that intuitionistic implication will coincide with conditional probability when probability is equal to 1.

Let us start with a standard possible-world model $M = \langle W, H, \mu, v \rangle$. We may define a preorder (reflexive and transitive relation) R on W by: uRw iff for every primitive proposition p, $v(u, p) = true$ implies $v(w, p) = true$. From this we may obtain a partial order in the usual way. First we introduce an equivalence relation $\sim$ defined by: $u \sim w$ iff uRw and wRu, and then we split W into equivalence classes: $C_u = \{w : u \sim w\}$. Now we may pick a selection $W' \subset W$ of representatives of equivalence classes (one for each class). So we have $(\forall u \in W)(\exists w \in W')(u \sim w)$. Obviously, R induces a partial order $\leq$ on W' such that $u \leq w$ iff $u R w$. Now we have a Kripke model with a partial ordering relation on worlds $\langle W', \leq, v \rangle$ which makes it a model for intuitionistic logic. Namely, we may define (semantically) a new propositional connective $\rightarrow$ by: $w \models \alpha \rightarrow \beta$ iff $(\forall w' \geq w)(w' \models \alpha$ implies $w' \models \beta)$. We may also define a new, intuitionistic, negation by: $-\alpha = \alpha \rightarrow \bot$. Therefore, even if we start with classical logic, when we come to models, we have an intuitionistic implication built in.

The interest in intuitionistic implication, besides the arguments proposed at the start of this section, comes from the fact that conditional probability, which is often used as the proper form of entailment in the context of probability logic, coincides in a sense with the intuitionistic implication:

Theorem 5.16

$$\mu(\alpha \rightarrow \beta) = 1 \;\textit{iff}\; \frac{\mu(\alpha \wedge \beta)}{\mu(\alpha)} = 1.$$

■

However, this symmetry holds only in the case when probability is equal to 1. It is possible to construct models in which conditional probability is high while the probability of (intuitionistic) implication is low and vice versa. The reason is that, despite the fact that both operators are defined globally (and not locally, in each world) the definitions are quite different. Conditional probability considers (i.e., counts) only worlds in which α is true, while intuitionistic implication takes into account also their predecessors. We may say, in a sense, that conditional probability disregards the development of events and regards only the final stages (with regard to the validity of α), i.e., the analysis starts with the worlds in which α is true and disregards the previous stages in which α may be "not yet true". Existence of long time-lines which end with worlds in which α is not true adds to the probability of $\alpha \rightarrow \beta$, while it is irrelevant for the conditional probability. On the other hand, a long sequence which has an ending in which α is true and β is not, reduces considerably the probability of $\alpha \rightarrow \beta$, while it may, in the presence of a relevant number of worlds in which both α and β are true, be insignificant for conditional probability.

5.7 Logics with Conditional Probability Operators

Our work on complete axiomatization of the notion of conditional probability involves both Kolmogorov and de Finetti concepts of developing probability theory [7, 8, 22–24]. While Kolmogorov derives conditional probability from the notion of probability, de Finetti postulates conditional probability as a primitive notion and from it derives the notion of probability.

Here we shall just outline the corresponding formalization in the case of logic $LPCP_2^{[0,1]_{\mathbb{Q}(\varepsilon)},\approx}$ with approximate conditional probabilities. Details can be found for instance in [26, 33].

5.7.1 A Logic $LPCP_2^{[0,1]_{\mathbb{Q}(\varepsilon)},\approx}$ with Approximate Conditional Probabilities

A Hardy field is a recursive non-Archimedean field which contains all rational functions of a fixed positive infinitesimal ε which belongs to $^*\mathbb{R}$. Some examples of infinitesimal are (in ascending order, if $\varepsilon > 0$): $\varepsilon^3 + \varepsilon^4$, $\varepsilon^2 - 5\varepsilon^6$, $\frac{\varepsilon}{100}$, 85ε, or negative infinitesimals: $-\varepsilon$, $-\varepsilon^2$, ...$\mathbb{Q}(\varepsilon)$ contains all rational numbers. The unit interval of $\mathbb{Q}(\varepsilon)$ is denoted by $[0, 1]_{\mathbb{Q}(\varepsilon)}$.

The basic probability operators are the following binary operators of the form:

$$CP_{\geq s},\ CP_{\leq s} \text{ and } CP_{\approx r},$$

where $s \in [0, 1]_{\mathbb{Q}(\varepsilon)}$ and $r \in [0, 1]_{\mathbb{Q}}$. The intended meaning is obvious: for instance, $CP_{\geq s}(\alpha, \beta)$ means that $Pr(\alpha|\beta) \geq s$, while $CP_{\approx r}(\alpha, \beta)$ means that $Pr(\alpha|\beta) - r$ is an infinitesimal.

Similarly to the case of the LPP_2-logic, probabilistic $LPCP_2^{[0,1]_{\mathbb{Q}(\varepsilon)},\approx}$-formulas are Boolean combinations of the basic (atomic) probabilistic (or probability) formulas. Moreover,

$$P_{*s}\alpha =_{\text{def}} CP_{*s}(\alpha, \rightarrow p),$$

where $* \in \{\geq, \leq, \approx\}$, introduces other probability operators.

Semantics is defined similarly as for the LPP_2-logic. The only difference is in definition of satisfiability of atomic probability formulas, which will be explicitly stated in Definition 5.13. Closely related to that is the following convention that extends the conditional probability to entire H^2: for any $[0, 1]_{\mathbb{Q}(\varepsilon)}$-valued probability measure μ and any pair of $X, Y \in H$ we assume that

$$\mu(X|Y) = 1 \text{ whenever } \mu(Y) = 0.$$

Definition 5.13 Let $\mathbf{M} = \langle W, H, \mu, v\rangle$ be an $LPCP_2^{[0,1]_{\mathbb{Q}(\varepsilon)},\approx}$ model. The satisfiability of the basic probabilistic formulas is defined as follows:

1. $\mathbf{M} \models CP_{\leq s}(\alpha, \beta)$ iff $\mu([\alpha]_{\mathbf{M}} \mid [\beta]_{\mathbf{M}}) \leq s$;
2. $\mathbf{M} \models CP_{\geq s}(\alpha, \beta)$ iff $\mu([\alpha]_{\mathbf{M}} \mid [\beta]_{\mathbf{M}}) \geq s$;
3. $\mathbf{M} \models CP_{\approx r}(\alpha, \beta)$ iff $\mu([\alpha]_{\mathbf{M}} \mid [\beta]_{\mathbf{M}}) - r$ is an infinitesimal. ■

5.7.2 Axiomatization

The axiom system $Ax_{LPCP_2^{[0,1]_{\mathbb{Q}(\varepsilon)},\approx}}$ which characterizes the set of all $LPCP^{[0,1]_{\mathbb{Q}(\varepsilon)},\approx}_{2,\text{Meas},\text{Neat}}$-valid formulas contains the following set of axiom schemata:

1. all For_C-instances of classical propositional tautologies
2. all For_P-instances of classical propositional tautologies
3. $CP_{\geq 0}(\alpha, \beta)$
4. $CP_{\leq s}(\alpha, \beta) \rightarrow CP_{<t}(\alpha, \beta)$, $t > s$
5. $CP_{<s}(\alpha, \beta) \rightarrow CP_{\leq s}(\alpha, \beta)$
6. $P_{\geq 1}(\alpha \leftrightarrow \beta) \rightarrow (P_{=s}\alpha \rightarrow P_{=s}\beta)$
7. $P_{\leq s}\alpha \leftrightarrow P_{\geq 1-s}\neg\alpha$
8. $(P_{=s}\alpha \wedge P_{=t}\beta \wedge P_{\geq 1}\neg(\alpha \wedge \beta)) \rightarrow P_{=\min(1,s+t)}(\alpha \vee \beta)$
9. $P_{=0}\beta \rightarrow CP_{=1}(\alpha, \beta)$
10. $(P_{=t}\beta \wedge P_{=s}(\alpha \wedge \beta)) \rightarrow CP_{=s/t}(\alpha, \beta)$, $t \neq 0$
11. $CP_{\approx r}(\alpha, \beta) \rightarrow CP_{\geq r_1}(\alpha, \beta)$, for every rational $r_1 \in [0, r)$
12. $CP_{\approx r}(\alpha, \beta) \rightarrow CP_{\leq r_1}(\alpha, \beta)$, for every rational $r_1 \in (r, 1]$

and inference rules:

1. From φ and $\varphi \rightarrow \psi$ infer ψ.
2. If $\alpha \in For_C$, from α infer $P_{\geq 1}\alpha$.
3. From $A \rightarrow P_{\neq s}\alpha$, for every $s \in [0, 1]_{\mathbb{Q}}$, infer $A \rightarrow \bot$.
4. For every $r \in [0, 1]_{\mathbb{Q}}$, from $A \rightarrow CP_{\geq r-1/n}(\alpha, \beta)$, for every integer $n \geq 1/r$, and $A \rightarrow CP_{\leq r+1/n}(\alpha, \beta)$ for every integer $n \geq 1/(1-r)$, infer $A \rightarrow CP_{\approx r}(\alpha, \beta)$.

It is easy to see (just put $\rightarrow p$ instead of β) that the axioms 3–5 generalize the corresponding axioms from the system Ax_{LPP_2}. Axiom 9 conforms with the useful practice of assuming conditional probability to be 1, whenever the condition has the probability 0. Axiom 10 expresses the standard definition of conditional probability, while the axioms 11 and 12, and Rule 4 describe the relationship between the standard conditional probability and the conditional probability infinitesimally close to some rational $r \in [0, 1]_{\mathbb{Q}}$. The rules 3 and 4 are infinitary. Rule 3 guarantees that the probability of a formula belongs to the set S.

The main difference in the completeness-proof technique is in the construction of a canonical model from a maximal consistent set of formulas T, namely the definition of a measure μ_T. Here μ_T is defined as follows:

$$\mu([\alpha]) = s \Leftrightarrow_{\text{def}} T \vdash P_{=s}\alpha.$$

The existence of such s is a consequence of Lemma 5.1, while the uniqueness is a consequence of the consistency of T. In particular, we have the following theorem:

Theorem 5.17 (Strong completeness for $LPCP_{2,\mathrm{Meas},\mathrm{Neat}}^{[0,1]_{\mathbb{Q}(\varepsilon)},\approx}$) *A set T of formulas is $Ax_{LPCP_2^{[0,1]_{\mathbb{Q}(\varepsilon)},\approx}}$-consistent iff it is $LPCP_{2,\mathrm{Meas},\mathrm{Neat}}^{[0,1]_{\mathbb{Q}(\varepsilon)},\approx}$-satisfiable.* ■

5.8 Polynomial Weight Formulas

In this section, we shall briefly outline the solution of the problem of finding a strongly complete propositional axiomatization of logics for polynomial weight formulas proposed by Fagin et al. [5]. In the same way, it is possible to address the related issue for logic with linear weight formulas from [5]. All technical details can be found in [2, 28–30].

First we introduce the notion of a polynomial weight term. The corresponding inductive definition is given below

$$w(\alpha) \mid \underline{r} \mid fg \mid f + g.$$

Here α is an arbitrary classical propositional formula, $r \in [0, 1]_{\mathbb{Q}}$ and f, g are variables for polynomial weight terms.

An atomic polynomial weight formula is a formula of the form

$$f \leq g,$$

where f, g are arbitrary polynomial weight terms. The set of polynomial weight formulas For_P contains Boolean combinations of atomic polynomial weight formulas, while the set of formulas $\mathrm{For}_{PWF} = \mathrm{For}_C \cup \mathrm{For}_P$.

The other types of the standard qualitative formulas $f \geq g, f < g, f > g$ and $f = g$ are introduced in the usual way. Note that the usual meaning of the atomic weight term $w(\alpha)$ is the weight of the formula α, which can be interpreted in terms of uncertain reasoning as the agent's measure (level of confidence) in α. The corresponding semantics for the logic PWF of polynomial weight formulas is virtually the same as for the LPP_2. A complete axiomatization Ax_{PWF} is given below

Axioms for propositional reasoning

1. all substitutional instances of the classical propositional tautologies.

Axioms for probabilistic reasoning

2. $P(\alpha) \geqslant \underline{0}$.
3. $P(\alpha \vee \beta) = P(\alpha) + P(\beta)$, for disjoint α and β.

Axioms about rational numbers

4. $\underline{r} \geqslant \underline{s}$, iff $r \geqslant s$.

Axioms about commutative ordered rings

5. $f + g = g + f$
6. $(f + g) + h = f + (g + h)$
7. $f + \underline{0} = f$
8. $f - f = \underline{0}$
9. $f \cdot g = g \cdot f$
10. $f \cdot (g \cdot h) = (f \cdot g) \cdot h$
11. $f \cdot \underline{1} = f$
12. $f \cdot (g + h) = (f \cdot g) + (f \cdot h)$
13. $f \geqslant g \vee g \geqslant f$
14. $(f \geqslant g \wedge g \geqslant h) \to f \geqslant h$
15. $f \geqslant g \to f + h \geqslant g + h$
16. $(f \geqslant g \wedge h > 0) \to f \cdot h \geqslant g \cdot h$

Inference rules

1. From Φ and $\Phi \to \Psi$ infer Ψ.
2. From α infer $P(\alpha) = \underline{1}$.
3. From $\{\phi \to f \geqslant \underline{-n^{-1}} \mid n = 1, 2, 3, \ldots\}$ infer $\phi \to f \geqslant \underline{0}$.

All relevant proof-theoretical notions are the same as for the LPP_2-logic. Using the same technique that is thoroughly elaborated in Chap. 3, we can show the corresponding strong completeness theorem. Decidability and complexity (PSPACE-completeness) of the PSAT problem are proved in [5].

5.9 Logics with Unordered or Partially Ordered Ranges

We conclude this chapter with a discussion on probability logics with unordered ranges of probability functions that are studied in [9, 11–14] (consider also Keynes' view described in Sect. 2.6.1). The particular ranges that we have in mind are the field of p-adic numbers $\mathbb{Q}_p$ for the given prime number p, the field of complex numbers and a lattice with the additional underlying structure (e.g., $[0, 1]_{\mathbb{Q}} \times [0, 1]_{\mathbb{Q}}$ with the product order).

Probability functions with such ranges naturally arose in various phenomena involving quantum physics, incomparability, and indeterminacy. Theoretical background can be found in so-called vector valued measure theory. The main idea is to replace the notion of an ordering with the notion of a metric space (Banach space, pre-Hilbert space, Hilbert space, etc.).

In terms of probability assertions, the atomic statements have the following two forms:

- $Pr(\alpha) \in B[c, r]$, i.e., the probability of α is in the closed ball with the center c and radius r;
- $Pr(\alpha) \in B(c, r)$, i.e., the probability of α is in the open ball with the center c and the radius r.

Due to the fact that any p-adic norm $|\ |_p$ generates an ultrametric space with respect to p-adic metric $d_p(x, y) = |x - y|_p$, we have a peculiar consequence that $B[c, r] = B(c, r)$ and $B[c, r] = B[x, r]$ for all $x \in B[c, r]$. In the case of complex valued probability functions, this will not be the case due to the well-known properties of the Euclidean metrics on . We shall illustrate the completeness-proof technique in the case of the logic $L_{\mathbb{Q}_p}$.

Given the prime number p, a p-adic norm $|\ |_p$ of any integer k is defined by

$$k = \frac{1}{p^m},$$

where $k = p^m \cdot r$ (r is not divisible by p) is the unique prime factorization of k. This notion is naturally extended on rational numbers by

$$\left|\frac{k}{l}\right|_p = \frac{|k|_p}{|l|_p}.$$

The field $\mathbb{Q}_p$ of p-adic numbers can be obtained as a completion (with respect to Cauchy sequences) of the field of rational numbers $\mathbb{Q}$ in p-adic norm $|\ |_p$ on $\mathbb{Q}$.

5.9.1 A Logic for Reasoning About p-adic Valued Probabilities

Let p be a fixed prime and $M \in \mathbb{N}$ be an arbitrary large but fixed positive integer. We introduce the following sets:

1. $\mathbb{Q}_M = \{r \in \mathbb{Q} : |r|_p \leq p^M\}$,
2. $\mathbb{Z}_M = \mathbb{Z}^- \cup \{0, 1, 2, \ldots, M\}$, where $\mathbb{Z}^-$ denotes the set of all negative integers, and
3. $R = \{p^{M-n} : n \in \mathbb{N}\} \cup \{0\} = \{p^n : n \in \mathbb{Z}_M\} \cup \{0\}$.

Suppose that ϕ is a countable set of propositional letters. By For_C we will denote the set of all propositional formulas over ϕ. Propositional formulas will be denoted by α, β, γ, etc. The set $\mathrm{For}_P^{L_{\mathbb{Q}_p}}$ of probabilistic formulas is defined as the least set satisfying

- If $\alpha \in For_C$, $r \in \mathbb{Q}_M$, $\rho \in R$ then $K_{r,\rho}\alpha$ is probabilistic formula.
- If φ, ϕ are probabilistic formulas then $(\neg\varphi)$, $(\varphi \wedge \phi)$ are probabilistic formulas

The set $For_{LPP_2}^{L_{\mathbb{Q}_p}}$ of all $L_{\mathbb{Q}_p}$-formulas is $For_C \cup For_P^{L_{\mathbb{Q}_p}}$.

Definition 5.14 An $L_{\mathbb{Q}_p}$-model is a structure $M = \langle W, H, \mu, v\rangle$ where W, H and v have the same meaning as before. The only difference is in the definition of μ

- $\mu : H \to B[0, p^M]$ is an additive function such that $\mu(W) = 1$. ■

The satisfiability relation is defined similarly as before. In the next definition, we emphasis the only difference, namely the satisfiability of the atomic formula $K_{r,\rho}$.

Definition 5.15 Let $M = \langle W, H, \mu, v\rangle$ be an $L_{\mathbb{Q}_p}$-model. The satisfiability relation on atomic formulas of the form $K_{r,\rho}$ is defined by

- If $\alpha \in For_C$ then $M \models K_{r,\rho}\alpha$ iff $|\mu([\alpha]) - r|_p \leq \rho$. ■

Let us pay attention on the definition of satisfiability for the formulas of the form $K_{r,\rho}\alpha$. For arbitrary ρ, $M \models K_{r,\rho}\alpha$ means that $\mu([\alpha])$ belongs to the p-adic ball with the center r and the radius ρ. In the special case when $\rho = 0$ and $M \models K_{r,0}\alpha$, according to Definition 5.15, $\mu([\alpha]) = r$.

Also, we can calculate the measure of tautology ($\top$) and contradiction ($\bot$). μ is a measure that satisfies conditions of normalization and additivity (see Definition 5.14) and since $[\top] = W$ and $[\bot] = \emptyset$ we conclude that $\mu([\top]) = 1$ and $\mu([\bot]) = 0$, i.e., for every model M, $M \models K_{1,0} \to p$ and $M \models K_{0,0}\bot$.

The axiom system $AX_{L_{\mathbb{Q}_p}}$ involves the next axiom schemas

1. Substitutional instances of tautologies.
2. $K_{r,\rho}\alpha \to K_{r,\rho'}\alpha$, whenever $\rho' \geq \rho$
3. $K_{r_1,\rho_1}\alpha \wedge K_{r_2,\rho_2}\beta \wedge K_{0,0}(\alpha \wedge \beta) \to K_{r_1+r_2,\max(\rho_1,\rho_2)}(\alpha \vee \beta)$.
4. $K_{r_1,\rho_1}\alpha \to \neg K_{r_2,\rho_2}\alpha$, if $|r_1 - r_2|_p > \max(\rho_1, \rho_2)$
5. $K_{r_1,\rho}\alpha \to K_{r_2,\rho}\alpha$, if $|r_1 - r_2|_p \leq \rho$

and inference rules

1. From φ and $\varphi \to \psi$ infer ψ.
2. From α infer $K_{1,0}\alpha$
3. If $n \in \mathbb{N}$, from $\varphi \to \neg K_{r,p^{M-n}}\alpha$ for every $r \in \mathbb{Q}_M$, infer $\varphi \to \bot$.
4. From $\alpha \to \bot$, infer $K_{0,0}\alpha$
5. If $r \in \mathbb{Q}_M$, from $\varphi \to K_{r,p^{M-n}}\alpha$ for every $n \in \mathbb{N}$, infer $\varphi \to K_{r,0}\alpha$.
6. From $\alpha \leftrightarrow \beta$ infer $(K_{r,\rho}\alpha \to K_{r,\rho}\beta)$.

Completeness technique is essentially the same as it was outlined in Sect. 5.1. On the other hand, the analysis of decidability involves a nontrivial adaptation of the corresponding argument for the LPP_2 logic. Specific technical details can be found in [13].

5.10 Other Extensions

Here we will briefly mention some other of our results related to probability logics. The paper [31] discusses the proof-theoretical and the model-theoretical approaches to a probabilistic logic which allows reasoning about the finitely additive probability measures on formulas generated by an arbitrary [0, 1]-valued evaluation of the set of propositional letters using Gödel's and product t-norms. The corresponding formal language enables efficient formalization of classification problems with criteria expressible as propositional formulas. Combinations of probability and justification logics are analyzed in [16, 17]. Doder and Ognjanović [3] presents an approach which allows reasoning about independence and probabilistic support in a probability logic. An infinitary strongly complete propositional logic with unary operators that speak about upper and lower probabilities is introduced in [34], while for some restricted fragments of the logic, finitary axiomatic systems are provided. Strong completeness for the class of measurable structures of an epistemic logic with probabilistic common knowledge and infinitely many agents is given in [35]. Ghilezan et al. [6] is a first step towards a formal system for probabilistic reasoning in λ-calculus with intersection types.

References

1. Bhaskara Rao, K.P.S., Bhaskara Rao, M.: Theory of Charges. Academic Press, Cambridge (1983)
2. Doder, D., Marinković, B., Maksimović, P., Perović, A.: A logic with conditional probability operators. Publications de L'Institut Mathematique (N.S.) **87**(101), 85–96 (2010). http://elib.mi.sanu.ac.rs/files/journals/publ/107/n101p085.pdf
3. Doder, D., Ognjanović, Z.: Probabilistic logics with independence and probabilistic support. progic 2015: The Seventh Workshop on Combining Probability and Logic, 22–24 April 2015, University of Kent, Canterbury, 25–26 (2015). https://blogs.kent.ac.uk/jonw/files/2015/04/ProgicAbstractsBooklet2015.pdf
4. Dordević, R., Rašković, M., Ognjanović, Z.: Completeness theorem for propositional probabilistic models whose measures have only finite ranges. Arch. Math. Logic **43**, 557–563 (2004)
5. Fagin, R., Halpern, J., Megiddo, N.: A logic for reasoning about probabilities. Inf. Comput. **87**(1–2), 78–128 (1990)
6. Ghilezan, S., Ivetić, J., Ognjanovic, Z., Savić, N.: Towards probabilistic reasoning about lambda terms with intersection types. In: Ghilezan, S., Ivetic, J. (ed.), Proceedings of the 22nd International Conference on Types for Proofs and Programs, TYPES 2016, pp. 23–26. Novi Sad, Serbia (2016)
7. Ikodinović, N.: Some Probability and Topological Logics. Phd thesis (in Serbian). University of Kragujevac (2005). http://elibrary.matf.bg.ac.rs/bitstream/handle/123456789/194/phdNebojsaIkodinovic.pdf?sequence=1
8. Ikodinović, N., Ognjanović, Z.: A logic with coherent conditional probabilities. In: Godo, L. (ed.), Proceedings of the 8th European Conference Symbolic and Quantitative Approaches to Reasoning with Uncertainty, ECSQARU 2005, Barcelona, Spain, July 6–8, 2005. Lecture Notes in Computer Science, vol. 3571, pp. 726–736. Springer, Heidelberg (2005)
9. Ikodinović, N., Rašković, M., Marković, Z., Ognjanović, Z.: Logics with generalized measure operators. J Multiple Valued Logic Soft Comput. **20**(5–6), 527–555 (2013)

10. Ikodinović, N., Ognjanović, Z., Perović, A., Rašković, M.: Hierarchies of probabilistic logics. Int. J. Approx. Reason. **55**(9), 1830–1842 (2014)
11. Ilić-Stepić, A., Ognjanović, Z.: Complex valued probability logics. Publications de l'Institut Mathematique, N.s. tome 95(109), 73–86 (2014). http://elib.mi.sanu.ac.rs/files/journals/publ/115/n109p073.pdf
12. Ilić-Stepić, A., Ognjanović, Z.: Logics for reasoning about processes of thinking with information coded by p-adic numbers. Studia Logica **103**, 145–174 (2015)
13. Ilić-Stepić, A., Ognjanović, Z., Ikodinović, N., Perović, A.: A p-adic probability logic. Math. Logic Q. **58**(4–5), 63–280 (2012)
14. Ilić-Stepić, A., Ognjanović, Z., Ikodinović, N.: Conditional p-adic probability logic. Int. J. Approx. Reason. **55**(9), 1843–1865 (2014)
15. Keisler, J.: Model Theory for Infinitary Logic. North-Holland, Amsterdam (1971)
16. Kokkinis, I., Maksimović, P., Ognjanović, Z., Studer, T.: First steps towards probabilistic justification logic. Logic J. IGPL **23**(4), 662–687 (2015)
17. Kokkinis, I., Ognjanović, Z., Studer, T.: Probabilistic justification logic. In: Artemov, S., Nerode, A. (eds.), Proceedings of Logical Foundations of Computer Science International Symposium, LFCS 2016. Lecture Notes in Computer Science, January 4–7, 2016, vol. 9537, pp. 174–186. Deerfield Beach, FL, USA (2016)
18. Kripke, S.: Semantical analysis of intuitionistic logic I. In: Formal Systems and Recursive Functions, pp. 93–130. North-Holland, Amsterdam (1965)
19. Marković, Z., Ognjanović, Z., Rašković, M.: A probabilistic extension of intuitionistic logic. Math. Logic Q. **49**, 415–424 (2003)
20. Marković, Z., Ognjanović, Z., Rašković, M.: An intuitionistic logic with probabilistic operators. Publications de L'Institute Matematique (N.S.) **73**(87): 31–38 (2003). http://elib.mi.sanu.ac.rs/files/journals/publ/93/n087p031.pdf
21. Marković, Z., Ognjanović, Z., Rašković, M.: What is the proper propositional base for probabilistic logic? IPMU **2004**, 443–450 (2004)
22. Milošević, M., Ognjanović, Z.: A first-order conditional probability logic. Logic J. Interest Gr. Pure Appl. Logics **20**(1), 235–253 (2012)
23. Milošević, M., Ognjanović, Z.: A first-order conditional probability logic with iterations. Publications de L'Institute Matematique, N.S. **93**(107), 19–27 (2013). http://elib.mi.sanu.ac.rs/files/journals/publ/113/n107p019.pdf
24. Ognjanović, Z., Ikodinović, N.: A logic with higher order conditional probabilities. Publications de L'Institute Matematique (Beograd), ns. **82**(96), 141–154 (2007). http://elib.mi.sanu.ac.rs/files/journals/publ/102/n096p141.pdf
25. Ognjanović, Z., Rašković, M.: Some probability logics with new types of probability operators. J. Logic Comput. **9**(2), 181–195 (1999)
26. Ognjanović, Z., Marković, Z., Rašković, M.: Completeness theorem for a Logic with imprecise and conditional probabilities. Publications de L'Institute Matematique Ns. **78**(92), 35–49 (2005). http://elib.mi.sanu.ac.rs/files/journals/publ/98/n092p035.pdf
27. Ognjanović, Z., Perović, A., Rašković, M.: Logics with the qualitative probability operator. Logic J. IGPL **16**(2), 105–120 (2008)
28. Perović, A.: Some applications of the formal method in set theory, model theory, probabilistic logics and fuzzy logics. Ph.d. thesis (in Serbian). University of Belgrade (2008). http://elibrary.matf.bg.ac.rs/bitstream/handle/123456789/100/phdAleksandarPerovic.pdf?sequence=1
29. Perović, A., Ognjanović, Z., Rašković, M., Marković, Z.: A probabilistic logic with polynomial weight formulas. In: Hartmann, S., Kern-Isberner, G. (eds.), 5th International Symposium on Foundations of Information and Knowledge Systems, FoIKS 2008, Proceedings. Lecture Notes in Computer Science, Pisa, Italy, February 11–15, 2008, vol. 4932, pp. 239–252. Springer, Heidelberg (2008)
30. Perović, A., Ognjanović, Z., Rašković, M., Marković, Z.: How to restore compactness into probabilistic logics? In: Hölldobler, S., Lutz, C., Wansing, H. (eds.), 11th European Conference on Logics in Artificial Intelligence, JELIA 2008, Proceedings. Lecture Notes in Computer Science, Dresden, Germany, September 28–October 1, 2008, vol. 5293, pp. 338–348. Springer, Heidelberg (2008)

31. Perović, A., Ognjanović, Z., Rašković, M., Radojević, D.: Finitely additive probability measures on classical propositional formulas definable by Godel's t-norm and product t-norm. Fuzzy Sets Syst. **169**, 65–90 (2011)
32. Rašković, M.: Completeness theorem for biprobability models. J. Symb. Logic **51**(3), 586–590 (1986)
33. Rašković, M., Marković, Z., Ognjanović, Z.: A logic with approximate conditional probabilities that can model default reasoning. Int. J. Approx. Reason. **49**(1), 52–66 (2008)
34. Savić, N., Doder, D., Ognjanović, Z.: A logic with upper and lower probability operators. In: Augustin, T., Doria, S., Miranda, E., Quaeghebeur, E. (eds.) Proceedings of the 9th International Symposium on Imprecise Probability: Theories and Applications, ISIPTA '15, Pescara, Italy, 2015, pp. 267–276. SIPTA Society for Imprecise Probability, Theories and Applications (2015)
35. Tomović, S., Ognjanović, Z., Doder, D.: Probabilistic common knowledge among infinite number of agents. In: Destercke, S., Denoeux, T. (eds.), Proceedings of the 13th European Conference Symbolic and Quantitative Approaches to Reasoning with Uncertainty ECSQARU 2015. Lecture Notes in Artificial Intelligence, Compiegne, France, July 15–17, 2015, vol. 9161, pp.496–505 (2015)

Chapter 6
Some Applications of Probability Logics

Dragan Doder, Aleksandar Perović and Angelina Stepić

Abstract In this chapter, we review some application areas of probability logics. The first part of the chapter is devoted to the theory of nonmonotonic inference and its deep connections with probability logic. We describe how our system $LPCP_2^{[0,1]_{\mathbb{Q}(\varepsilon)},\approx}$ (from Sect. 5.7) can be used to model default reasoning. Then, using the techniques from the Chaps. 4 and 5, we present a solution for an open problem of construction a propositional logic for reasoning about evidence (Halpern and Pucella, J Artif Intell Res 26:1–34, 2006 [15]). In the third section, we discuss a method for logical modeling of the process of human thinking based on p-adic numbers. The fourth section briefly describes some additional papers in the application area of probability logics. This chapter covers some results from Doder, Publications de l'Institut Mathématique, ns. 90(104):13–22, 2011 [5], Doder and Woltran, Scalable uncertainty management SUM 2014. Lecture notes in artificial intelligence, vol 8720, pp 134–147, 2014 [6], Doder et al., A probabilistic temporal logic that can model reasoning about evidence, vol 5956, pp 9–24, 2010 [7], Doder et al. Int J Approx Reason 51:832–845, 2010 [8], Doder et al., Symbolic and quantitative approaches to reasoning with uncertainty ECSQARU 2011. Lecture notes in artificial intelligence, vol 6717, pp 459–471, 2011 [9], Doder et al., J Log Comput 23(3):487–515, 2013 [10], Ilić-Stepić and Ognjanović, Studia Logica 103:145–174, 2015 [17], Ognjanović et al., Ann Math Artif Intell 65(2–3):217–243, 2012, [25], Rašković et al., Int J Approx Reason 49(1):52–66, 2008 [28].

6.1 Nonmonotonic Reasoning and Probability Logics

6.1.1 System P and Rational Monotonicity

Nonmonotonic reasoning is a field of artificial intelligence that studies behavior of the so-called common sense reasoning from available, but incomplete data. Often, an expert possesses incomplete knowledge and use it to infer further information in order to make decisions and plan actions. Thus, nonmonotonic logics deal with principled reasoning about normal or typical situations where the conclusion might be retracted after a new information is added. In [13], Gabbay suggested that the study

Z. Ognjanović et al., *Probability Logics*,
DOI 10.1007/978-3-319-47012-2_6

of default reasoning should be focused on the corresponding consequence relations $\mid\!\sim$ on formulas, where a default rule $\alpha \mid\!\sim \beta$ can be read as "if α, then generally β." Soon after, Kraus, Lehmann and Magidor proposed in [20] a set of properties, named System P (P stands for preferential), that every nonmonotonic consequence relation should satisfy. Those properties are widely accepted as the core of nonmonotonic reasoning (see, for example [12]).

A preferential relation [20] is a binary relation $\mid\!\sim$ on the set of propositional formulas, which satisfies the following properties of so-called System P (*REF*–Reflexivity, *LLE*–Left logical equivalence, *RW*–Right weakening, *CM*–Cautious monotonicity):

$$REF : \frac{}{\alpha\mid\!\sim\alpha}; \qquad LLE : \frac{\vdash \alpha \leftrightarrow \beta,\ \alpha\mid\!\sim\gamma}{\beta\mid\!\sim\gamma};$$

$$RW : \frac{\vdash \alpha \rightarrow \beta,\ \gamma\mid\!\sim\alpha}{\gamma\mid\!\sim\beta}; \qquad AND : \frac{\alpha\mid\!\sim\beta,\ \alpha\mid\!\sim\gamma}{\alpha\mid\!\sim\beta\wedge\gamma};$$

$$OR : \frac{\alpha\mid\!\sim\gamma,\ \beta\mid\!\sim\gamma}{\alpha\vee\beta\mid\!\sim\gamma}; \qquad CM : \frac{\alpha\mid\!\sim\beta,\ \alpha\mid\!\sim\gamma}{\alpha\wedge\beta\mid\!\sim\gamma}.$$

For a set of defaults Δ, we write $\Delta \vdash_P \alpha \mid\!\sim \beta$ if the default $\alpha \mid\!\sim \beta$ is deducible from Δ using System P. In the paper [21] of Lehmann and Magidor, the additional rule of Rational monotonicity is considered:

$$RM : \frac{\alpha \mid\!\sim \gamma,\ \ \alpha \not\mid\!\sim \neg\beta}{\alpha\wedge\beta \mid\!\sim \gamma}$$

and nonstandard probabilistic semantics is proposed. A preferential relation which satisfies the rule *RM* is called a rational relation. Let μ be a finitely additive nonstandard probability measure on formulas and let the binary relation $\mid\!\sim_\mu$ be defined as

$$\alpha \mid\!\sim_\mu \beta \ \text{ iff } \ \mu(\beta|\alpha) \approx 1 \text{ or } \mu(\alpha) = 0.$$

Lehmann and Magidor have proved that each rational relation is generated by some neat finitely additive probability measure, i.e., for each rational relation $\mid\!\sim$ there is a finitely additive hyperreal-valued probability measure μ such that $\mid\!\sim=\mid\!\sim_\mu$.

The classes of preferential and rational relations are not distinguishable using the language of defaults (that contains $\mid\!\sim$, but not $\not\mid\!\sim$).

6.1.2 Modeling Defaults in $LPCP_2^{[0,1]_{\mathbb{Q}(\varepsilon)},\approx}$

We can use $CP_{\approx 1}(\beta,\alpha)$ to syntactically represent the default $\alpha \mid\!\sim \beta$. In the sequel, we will use $\alpha \mid\!\sim \beta$ both in the original context of the system P and to denote the corresponding translation $CP_{\approx 1}(\beta,\alpha)$ in $LPCP_2^{[0,1]_{\mathbb{Q}(\varepsilon)},\approx}$ (Sect. 5). In the case of a finite default base our approach produces the same result as the other mentioned approaches, namely it is equivalent to P.

Theorem 6.1 *For every finite default base Δ and for every default $\alpha \mid\sim \beta$*

$$\Delta \vdash_P \alpha \mid\sim \beta \ \text{ iff } \ \Delta \vdash_{Ax_{LPCP_2^{[0,1]_{\mathbb{Q}(\varepsilon)},\approx}}} \alpha \mid\sim \beta.$$

Proof Since (the corresponding translation of) all axioms and rules (e.g. $(CP_{\approx 1}(\beta, \alpha) \wedge CP_{\approx 1}(\gamma, \alpha)) \rightarrow CP_{\approx 1}(\beta \wedge \gamma, \alpha)$ corresponds to AND rule) of the system P are valid in the class of nonstandard probability models from [21], and $LPCP^{[0,1]_{\mathbb{Q}(\varepsilon)},\approx}_{2,\text{Meas},\text{Neat}}$ is a subclass of that class, P is sound with respect to $LPCP^{[0,1]_{\mathbb{Q}(\varepsilon)},\approx}_{2,\text{Meas},\text{Neat}}$. On the other hand, following the ideas from [21, Lemma 4.9], we can show that for every finite default base Δ and for every default $\alpha \mid\sim \beta$, if $\Delta \nvdash_P \alpha \mid\sim \beta$ then $\Delta \nvdash_{Ax_{LPCP_2^{[0,1]_{\mathbb{Q}(\varepsilon)},\approx}}} \alpha \mid\sim \beta$. The key step in the proof is that there is a finite rational model $\mathbf{M}$ which satisfies Δ and does not satisfy $\alpha \mid\sim \beta$. $\mathbf{M}$ can be transformed to an $LPCP^{[0,1]_{\mathbb{Q}(\varepsilon)},\approx}_{2,\text{Meas},\text{Neat}}$-model $\mathbf{M}'$ such that for every default d, $\mathbf{M} \models d$ iff $\mathbf{M}' \models d$. The transformation can be as follows. For an arbitrary infinitesimal $\varepsilon' \in [0, 1]_{\mathbb{Q}(\varepsilon)}$ a probability distribution μ on W can be defined so that

- $\frac{\mu(w_{n+1})}{\mu(w_n)} = \varepsilon'$, where w_n and w_{n+1} are the sets of all states of the rank n and $n + 1$ respectively, and
- all states of the same rank have equal probabilities.

Since $\mathbf{M} \models \Delta$ and $\mathbf{M} \not\models \alpha \mid\sim \beta$, the same holds for $\mathbf{M}'$, and from the completeness of $LPCP_2^{[0,1]_{\mathbb{Q}(\varepsilon)},\approx}$ we obtain that $\Delta \nvdash_{Ax_{LPCP_2^{[0,1]_{\mathbb{Q}(\varepsilon)},\approx}}} \alpha \mid\sim \beta$. ■

Theorem 6.1 cannot be generalized to an arbitrary default base Δ, as it is illustrated by the following example:

Example 6.1 It is proved in [21, Lemma 2.7] that the infinite set of defaults $T = \{p_i \mid\sim p_{i+1}, p_{i+1} \mid\sim \neg p_i\}$, where p_i's are propositional letters for every integer $i \geq 0$, has only non well-founded preferential models (a preferential model containing an infinite descending chain of states) in which $p_0 \not\mid\sim \bot$, i.e., p_0 is consistent. It means that $T \nvdash_P p_0 \mid\sim \bot$. On the other hand, $T \vdash_{Ax_{LPCP_2^{[0,1]_{\mathbb{Q}(\varepsilon)},\approx}}} p_0 \mid\sim \bot$ since the following holds. Let an $LPCP^{[0,1]_{\mathbb{Q}(\varepsilon)},\approx}_{2,\text{Meas},\text{Neat}}$-model $\mathbf{M} = \langle W, H, \mu, v\rangle$ satisfy the set T. If $\mu([p_i]) = 0$, for some $i > 0$, then it must be $\mu([p_0]) = 0$, and $\mathbf{M} \models p_0 \mid\sim \bot$. Thus, suppose that $\mu([p_i]) \neq 0$, for every $i > 0$. Then, for every $i \geq 0$: $\frac{\mu([p_i \wedge p_{i+1}])}{\mu([p_i])} \approx 1$ and $\frac{\mu([\neg p_i \wedge p_{i+1}])}{\mu([p_{i+1}])} \approx 1$, i.e., $\frac{\mu([p_i \wedge p_{i+1}])}{\mu([p_i])} = 1 - \varepsilon_1$ and $\frac{\mu([\neg p_i \wedge p_{i+1}])}{\mu([p_{i+1}])} = 1 - \varepsilon_2$, for some infinitesimals ε_1 and ε_2. A simple calculation shows that which means that $\mu([p_i]) \leq \varepsilon_0 \mu([p_{i+1}])$ for some infinitesimal ε_0. Since, for some c and k, $\varepsilon_0 \leq c\varepsilon^k$, it follows that for every $i > 0$, $0 \leq \mu([p_0]) \leq \varepsilon^i$. Since $\mu([p_0]) \in S$ and there is no positive element of S with such property, it follows that

$$\mu([p_0]) = 0, \ [p_0] = \emptyset \text{ and } M \models p_0 \mid\sim \bot.$$

Since $\mathbf{M}$ is an arbitrary $LPCP^{[0,1]_{\mathbb{Q}(\varepsilon)},\approx}_{2,\text{Meas},\text{Neat}}$-model, $T \vdash_{LPCP_2^{[0,1]_{\mathbb{Q}(\varepsilon)},\approx}} p_0 \mid\sim \bot$. ■

Note that the above proof of $\mu([p_0]) = 0$, does not hold in the case when the range of the probability is the unit interval of $^*R$ because $^*R$ is ω_1-saturated (which means that the intersection of any countable decreasing sequence of nonempty internal sets must be nonempty). As a consequence, thanks to the restricted ranges of probabilities that are allowed in $LPCP_{2,\mathrm{Meas},\mathrm{Neat}}^{[0,1]_{\mathbb{Q}(\varepsilon)},\approx}$-class of models, our system goes beyond the system P, when we consider infinite default bases.

$LPCP_2^{[0,1]_{\mathbb{Q}(\varepsilon)},\approx}$ is rich enough not only to express formulas that represents defaults but also to describe more: probabilities of formulas, negations of defaults, combinations of defaults with the other (probabilistic) formulas, etc. Let us now consider some situations where these possibilities allow us to obtain more conclusions than in the framework of the language of defaults.

Example 6.2 The translation of rational monotonicity, $((\alpha \mid\!\sim \beta) \wedge \neg(\alpha \mid\!\sim \neg\gamma)) \rightarrow ((\alpha \wedge \gamma) \mid\!\sim \beta)$, is $LPCP_{2,\mathrm{Meas},\mathrm{Neat}}^{[0,1]_{\mathbb{Q}(\varepsilon)},\approx}$-valid since rational monotonicity is satisfied in every $^*R$-probabilistic model, and $LPCP_{2,\mathrm{Meas},\mathrm{Neat}}^{[0,1]_{\mathbb{Q}(\varepsilon)},\approx}$ is a subclass of that class of models. The same holds for the formula $\neg(true \mid\!\sim false)$ corresponding to another property called normality in [12]. ■

Note that in the above example we use negated defaults that are not expressible in P.

Example 6.3 Let the default base consist of the following two defaults $s \mid\!\sim b$ and $s \mid\!\sim t$, where s, b and t means Swedes, blond and tall, respectively [3]. Because of the inheritance blocking problem, in some systems (for example in P) it is not possible to conclude that Swedes who are not tall are blond $((s \wedge \neg t) \mid\!\sim b)$. Since our system and P coincide if the default base is finite, the same holds in our framework. In fact, there are some $LPCP_{2,\mathrm{Meas},\mathrm{Neat}}^{[0,1]_{\mathbb{Q}(\varepsilon)},\approx}$-models in which the previous formula is not satisfied. Avoiding a discussion of intuitive acceptability of the above conclusion, we point out that by adding some additional assumptions ($CP_{=1-\varepsilon}(t, s)$ and $CP_{=1-\varepsilon^2}(b, s)$) to the default base we can entail that conclusion too. First, note that the assumptions are compatible with defaults $s \mid\!\sim t$ and $s \mid\!\sim b$. Then, an easy calculation shows that $\frac{P(s\wedge\neg t)}{P(s)} = \frac{P(s)-P(s\wedge t)}{P(s)} = \frac{P(s)-P(s)+P(s)\varepsilon}{P(s)} = \varepsilon$, and similarly $\frac{P(s\wedge\neg b)}{P(s)} = \varepsilon^2$. Finally, we can estimate the conditional probability of b given $s \wedge \neg t$:

$$\frac{P(s \wedge \neg t \wedge b)}{P(s \wedge \neg t)} = \frac{P(s \wedge \neg t) - P(s \wedge \neg t \wedge \neg b)}{P(s \wedge \neg t)} \geq \frac{\varepsilon P(s) - \varepsilon^2 P(s)}{\varepsilon P(s)} = 1 - \varepsilon.$$

It follows that $(s \wedge \neg t) \mid\!\sim b$. ■

6.1.2.1 Decidability of $LPCP_2^{[0,1]_{\mathbb{Q}(\varepsilon)},\approx}$

The decision method presented in this section extends the procedure from Sect. 3.5. We can show that every probability formula A is equivalent to a disjunctive normal form of A: $DNF(A) = \bigvee_{i=1}^{m} \bigwedge_{j=1}^{k_i} \pm X_{i,j}(p_1, \ldots, p_n)$, where n is the number of propositional letters from A, and

- $X_{i,j} \in \{CP_{\geq s}, CP_{\leq s}\}_{s \in [0,1]_{\mathbb{Q}(\varepsilon)}} \cup \{CP_{\approx r}\}_{r \in [0,1]_{\mathbb{Q}}}$,
- $X_{i,j}(p_1, \ldots, p_n)$ denotes that the propositional formula which is in the scope of the probability operator $X_{i,j}$ is in the complete disjunctive normal form, i.e., the propositional formula is a disjunction of the atoms of A.

Obviously, to prove decidability of our logic, it is enough to show that satisfiability of probability formulas of the form $\bigwedge_{i=1}^{k} \pm X_i(p_1, \ldots, p_n)$ is decidable.

For every conditional probability formula ($\pm CP_{\geq s}(\alpha, \beta)$, $\pm CP_{\leq s}(\alpha, \beta)$, and $\pm CP_{\approx r}(\alpha, \beta)$) we can distinguish two cases

1. the probability of β is zero, in which case
   - $CP_{\geq s}(\alpha, \beta)$, for $s \in [0, 1]_{\mathbb{Q}(\varepsilon)}$, $\neg CP_{\leq s}(\alpha, \beta)$, for $s \in [0, 1]_{\mathbb{Q}(\varepsilon)} \setminus \{1\}$, $CP_{\leq 1}(\alpha, \beta)$, and $CP_{\approx 1}(\alpha, \beta)$ hold—and can be deleted from the formula, while
   - $\neg CP_{\geq s}(\alpha, \beta)$, for $s \in [0, 1]_{\mathbb{Q}(\varepsilon)}$, $CP_{\leq s}(\alpha, \beta)$, $CP_{\approx r}(\alpha, \beta)$, for $s \in [0, 1]_{\mathbb{Q}(\varepsilon)} \setminus \{1\}$, $r \in [0, 1]_{\mathbb{Q}} \setminus \{1\}$ $\neg CP_{\leq 1}(\alpha, \beta)$, and $\neg CP_{\approx 1}(\alpha, \beta)$ do not hold—and the whole conjunction is not satisfiable,
2. the probability of β is greater than zero.

As a consequence, to prove decidability of our logic it is enough to prove decidability of satisfiability of formulas which are conjunctions of conditional probabilistic formulas of the forms: $\pm CP_{\geq s}(\alpha, \beta)$, $\pm CP_{\leq s}(\alpha, \beta)$, and $\pm CP_{\approx r}(\alpha, \beta)$, such that the probability of β is greater than 0.

In the next step, we will reduce the satisfiability problem to linear programming problem. However, in the logic we discuss here, the range of probabilities is recursive and contains nonstandard values, and there are operators of the form $CP_{\approx r}$ that do not appear in Sect. 3.5. Thus, we should perform the reduction carefully to obtain linear systems that are suitable for establishing decidability, which, in our approach, means that Fourier–Motzkin elimination can be applied to them. The idea is to eliminate $\approx$ and $\not\approx$ signs and to try to solve linear systems in an extension of $\mathbb{Q}(\varepsilon)$. We will use the following abbreviations:

- x_i denotes the measure of the atom $a_i \in At(A)$, $i = 1, \ldots, 2^n$,
- $a_i \models \alpha$ means that the atom a_i appears in the complete disjunctive normal form of a classical propositional formula α,
- $\sum(\alpha)$ denotes $\sum_{a_i \in At(A): a_i \models \alpha} x_i$, and
- $\mathrm{C}\sum(\alpha, \beta)$ denotes $\frac{\sum(\alpha \wedge \beta)}{\sum(\beta)}$.

Recall that $[\alpha]_M$ denotes the set of all worlds of an $LPCP^{[0,1]_{\mathbb{Q}(\varepsilon)},\approx}_{2,\mathrm{Meas},\mathrm{Neat}}$-model M that satisfy α. Since $[\alpha]_M = \cup_{a_i \in At(A): a_i \models \alpha}[a_i]_M$, and different atoms are mutually exclusive, i.e., $[a_i]_M \cap [a_j]_M = \emptyset$ for $i \neq j$, $CP_{\geq s}(\alpha, \beta)$ holds in M iff $\sum(\beta) = 0$, or $\sum(\beta) > 0$ and $\mathrm{C}\sum(\alpha, \beta) \geq s$ (and similarly for $CP_{\leq s}$, and $CP_{\approx r}$).

Let us consider a formula A of the form

$$(\wedge_{i=1,I} \pm CP_{\geq s_i}(\alpha_i, \beta_i)) \wedge (\wedge_{j=1,J} \pm CP_{\leq s_j}(\alpha_j, \beta_j)) \wedge (\wedge_{l=1,L} \pm CP_{\approx r_l}(\alpha_l, \beta_l)).$$

Then, A is satisfiable iff the following system is satisfiable:

$$
\begin{array}{ll}
\sum_{i=1}^{2^n} x_i = 1 & \\
x_i \geq 0 & \text{for } i = 1, 2^n \\
\sum(\beta) > 0 & \text{for every formula } \beta \text{ appearing in the formulas of the} \\
 & \text{form } \pm CP_\diamond(\alpha, \beta) \text{ from } A, \text{ and } \diamond \in \{\geq s_i, \leq s_j, \approx r_l\} \\
\mathrm{C}\sum(\alpha_i, \beta_i) \geq s_i & \text{for every formula } CP_{\geq s_i}(\alpha_i, \beta_i) \text{ from } A \\
\mathrm{C}\sum(\alpha_i, \beta_i) < s_i & \text{for every formula } \neg CP_{\geq s_i}(\alpha_i, \beta_i) \text{ from } A \\
\mathrm{C}\sum(\alpha_j, \beta_j) \leq s_j & \text{for every formula } CP_{\leq s_j}(\alpha_j, \beta_j) \text{ from } A \\
\mathrm{C}\sum(\alpha_j, \beta_j) > s_j & \text{for every formula } \neg CP_{\leq s_j}(\alpha_j, \beta_j) \text{ from } A \\
\mathrm{C}\sum(\alpha_l, \beta_l) \approx r_l & \text{for every formula } CP_{\approx r_l}(\alpha_l, \beta_l) \text{ from } A \\
\mathrm{C}\sum(\alpha_l, \beta_l) \not\approx r_l & \text{for every formula } \neg CP_{\approx r_l}(\alpha_l, \beta_l) \text{ from } A.
\end{array}
$$

We can further simplify the above system by observing that every expression of the form $\mathrm{C}\sum(\alpha_l, \beta_l) \approx r_l$ can be seen as

$$\mathrm{C}\sum(\alpha_l, \beta_l) - r_l \approx 0 \text{ and } \mathrm{C}\sum(\alpha_l, \beta_l) - r_l \geq 0 \tag{6.1}$$

or

$$\mathrm{C}\sum(\alpha_l, \beta_l) - r_l \approx 0 \text{ and } r_l - \mathrm{C}\sum(\alpha_l, \beta_l) \geq 0. \tag{6.2}$$

Similarly, every expression of the form $\mathrm{C}\sum(\alpha_l, \beta_l) \not\approx r_l$ can be seen as

$$\mathrm{C}\sum(\alpha_l, \beta_l) - r_l \not\approx 0 \text{ and } \mathrm{C}\sum(\alpha_l, \beta_l) - r_l > 0 \tag{6.3}$$

or

$$\mathrm{C}\sum(\alpha_l, \beta_l) - r_l \not\approx 0 \text{ and } r_l - \mathrm{C}\sum(\alpha_l, \beta_l) > 0. \tag{6.4}$$

Thus, we will consider systems containing expressions of the forms (6.1)–(6.4) instead of $\mathrm{C}\sum(\alpha_l, \beta_l) \approx r_l$, and $\mathrm{C}\sum(\alpha_l, \beta_l) \not\approx r_l$, respectively. Let us use $S(\overrightarrow{x}, \varepsilon)$ to denote a system of that form. Note that

$$\mathrm{C}\sum(\alpha_l, \beta_l) - r_l \approx 0 \text{ and } \mathrm{C}\sum(\alpha_l, \beta_l) - r_l \geq 0 \tag{6.5}$$
$$\text{is equivalent to } (\exists n_l \in \mathbb{N})\, 0 \leq \mathrm{C}\sum(\alpha_l, \beta_l) - r_l < n_l \cdot \varepsilon,$$

$$\mathrm{C}\sum(\alpha_l, \beta_l) - r_l \approx 0 \text{ and } r_l - \mathrm{C}\sum(\alpha_l, \beta_l) \geq 0 \tag{6.6}$$
$$\text{is equivalent to } (\exists n_l \in \mathbb{N})\, 0 \geq \mathrm{C}\sum(\alpha_l, \beta_l) - r_l > -n_l \cdot \varepsilon,$$

$$C\sum(\alpha_l, \beta_l) - r_l \not\approx 0 \text{ and } C\sum(\alpha_l, \beta_l) - r_l > 0 \tag{6.7}$$
$$\text{is equivalent} \quad \text{to} \quad (\exists n_l \in \mathbb{N})C\sum(\alpha_l, \beta_l) - r_l > \frac{1}{n_l},$$

$$C\sum(\alpha_l, \beta_l) - r_l \not\approx 0 \text{ and } r_l - C\sum(\alpha_l, \beta_l) > 0 \tag{6.8}$$
$$\text{is equivalent} \quad \text{to} \quad (\exists n_l \in \mathbb{N})C\sum(\alpha_l, \beta_l) - r_l < -\frac{1}{n_l}.$$

Since we have only finitely many expressions of the forms (6.1)–(6.4) in our system, we can use a unique $n_0 \in \mathbb{N}$ instead of many n_l's in expressions (6.5)–(6.8) (we can choose any n_0 greater than all n_l's). In other words, if we denote by $S(\overrightarrow{x}, \varepsilon, n_0)$ the conjunction of all the formulas that appear in (6.5)–(6.8) under the scope of existential quantifier (with n_l's replaced by n_0), then

$$S(\overrightarrow{x}, \varepsilon) \text{ has a solution in } \mathbb{Q}(\varepsilon) \text{ iff } (\exists n_0 \in \mathbb{N})S(\overrightarrow{x}, \varepsilon, n_0) \text{ is satisfiable in } \mathbb{Q}(\varepsilon). \tag{6.9}$$

Note that n_0 is not determined in the (6.9). Now, we will replace n_0 with another, infinite but fixed, parameter K which will also have some suitable characteristics in relation to ε. The role of K is to help us to avoid the standard approach to the analysis of inequalities, where we have very often to discuss arguments of the form "it holds for all, large enough integers." Since K is a positive infinite integer, if an inequality holds for every $n \in \mathbb{N}$ greater than some fixed finite n_0, by the overspill principle it also holds for K. The other direction is a consequence of the underspill principle which says that if an inequality holds for every infinite number less than K, it also holds for some finite positive integer. Thus, let us consider the following set

$$O = \{n \in {}^*\mathbb{N} : S(\overrightarrow{x}, n, \varepsilon) \text{ has a solution in } {}^*\mathbb{R}\}.$$

O is an internal set. If O is nonempty, it contains all natural numbers greater than some fixed natural number n'. Using the overspill and underspill principles, we conclude that O contains all infinite numbers from ${}^*\mathbb{N}$ which are less than a fixed infinite natural number K, i.e., for some $n' \in \mathbb{N}$, and $K \in {}^*\mathbb{N} \setminus \mathbb{N}$, $[n', K] = \{n \in {}^*\mathbb{N} : n' \leq n \leq K\} \subset O$. Then,

$$S(\overrightarrow{x}, n_0, \varepsilon) \text{ has a solution in } {}^*\mathbb{R} \text{ iff } S(\overrightarrow{x}, K, \varepsilon) \text{ has a solution in } {}^*\mathbb{R}. \tag{6.10}$$

We can choose K so that for every $k \in \mathbb{N}$, $K^k \cdot \varepsilon \approx 0$. That can be explained as follows. Let us consider the internal set $O' = \{n \in {}^*\mathbb{N} : n^n < \frac{1}{\sqrt{\varepsilon}}\}$. Obviously, $\mathbb{N} \subset O'$. Using the overspill principle, there is some $K \in {}^*\mathbb{N} \setminus \mathbb{N}$ such that

$$0 < K^K < \frac{1}{\sqrt{\varepsilon}} \text{ and } 0 < K^K \cdot \varepsilon < \sqrt{\varepsilon}.$$

Thus, for every $k \in \mathbb{N}$,

$$0 < K^k \cdot \varepsilon < \sqrt{\varepsilon} \text{ and } K^k \cdot \varepsilon \approx 0.$$

Note that $\approx$ and $\not\approx$ do not appear in the system $S(\overrightarrow{x}, K, \varepsilon)$. Thus, we can freely multiply (in)equalities by the denominators of the expressions of the form $\mathrm{C}\sum(\alpha, \beta)$ and in that way obtain linear (in)equalities of the form

$$\sum(\alpha \wedge \beta) - s\sum(\beta)\ \rho\ 0,$$

where s is a rational function in ε and K, and $\rho \in \{\geq, >, =, <, \leq\}$.

Next, we perform Fourier–Motzkin elimination, which iteratively rewrites the starting system into a new system without a variable x_i such that two systems are equisatisfiable. During the procedure, numerators and denominators of coefficients in (in)equalities remain polynomials in ε and K. When no variables are left, we have to check satisfiability of relations between numerical expressions with the parameter ε and K which is decidable since K is chosen so that for every $k \in \mathbb{N}$, $K^k \cdot \varepsilon \approx 0$. Namely, we compare two polynomials $Q_1(\varepsilon, K)$ and $Q_2(\varepsilon, K)$ in ε and K of the forms: $Q_1(\varepsilon, K) = q_{1,0}Q_{1,0}(K)\varepsilon^0 + q_{1,1}Q_{1,1}(K)\varepsilon^1 + \cdots + q_{1,n_1}Q_{1,n_1}(K)\varepsilon^{n_1}$, and $Q_2(\varepsilon, K) = q_{2,0}Q_{2,0}(K)\varepsilon^0 + q_{2,1}Q_{2,1}(K)\varepsilon^1 + \cdots + q_{2,n_2}Q_{2,n_2}(K)\varepsilon^{n_2}$, where $q_{i,j}$'s are rationals, and $Q_{i,j}(K)$'s are polynomials in K with rational coefficients. Comparison of $Q_1(\varepsilon, K)$ and $Q_2(\varepsilon, K)$ starts by examining $q_{1,0}Q_{1,0}(K)$ and $q_{2,0}Q_{2,0}(K)$ in the standard way. If they are equal, we have to examine $q_{1,1}Q_{1,1}(K)$ and $q_{2,1}Q_{2,1}(K)$ and so on. Since ε is an infinitesimal, the above examination of expressions sharing the same powers of ε is done in a reverse order with respect to the standard procedure of comparison of polynomials.

It follows that the problem of solving whether $S(\overrightarrow{x}, K, \varepsilon)$ has a solution in $^*\mathbb{R}$ is decidable, i.e., it is decidable whether the above defined set O is nonempty. As we noted above, in the case $O \neq \emptyset$, there is a nonempty $[n', K] \subset O$. In particular, if we replace K with an arbitrary finite $n \in [n', K]$ in the inequalities resulting from Fourier–Motzkin elimination applied to $S(\overrightarrow{x}, K, \varepsilon)$, we obtain the inequalities solvable in $\mathbb{Q}(\varepsilon)$ iff the original inequalities are solvable in $^*\mathbb{R}$ which is decidable.

If $S(\overrightarrow{x}, \varepsilon)$ is solvable, we can define an $LPCP_2^{[0,1]_{\mathbb{Q}(\varepsilon)},\approx}$-model $\mathbf{M} = \langle W, H, \mu, v\rangle$ such that

- $W = \text{Atoms}(A)$,
- $H = \mathbb{P}(W)$,
- μ is defined according to the solutions of $S(\overrightarrow{x}, \varepsilon)$, and
- $v(a)(p) = true$ iff p (and not $\neg p$) appears in the conjunction which constitutes the atom a.

Obviously, $\mathbf{M} \models A$. However, even if $S(\overrightarrow{x}, \varepsilon)$ has a solution, some of x_i's might be 0. It means that $\mathbf{M}$ does not satisfy the neatness condition, i.e., that some nonempty sets of worlds (represented by the corresponding atoms that hold in those worlds) have the zero probabilities. In that case, we can simply remove those worlds and denote the obtained model by $\mathbf{M}'$. It is easy to see that for every formula A, $\mathbf{M} \models A$ iff $\mathbf{M}' \models A$. Thus, we have

Theorem 6.2 *The problem of $LPCP_{2,\mathrm{Meas},\mathrm{Neat}}^{[0,1]_{\mathbb{Q}(\varepsilon)},\approx}$-satisfiability is decidable.* ■

Example 6.4 Let us consider the formula $A = C \wedge ((D \vee B) \rightarrow (D \wedge B))$, where B, C and D denote $CP_{\approx 0}(q, \top)$, $CP_{\approx 1}(\neg p \wedge \neg q, \neg q)$ and $CP_{\approx 0.4}(p \wedge q, q)$, respectively. The set of atoms, Atoms(A), contains $a_1 = p \wedge q$, $a_2 = p \wedge \neg q$, $a_3 = \neg p \wedge q$ and $a_4 = \neg p \wedge \neg q$. Let x_i denote the measure of atom a_i. The formula A is equivalent to $(B \wedge C \wedge D) \vee (\neg B \wedge C \wedge \neg D)$. We start with the first conjunct $B \wedge C \wedge D$. According to the above procedure suppose that the measures of q and $\neg q$ are greater than zero, i.e., that $x_1 + x_3 > 0$, and $x_2 + x_4 > 0$. $B \wedge C \wedge D$ is satisfiable iff the same holds for the following system:

$$\begin{array}{ll} x_1 + x_2 + x_3 + x_4 = 1\ , x_i \geq 0 & \text{for } i = 1, 4 \\ x_1 + x_3 > 0 & x_2 + x_4 > 0 \\ x_1 + x_3 \approx 0 & \\ x_2/(x_2 + x_4) \approx 1 & \\ x_1/(x_1 + x_3) \approx 0.4 & \end{array}$$

which is equivalent to

$$\begin{array}{ll} x_1 + x_2 + x_3 + x_4 = 1, x_i \geq 0 & \text{for } i = 1, 4 \\ x_1 + x_3 > 0 & x_2 + x_4 > 0 \\ 0 < x_1 + x_3 < n_1\varepsilon & \\ x_4/(x_2 + x_4) < 1/n_2 & \\ 0.4 - n_3\varepsilon < x_1/(x_1 + x_3) < 0.4 + n_3\varepsilon & \end{array}$$

for some $n_1, n_2, n_3 \in \mathbb{N}$. If we replace n_1, n_2, n_3 by their maximum denoted by n, we obtain an equivalent system. Since $\approx$ does not appear in the last system, Fourier–Motzkin elimination can be performed in the standard way. The procedure finishes with the true condition

$$\frac{1 - n\varepsilon}{n} < 1$$

which means that the considered formula is satisfiable. ■

6.1.3 Approximate Defaults and $LPCP_2^{[0,1]_{\mathbb{Q}(\varepsilon)},\approx}$

Now we turn to the issue of combining default knowledge and probabilistic knowledge. We are interested in probabilistic approximations of defaults of the form $\alpha \mid\!\sim_n \beta$, where the corresponding conditional probability is not approximately 1, but at least $1 - \frac{1}{n}$ [8]. We are interested in the following question: if we use both rational relation $\mid\!\sim$ and $\mid\!\sim_n$ in one of the nonmonotonic rules, does the strength of a premise (the number n) transfer to the conclusion? Consider the following example, which is essentially a modification of examples from [23].

Example 6.5 Suppose that the statistical knowledge "more than 95 % of birds fly" is available, and that we accept the default rule "generally, birds have wings." (The former can be expressed in our terminology with $b \mid\!\sim_{20} f$, while the later is expressible by default rule $b \mid\!\sim w$, usually interpreted as "conditional probability of w knowing b is approximately 1.")

What can we say about the birds with the wings? Intuitively, the conclusion that they fly with the probability greater than 95 % is quite acceptable. On the other hand, the best we can calculate is that the probability is either greater than or infinitely close to 95 %. ■

We overcome the above difficulty by slightly changing the notion of $\mid\!\sim_n$.

Theorem 6.3 *Let $\mid\!\sim$ be a rational relation, and let μ be a corresponding neat nonstandard probability measure. If the binary relations $\mid\!\sim_n$ are defined by $\alpha \mid\!\sim_n \beta$ iff $\mu(\beta|\alpha) > 1 - \frac{1}{n}$ or $\mu(\beta|\alpha) \approx 1 - \frac{1}{n}$, then the following rules hold:*

$$LLE_n^{\approx} : \frac{\vdash \alpha \leftrightarrow \beta, \alpha \mid\!\sim_n \gamma}{\beta \mid\!\sim_n \gamma}; \qquad RW_n^{\approx} : \frac{\vdash \alpha \rightarrow \beta, \gamma \mid\!\sim_n \alpha}{\gamma \mid\!\sim_n \beta};$$

$$OR_n^{\approx} : \frac{\alpha \mid\!\sim \gamma, \beta \mid\!\sim_n \gamma}{\alpha \vee \beta \mid\!\sim_n \gamma}; \qquad AND_n^{\approx} : \frac{\alpha \mid\!\sim \beta, \alpha \mid\!\sim_n \gamma}{\alpha \mid\!\sim_n \beta \wedge \gamma};$$

$$CM1_n^{\approx} : \frac{\alpha \mid\!\sim_n \beta, \alpha \mid\!\sim \gamma}{\alpha \wedge \beta \mid\!\sim_n \gamma}; \qquad CM2_n^{\approx} : \frac{\alpha \mid\!\sim \beta, \alpha \mid\!\sim_n \gamma}{\alpha \wedge \beta \mid\!\sim_n \gamma}.$$

Proof As an illustration, we will prove $OR_n^{\approx}$. Let us suppose $\mu(\gamma|\alpha) = 1 - \varepsilon_1$ and $\mu(\gamma|\beta) > 1 - \frac{1}{n} - \varepsilon_2$ ($\varepsilon_1, \varepsilon_2 \approx 0$). As in the proof of OR_n, we can obtain $\mu(\gamma|\alpha \vee \beta) \geq 1 - \frac{\mu(\alpha_0 \vee \beta_0)}{\mu(\alpha \vee \beta)}$, where α_0 is $\alpha \wedge \neg\gamma$ and β_0 is $\beta \wedge \neg\gamma$. From $\mu(\alpha_0) \leq \varepsilon_1 \mu(\alpha \vee \beta)$ and $\mu(\beta_0) < (\varepsilon_2 + \frac{1}{n})\mu(\alpha \vee \beta)$ we conclude $\mu(\alpha_0 \vee \beta_0) \leq (\varepsilon_1 + \varepsilon_2 + \frac{1}{n})\mu(\alpha \vee \beta)$. Thus, $\mu(\gamma|\alpha \vee \beta) \geq 1 - \frac{1}{n} - \varepsilon_1 + \varepsilon_2 \approx 1 - \frac{1}{n}$. ■

The above statement is in spirit of [23, Theorem 5.1] where some rules with similar combinations of default knowledge and probabilistic knowledge are presented.

The $LPCP_2^{[0,1]_{\mathbb{Q}(\varepsilon)},\approx}$-logic is a suitable syntactic framework for modeling default reasoning. For example, the above rule $OR_n^{\approx}$ can be written as

$$CP_{\approx 1}(\gamma, \alpha) \wedge (CP_{\geq 1-\frac{1}{n}}(\gamma, \beta) \vee CP_{\approx 1-\frac{1}{n}}(\gamma, \beta)) \rightarrow$$
$$(CP_{\geq 1-\frac{1}{n}}(\gamma, \alpha \vee \beta) \vee CP_{\approx 1-\frac{1}{n}}(\gamma, \alpha \vee \beta)).$$

Section 5.7.1 provides a complete axiomatization of the logic $LPCP_2^{[0,1]_{\mathbb{Q}(\varepsilon)},\approx}$, so that (the $LPCP_2^{[0,1]_{\mathbb{Q}(\varepsilon)},\approx}$-translations) of all the rules from Theorem 6.3 are theorems of $LPCP_2^{[0,1]_{\mathbb{Q}(\varepsilon)},\approx}$. Obviously, if a formula (representing the $LPCP_2^{[0,1]_{\mathbb{Q}(\varepsilon)},\approx}$-translation of defaults and/or approximate defaults) is not an $LPCP_2^{[0,1]_{\mathbb{Q}(\varepsilon)},\approx}$-theorem, then it is not a consequence of (the $LPCP_2^{[0,1]_{\mathbb{Q}(\varepsilon)},\approx}$-translations) of the above rules.

6.2 Logic for Reasoning About Evidence

6.2.1 Evidence

As a noun, evidence can be defined as something that proves or disproves certain claims—hypotheses. In the philosophy of science the notion of evidence is one of the main concepts in confirmation theory and evidence theory. Let us first consider the introductory example from [15].

Example 6.6 Suppose that a box with fair and double-headed coins is given and that one coin is randomly chosen from that box. After finitely many (say 10) tosses of the coin and the observations of the outcomes of each toss, what can we say about the likelihood of the hypothesis that the coin is double-headed without the prior knowledge of the exact number of fair coins and the exact number of the double-headed coins in the box?

If the coin lands tails in at least one toss, then the hypothesis that the coin is double-headed is obviously false. On the other hand, if the coin lands heads in all tosses, than we can say that our experiment favors the mentioned hypothesis (the coin is double-headed). In this particular case, the amount of evidence increases with the increment of the number of tosses in the experiment.

Of course, the relative frequencies of double-headed and fair coins in the box are sufficient for the computation of the probability of hypotheses after experiment. However, the prior probabilities are usually unknown, so we cannot compute posterior probabilities. Still, experiments do provide some evidence in favor of one or several possible hypotheses. ■

Intuitively, the probability of a hypothesis depends on

- the prior probabilities of the hypothesis (i.e., the proportion of double-headed coins in the box);
- to what extent the observations support the hypothesis.

The second item is formalized by the weight of evidence—the function which assigns a number from the unit interval to every observation and hypothesis. Let us give a short overview of how evidence is modeled and formalized in the case of one or more observations.

Let $H = \{h_1, \ldots, h_m\}$ be the set that represents mutually exclusive and exhaustive hypotheses and $O = \{o_1, \ldots, o_n\}$ be the set of possible observations.

For the hypothesis h_i, let μ_i be a likelihood function on O, i.e., the function which satisfies

- $\mu_i : O \longrightarrow [0, 1]$;
- $\mu_i(o_1) + \cdots + \mu_i(o_n) = 1$.

We assume that for every observation $o \in O$ there is $i \in \{1, \ldots, m\}$ such that $\mu_i(o) > 0$. An evidence space is a tuple $\mathrm{E} = \langle H, O, \mu_1, \ldots, \mu_m \rangle$. For an evidence space E, we define a weight function w_{E} by the following conditions:

- $w_{\mathrm{E}} : O \times H \longrightarrow [0, 1]$;
- $w_{\mathrm{E}}(o_i, h_j) = \frac{\mu_j(o_i)}{\mu_1(o_i)+\cdots+\mu_m(o_i)}$.

Intuitively, $w_{\mathrm{E}}(o, h)$ is the likelihood that the hypothesis h holds, if o is observed. Specially, $w_{\mathrm{E}}(o, h) = 0$ means that h is certainly false if o is observed, while $w_{\mathrm{E}}(o, h) = 1$ means that o fully confirms h.

Similarly as in the case of probability measures, we need a list of properties that will fully capture the notion of a weight function. In addition, those properties should be stated in such a way that will enable purely syntactical reformulation.

Reference [15] provides one such characterization of weight functions

Theorem 6.4 *Let* $H = \{h_1, \ldots, h_m\}$ *and* $O = \{o_1, \ldots, o_n\}$, *and let* f *be a real-valued function,* $f : O \times H \longrightarrow [0, 1]$. *Then there exists an evidence space*

$$\mathrm{E} = \langle H, O, \mu_1, \ldots, \mu_m \rangle$$

such that $f = w_{\mathrm{E}}$ *iff* f *satisfies the following properties:*

1. $f(o_i, h_1) + \cdots + f(o_i, h_m) = 1$, *for every* $i \in \{1, \ldots, n\}$.
2. *There exist* $x_1, \ldots, x_n > 0$ *such that, for all* $j \in \{1, \ldots, m\}$, $x_1 f(o_1, h_j) + \cdots + x_n f(o_n, h_j) = 1$. ■

Moreover, if 1. and 2. are satisfied, then the likelihood functions μ_i, $i \in \{1, \ldots, m\}$ are defined by

$$\mu_j(o_i) = \frac{f(o_i, h_j)}{x_j}.$$

A weight function can be seen as a qualitative assessment of the evidence in favor of one of the hypotheses. As we have mentioned earlier, we cannot determine the probability of the hypotheses after an observation without knowing the prior probabilities. On the other hand, if we know the prior probabilities, then we can use the Dempster's rule of combination to calculate the posterior probabilities.

Dempster's rule of combination combines probability distributions ν_1 and ν_2 on $\mathcal{G}$ in the following way: for every measurable $G \subseteq \mathcal{G}$

$$(\nu_1 \oplus \nu_2)(G) = \frac{\sum_{h \in G} \nu_1(h)\nu_2(h)}{\sum_{h \in \mathcal{G}} \nu_1(h)\nu_2(h)}.$$

Let μ be a probability measure on the set $For(H)$ of propositional formulas over $H = \{h_1, \ldots, h_m\}$, which satisfies

$$\mu(h_1) + \cdots + \mu(h_m) = 1.$$

Since hypotheses are mutually exclusive, μ should satisfy $\mu(h_i \wedge h_j) = 0$, for $i \neq j$. Then $\mu(h_1 \vee \ldots \vee h_m) = \mu(h_1) + \cdots + \mu(h_m)$, so the fact that hypotheses are exhaustive may be expressed by the equality $\mu(h_1) + \cdots + \mu(h_m) = 1$.

Note that for any $\phi \in For(H)$, there exists $\phi' \in For(H)$ of the form $\bigvee_{i \in I} h_i$, for some $I \subseteq \{1, \ldots, m\}$, such that $\mu(\phi) = \mu(\phi')$.

Indeed, it is obvious if $\phi \in H$; suppose that $\mu(\phi_1) = \mu(\phi_1')$ and $\mu(\phi_2) = \mu(\phi_2')$, where ϕ_1' is of the form $\bigvee_{i \in I_1} h_i$ and ϕ_2' is of the form $\bigvee_{i \in I_2} h_i$.

Then, $\mu(\phi_1 \wedge \phi_2) = \mu((\phi_1 \wedge \phi_2)')$ and $\mu(\neg\phi_1) = \mu((\neg\phi_1)')$ for $(\phi_1 \wedge \phi_2)' = \bigvee_{i \in I_1 \cap I_2} h_i$ and $(\neg\phi_1)' = \bigvee_{i \in \{1, \ldots, m\} \setminus I_1} h_i$.

Since for each observation o for which $\mu_1(o) + \cdots + \mu_m(o) > 0$, $w_{\mathrm{E}}(o, h_1) + \cdots + w_{\mathrm{E}}(o, h_m) = 1$ holds, there is a unique probability measure on $For(H)$ which is an extension of $w_{\mathrm{E}}(o, \cdot)$, such that hypotheses are mutually exclusive. Hence, we will also denote that measure with $w_{\mathrm{E}}(o, \cdot)$. Informally, we may assume that elements of $For(H)$ are subsets of H.

If we let μ be a prior probability on hypotheses, then we can calculate the probability of hypotheses after observing o in the following way:

$$\mu_o = \mu \oplus w_{\mathrm{E}}(o, \cdot).$$

If $\mathrm{E} = \langle H, O, \mu_1, \ldots, \mu_m \rangle$ is an evidence space, we define

$$\mathrm{E}^* = \langle H, O^*, \mu_1^*, \ldots, \mu_m^* \rangle$$

as follows:

- $O^* = \{\langle o^1 \ldots, o^k \rangle \mid k \in \omega, o^i \in O\}$.
- $\mu_i^* : O^* \longrightarrow [0, 1]$ is defined by

$$\mu_i^*(\langle o^1, \ldots, o^k \rangle) = \mu_i(o^1) \cdots \mu_i(o^k).$$

The sequence $\langle o^1, \ldots, o^k \rangle$ can be seen as a conjunction of its members, so the previous formula implicitly reflects independence of observations $o^1, \ldots, o^k$. It is shown in [15] that

$$w_{\mathrm{E}^*}(\langle o^1, \ldots, o^k \rangle, \cdot) = w_{\mathrm{E}}(o^1, \cdot) \oplus \cdots \oplus w_{\mathrm{E}}(o^k, \cdot).$$

Informally, $w_{\mathrm{E}^*}(\langle o^1, \ldots, o^k \rangle, h)$ is the weight that hypothesis h is true, after observing $o^1 \ldots, o^k$. We will also use the following equality in an axiomatization of our temporal logic:

$$w_{\mathrm{E}^*}(\langle o^1, \ldots, o^k \rangle, h_i) = \frac{w_{\mathrm{E}^*}(o^1, h_i) \cdots w_{\mathrm{E}^*}(o^k, h_i)}{w_{\mathrm{E}^*}(o^1, h_1) \cdots w_{\mathrm{E}^*}(o^k, h_1) + \cdots + w_{\mathrm{E}^*}(o^1, h_m) \cdots w_{\mathrm{E}^*}(o^k, h_m)}.$$

If the prior probability μ on the set of hypotheses H is known, then we can calculate the probability of hypotheses after observing $\langle o^1, \ldots, o^k \rangle$ in the following way:

$$\mu_{\langle o^1, \ldots, o^k \rangle} = \mu \oplus w_{\mathrm{E}}(\langle o^1, \ldots, o^k \rangle, \cdot).$$

As an illustration, we will slightly modify Example 2.5 from [15]

Example 6.7 Similarly as before, one coin is taken from the box that contains fair and double-headed coins. The coin is tossed 10 times. Each toss of the coin yields a different observation. The set of all possible outcomes (observations) is $O = \{h, t\}$, where h stands for "the coin lands heads," while t stands for "the coin lands tails." The set of hypotheses is $H = \{f, d\}$, where f stands for "the coin is fair," while d stands for "the coin is double-headed." In order to complete our evidence space, we need to define the likelihood functions μ_f and μ_d

- $\mu_f(h) = \mu_f(t) = 0.5$;
- $\mu_d(h) = 1$, $\mu_d(t) = 0$.

Using likelihood functions μ_f and μ_d we can easily compute the values of $\mu_f^*(\langle o^1, \ldots, o^k\rangle)$ $\mu_d^*(\langle o^1, \ldots, o^k\rangle)$ for any sequence of observations $\langle o^1, \ldots, o^k\rangle$, where $1 \leqslant k \leqslant 10$. For instance, let us consider the sequences

$$a = \langle h, h, h, t, h, h, h, h, h, h\rangle \text{ and } b = \langle h, h, h, h, h, h, h, h, h, h\rangle.$$

The corresponding μ^*-values are

- $\mu_f^*(a) = \mu_f^*(b) = 2^{-10}$;
- $\mu_d^*(a) = 0$, $\mu_d^*(b) = 1$.

From the μ^*-values we can calculate the respective values of evidence weights as follows:

- $w_{\mathrm{E}^*}(a, f) = 1$, $w_{\mathrm{E}^*}(a, d) = 0$;
- $w_{\mathrm{E}^*}(b, f) = \dfrac{2^{-10}}{1 + 2^{-10}}$, $w_{\mathrm{E}^*}(b, d) = \dfrac{1}{1 + 2^{-10}}$. ■

6.2.2 Axiomatizing Evidence

Now we introduce a temporal logic that can deal with sequences of observations made over the time. Let $H = \{h_1, \ldots, h_m\}$, $O = \{o_1, \ldots, o_n\}$ and $C = \{c_1, \ldots, c_n\}$. The elements of H will be called hypotheses and the elements of O will be called observations observations. The elements of C are constants occurring in the representation of the likelihood function. Let $For(H)$ be the set of propositional formulas over H.

We define the set *Term* of all probabilistic terms recursively as follows:

- $Term(0) = \{P(\alpha) \mid \alpha \in For(H)\} \cup \{w(o, h) \mid o \in O, h \in H\} \cup C \cup \{0, 1\}$.
- $Term(n + 1) = Term(n) \cup \{(\mathtt{f} + \mathtt{g}), (\mathtt{f} \cdot \mathtt{g}), (-\mathtt{f}) \mid \mathtt{f}, \mathtt{g} \in Term(n)\}$.
- $Term = \bigcup_{n=0}^{\infty} Term(n)$.

The set *For* of formulas is defined recursively as the smallest set that contains expressions of the form $\mathtt{f} \geqslant 0$, $\mathtt{f} \in \mathit{Term}$ (basic probability formulas), observations and hypotheses, and is closed under Boolean connectives and the temporal operators $\bigcirc$ and U. An example of the formula is

$$(o \wedge w(o,h) \geqslant r \wedge G(P(h) > 0)) \rightarrow F(o \rightarrow P(h) \geqslant s)$$

which can be read as "if o is observed, the weight of evidence of o for h is at least r and if the probability of h is always positive, then, sometimes in the future, probability of h will be at least s, if o is observed."

We define a model $\overline{\mathbf{M}}$ as an infinite sequence $\langle \mathbf{M}_0, \mathbf{M}_1, \mathbf{M}_2, \ldots \rangle$ such that

$$\mathbf{M}_k = \langle \mathrm{E}^*, \mu, h, d_1, \ldots, d_n, o^1, o^2, \ldots, o^k \rangle$$

for all $k \geqslant 1$ and $\mathbf{M}_0 = \langle \mathrm{E}^*, \mu, h, d_1, \ldots, d_n \rangle$. Notice that E^*, μ, $d_1, \ldots, d_n$ are the same in all $\mathbf{M}_k$'s and that with the increment of k we just add one new observation.

For a model $\overline{\mathbf{M}} = \langle \mathbf{M}_0, \mathbf{M}_1, \mathbf{M}_2, \ldots \rangle$ we define the satisfiability relation $\models$ recursively

- $\mathbf{M}_k \models h'$ if $h' = h$.
- $\mathbf{M}_k \models o'$ if $o' = o^k$.
- $\mathbf{M}_k \models \mathtt{f} \geqslant 0$ if $\mathtt{f}^{\mathbf{M}_k} \geqslant 0$, where $\mathtt{f}^{\mathbf{M}_k}$ is recursively defined in the following way:
  - $0^{\mathbf{M}_k} = 0$, $1^{\mathbf{M}_k} = 1$, $c_i^{\mathbf{M}_k} = d_i$.
  - $P(\phi)^{\mathbf{M}_k} = \mu \oplus w_{\mathrm{E}^*}(\langle o^1, \ldots, o^k \rangle, \cdot)(\phi)$, $\phi \in For(H)$.
  - $w(\langle o^{i_1}, \ldots, o^{i_k} \rangle, h')^{\mathbf{M}_k} = w_{\mathrm{E}^*}(\langle o^{i_1}, \ldots, o^{i_k} \rangle, h')$.
  - $(\mathtt{f} + \mathtt{g})^{\mathbf{M}_k} = \mathtt{f}^{\mathbf{M}_k} + \mathtt{g}^{\mathbf{M}_k}$.
  - $(\mathtt{f} \cdot \mathtt{g})^{\mathbf{M}_k} = \mathtt{f}^{\mathbf{M}_k} \cdot \mathtt{g}^{\mathbf{M}_k}$.
  - $(-\mathtt{f})^{\mathbf{M}_k} = -(\mathtt{f}^{\mathbf{M}_k})$.
- the conditions for Boolean operators and temporal operators are as usual (see Sect. 4.7).

Axiomatization of the logic extends the axiomatization of polynomial weight formulas from [27] (see also Sect. 5.8) with the temporal axioms 8–11 and inference rules 2 and 3 from $Ax_{LPP_1^{LTL}}$ (Sect. 4.7) and the following axioms:

Te1 $\phi \leftrightarrow \bigcirc\phi$, $\phi \in For(H)$.
Te2 $\mathtt{f} \geqslant 0 \leftrightarrow \bigcirc(\mathtt{f} \geqslant 0)$, if f does not contain an occurrence of P.
Ev1 $w(o,h) \geqslant 0$, $o \in O$, $h \in H$.
Ev2 $w(o,h_1) + \cdots + w(o,h_m) = 1$, $o \in O$.
Ev3 $\bigcirc(P(h) \geqslant r) \rightarrow P(h)w(o,h) \geqslant r(P(h_1)w(o,h_1) + \cdots + P(h_m)w(o,h_m))$, $o \in O$, $h \in H$, $r \in [0,1] \cap \mathbb{Q}$.
Ev4 $c_1 > 0 \wedge \ldots \wedge c_n > 0 \wedge c_1 w(o_1,h_1) + \cdots + c_n w(o_n,h_1) = 1 \wedge \ldots \wedge c_1 w(o_1,h_m) + \cdots + c_n w(o_n,h_m) = 1$.
Ev5 $w(o^1,h) \cdots w(o^k,h) =$
$w(\langle o^1, \ldots, o^k \rangle, h)(w(o^1,h_1) \cdots w(o^k,h_1) + \cdots + w(o^1,h_m) \cdots w(o^k,h_m))$.

Axioms Ev2 and Ev4 are counterparts of the items 1 and 2 of Theorem 6.4 respectively. In particular, Axiom Ev4 eliminates quantifiers in Axiom E4 of [15], which is an important part of the solution of the axiomatization problem posed in the same paper. Axiom Ev3 is a reformulation of Dempster's rule of combination in temporal settings: it gives the connection between probabilities in the current state (time instance) and its immediate future. Axiom Ev5 is formal representation of

$$w_{\mathrm{E}^*}(\langle o^1, \ldots, o^k \rangle, h_i) = \frac{w_{\mathrm{E}^*}(o^1, h_i) \cdots w_{\mathrm{E}^*}(o^k, h_i)}{w_{\mathrm{E}^*}(o^1, h_1) \cdots w_{\mathrm{E}^*}(o^k, h_1) + \cdots + w_{\mathrm{E}^*}(o^1, h_m) \cdots w_{\mathrm{E}^*}(o^k, h_m)}.$$

All the proof theoretical notions are the same as for the LPP_1^{LTL}-logic. Combining the techniques from the Sects. 4.7 and 5.8, we can prove that this axiomatization is strongly complete for the presented class of models.

6.3 Formalization of Human Thinking Processes in $L_{\mathbb{Q}_p}$

A mathematical model of the process of human thinking based on the dynamical systems over a field of p-adic numbers was presented in [1, 18, 19]. Here we briefly describe only the key parts of this model and for details we refer to the cited literature. Then we show how this process of thinking can be formalized in a version of the logic $L_{\mathbb{Q}_p}$ where the appropriate measure μ can be seen as a coding function that associates a p-adic number to each peace of information.

The process of human thinking can be seen as a nonlinear function $x_{n+1} = f(x_n)$, $x_n \in X_I$ where X_I is a space of information. In the biological model the space of information is defined as follows. Elementary units for processing information are neurons. All neurons have the same number of possible states (levels), say m. Each information I is represented by chains of neurons, $\mathcal{N} = (n_0, n_1 \ldots n_N)$, so every such chain of neurons must be found in some of the m^N different I-states $x = (\alpha_0, \alpha_1 \ldots \alpha_N), \alpha_i \in \{0, 1, \ldots, m-1\}$. Then, the space of information X_I is exactly the set of all possible I-states. Neurons have a hierarchy structure: the most important is the neuron n_0, the neuron n_1 is less important then n_0, but it is more important then $n_2, n_3 \ldots n_N$, and so on. Thus, because of this hierarchy, if n_j ignites, that can cause ignition of the neurons $n_{j+1}, n_{j+2} \ldots n_N$.

More formally, if n_j is in the highest state, $\alpha_j = m-1$, then ignition of this neuron causes that $\alpha_j = 0$, and n_{j+1} exceeds to higher level, $\alpha_{j+1} := \alpha_{j+1} + 1$. Note that the same holds when we increase by 1 any digit in p-adic representation of some number (when m is actually some prime number). This model assumes a finite number of neurons but in order to apply tools developed for p-adic analysis to examine the proposed model, it is useful to consider an ideal - infinite states: $x = (\alpha_0, \alpha_1, \ldots, \alpha_N \ldots), \alpha_i = 0, 1, \ldots, m-1$.

Note that in this model two pieces of information are close if they have a sufficiently long common prefix. The p-adic distance $d(x, y) = |x - y|_p$ allows us to measure nearness of information in this way. Therefore the set of p-adic integers $\mathbb{Z}_{\mathbb{Q}_p}$ is suitable to be chosen as the space of information, i.e., $X_I = \mathbb{Z}_{\mathbb{Q}_p}$. We consider dynamical system $f : \mathbb{Z}_{\mathbb{Q}_p} \to \mathbb{Z}_{\mathbb{Q}_p}$ where f is analytic function. If $f(x_0) = x_0$ then x_0 is a *fixed point* of the function f, i.e., fixed point of the dynamical system.

A fixed point x_0 of the function f is called attractor if there is a neighborhood $V(x_0)$ such that for every $y \in V(x_0)$, $\lim_{n\to\infty} f^n(y) = x_0$. For an attractor x_0, we define a *basin of attraction* $A(x_0) = \{y \in \mathbb{Z}_{\mathbb{Q}_p} : f^n(y) \to x_0, n \to \infty\}$.

The clopen ball $B[a, \frac{1}{p^n}]$, $n \in \mathbb{Z}$ is called Siegel disk, if each sphere $S(a, \frac{1}{p^l})$, $l \geq n$ is invariant sphere of f.

There are several results concerning dynamical functions of the forms $f_c(x) = x^2 + c$, $c \in \mathbb{Z}_{\mathbb{Q}_p}$ and $f_n(x) = x^n$, $n \in \mathbb{N}$, see for instance [1, 18, 19]. Here we give one of these results needed for the later discussion.

Theorem 6.5 *The dynamical system $f_n(x) = x^n$ has $m = n - 1$ fixed points $a_j, j = 1 \ldots m$ on the sphere $S[0, 1]$. All fixed points $a_j \neq 1$ belong to sphere $S[1, 1]$ and*

1. *If $GCD(n, p) = 1$ then all these points are centers of Siegel disks.*
2. *If $GCD(n, p) \neq 1$ then all these points are attractors and $A(a_j) = B[a_j, \frac{1}{p}]$* ■

The proposed model uses the idea that thinking is carried out on two levels: conscious and subconscious. The conscious formulates problems and then sends them to subconscious, i.e., conscious sends an initial information x_0 and a function f- the regime of work to subconscious. Subconscious works with the obtained data, i.e., starting with x_0, and using f generates the set of attractors, which conscious sees as possible solutions. Below we present an example similar to examples from [18] which illustrates thinking in $\mathbb{Q}_2$ with the function $f(x) = x^2$.

Example 6.8 A tourists must decide which of the given n countries she will visit based on the given conditions: each of these country has one of the labels-0 if tourist cannot go to that country and 1 if she can. The set of countries is ordered by their importance $B_0, B_1 \ldots B_{n-1}$, where B_0 is the country where she would most wish to go, B_1 is less desirable than B_0 etc. Her state of mind about countries can be described by a number $x = \alpha_0 + \alpha_1 \cdot 2 + \cdots \alpha_{n-1} \cdot 2^{n-1}$, where α_j is the label of B_j.

The tourist makes a decision which country she will visit, using some initial state of mind x_0. First, we consider the situation when her subconscious is described by the dynamical system $f(x) = x^2$

(a) If $x_0 \in B[0, \frac{1}{2}]$, i.e., $\alpha_0 = 0$, then, according to the previous theorem, the sequence $x_0, x_1, x_2 \ldots$ converges to the attractor $a_0 = 0 + 0 \cdot 2 + \cdots + 0 \cdot 2^{m-1}$, which can be interpreted as that if she can not go her favorite country, she would decide not to go anywhere.
(b) If $x_0 \in S[0, 1] = B[1, \frac{1}{2}]$, i.e., $\alpha_0 = 1$, then the sequence $x_0, x_1, x_2 \ldots$ converges to the attractor $a_1 = 1 + 0 \cdot 2 + \cdots + 0 \cdot 2^{n-1}$. Therefore, if she can go to B_0, she reject all other possibilities. ■

In this model, each piece of information corresponds to a p-adic number. From the logical point of view each piece of information is a proposition, and we want to represent it by a classical propositional formula. Then we have to define a coding function (or functions) which associates a p-adic number to that formula(information). It turns out that the logic $L_{\mathbb{Q}_p}$ is suitable for formalization of this coding function because axiomatic system $AX_{L_{\mathbb{Q}_p}}$ "obligates" this function to behave according to the topology of space $\mathbb{Z}_{\mathbb{Q}_p}$.

There are some changes that we need to make in order to adapt the logic for this new task. Thus, in this section we propose a new logic, denoted $L^{thinking}_{\mathbb{Z}_{\mathbb{Q}_p}}$, for reasoning about arbitrary coding p-adic function μ. The only two differences between this logic and $L_{\mathbb{Q}_p}$ are

- the additivity condition (Axiom 3) is withdrawn, since the corresponding feature is not required for coding functions, and
- the range of coding functions is $\mathbb{Z}_{\mathbb{Q}_p} = B[0, 1]$, and not $B[0, p^M] \subset \mathbb{Q}_p$ as above.

Special properties (if any) of coding functions could be described using particular theories in the logic. We illustrate our approach by describing a theory which formalizes Example 6.8.

The language of the logic $L^{thinking}_{\mathbb{Z}_{\mathbb{Q}_p}}$ is the same as the language of the logic L_{Q_p}, except that for the operators $K_{r,\rho}$ it must be $r \in \mathbb{Q}_0 = \{r \in \mathbb{Q} \ : \ |r|_p \leq 1\}$, $\rho \in R_0 = \{p^{-n} \ : \ n \in \mathbb{N}\} \cup \{0\}$. Using these operators we define the set of *cognitive formulas* For_{cgn} in the same way as we construct $L_{\mathbb{Q}_p}$-formulas. The fundamental difference between theses logics lies in the meaning of the formulas $K_{r,\rho}\alpha$. In the new formalism $K_{r,\rho}\alpha$ means that the code of α belongs to the p-adic ball with the center r and the radius ρ.

Finally, we use logic $L^{thinking}_{\mathbb{Z}_{\mathbb{Q}_p}}$ to formalize Example 6.8. Let the proposition α_i represent that the tourist will go to the country B_i and let $\alpha = \alpha_0 \vee \alpha_1 \vee \ldots \vee \alpha_{m-1}$. In this example the coding function satisfies $\mu([\alpha]) = \mu([\alpha_0]) + \mu([\alpha_1]) \cdot 2 + \cdots \mu([\alpha_{m-1}]) \cdot 2^{m-1}$. As in the original example, we suppose that the code of α_0 is given, and that the code of α should be obtained as a deductive consequence of the corresponding $L^{thinking}_{\mathbb{Z}_{\mathbb{Q}_p}}$-theory. First, we consider the case $f(x) = x^2$. Let

- $T_1 = \{K_{x_0, \frac{1}{2}}\alpha \Leftrightarrow K_{x_0, 0}\alpha_0\}$
- $T_2 = \{K_{x_0^{2^n}, \frac{1}{2^{n+1}}}\alpha \Rightarrow K_{x_0^{2^{n+1}}, \frac{1}{2^{n+2}}}\alpha | n \in \mathbb{N}\}$
- $T = T_1 \cup T_2$

Since $\mu([\alpha]) = \mu([\alpha_0]) + \mu([\alpha_1]) \cdot 2 + \cdots \mu([\alpha_{m-1}]) \cdot 2^{m-1}$, the theory T_1 provides that the code of α belongs to the p-adic ball with the center x_0 and the radius $\frac{1}{2}$, where $x_0 = \mu([\alpha_0])$. The theory T_2 asserts that thinking process consists of iterative applications of the function x^2 on the code of α. Precisely, T_2 says: if after the n-th iteration we know that $\mu([\alpha])$ belongs to the ball with the center $(\mu([\alpha_0]))^{2^n}$ and the radius $\frac{1}{2^{n+1}}$, then we conclude that after the next iteration, $\mu([\alpha])$ belongs to the ball with the center $(\mu([\alpha_0]))^{2^{n+1}}$ and the radius $\frac{1}{2^{n+2}}$. Furthermore, $K_{1,0}\alpha_0$ and $K_{0,0}\alpha_0$ mean that the tourist can (cannot) go to her favorite country.

Now, if she can go to her favorite country, we have

- $T, K_{1,0}\alpha_0 \vdash K_{1,\frac{1}{2}}\alpha$; i.e. if $\mu([\alpha_0]) = 1$, then according to T_1, $\mu([\alpha])$ belongs to the ball with the center 1 and the radius $\frac{1}{2}$.

Then using T_2 we obtain

- $T, K_{1,0}\alpha_0 \vdash K_{1,\frac{1}{2^n}}\alpha$ for every $n \in \mathbb{N}$; i.e., for every $n \in \mathbb{N}$, $\mu([\alpha])$ belongs to the ball with the center 1 and the radius $\frac{1}{2^n}$,

and therefore, according to Rule 5

- $T, K_{1,0}\alpha_0 \vdash K_{1,0}\alpha$.

Thus, $\mu([\alpha]) = 1 + 0 \cdot 2 + \cdots + 0 \cdot 2^{m-1}$, that is, she will go only to the country B_0.

In the same way we obtain

- $T, K_{0,0}\alpha_0 \vdash K_{0,0}\alpha$,

i.e., if she cannot go to B_0, she will go nowhere.

6.4 Other Applications

It is shown in [2] that the standard probabilistic semantics for System P is given by so-called big-stepped probabilities, i.e., the neat probability measures that satisfy the conditions

- $\sum_{at' \in \text{Atoms}, \mu(at') < \mu(at)} \mu(at') < \mu(at)$, for all $at \in$ Atoms, and
- $\mu(at) = \mu(at')$ iff $\vdash at \leftrightarrow at'$, for all $at, at' \in$ Atoms,

where Atoms $=$ Atoms$(\{p_1, p_2, \ldots, p_n\})$ is the set of atoms. Then we can interpret a default rule of the form $\alpha \mathrel{|\!\sim} \beta$ as $\mu(\beta|\alpha) > \mu(\neg\beta|\alpha)$. In [5] we showed that we can express this condition using the language of $LPP_{2,\preceq}$, and that we can extend the axiomatization of $LPP_{2,\preceq}$ in order to characterize big-stepped probabilities.

In [9] we provided probabilistic representations for three specific classes of preferential relations, obtained by adding to P one of the following rules: Determinacy preservation [24]

$$DP : \frac{\alpha \mathrel{|\!\sim} \beta,\ \alpha \wedge \gamma \mathrel{|\!\not\sim} \neg\beta}{\alpha \wedge \gamma \mathrel{|\!\sim} \beta},$$

Fragmented disjunction and conditional excluding middle [4]

$$FD : \frac{\alpha \mathrel{|\!\sim} \beta \vee \gamma,\ \alpha \mathrel{|\!\not\sim} \beta,\ \alpha \mathrel{|\!\not\sim} \gamma}{\neg\beta \mathrel{|\!\sim} \gamma}; \qquad CEM : \frac{\alpha \mathrel{|\!\not\sim} \beta}{\alpha \mathrel{|\!\sim} \neg\beta}.$$

Each of those rules imply Rational monotonicity [4]. For the three classes of relations we have identified the corresponding subclasses of nonstandard finitely additive probability measures that induces them, and proved the corresponding characterization theorems.

A few years ago a new formalism was introduced to represent spatiotemporal information in the presence of uncertainty [14, 26]. In [10], several probabilistic logics are developed for modeling such information. An atomic formula of the logics is of the form $loc(id, r, t)[\ell, u]$ and has the meaning that a particular object id is in a particular region r at a particular time t with a probability that is in the probability interval $[\ell, u]$. For all the logics, sound and strongly complete axiomatizations are presented, and decidability issues are discussed.

The paper [6] presents an application of probability logic in the field of abstract argumentation, nowadays a vivid field within artificial intelligence. Several enrichments of the standard Dung's argumentation frameworks [11] have been proposed in order to model scenarios where probabilities have to be expressed [16, 22]. In [6], the semantics from [22] are characterized in terms of probability logic (similar to LPP_2). This not only provides a uniform logical formalization but also might pave the way for future implementations.

References

1. Albeverio, S., Khrennikov, A., Kloeden, P.E.: Memory retrieval as p-adic dynamical system. BioSystems **49**, 105–115 (1999)
2. Benferhat, S., Dubois, D., Prade, H.: Possibilistic and standard probabilistic semantics of conditional knowledge bases. J. Log. Comput. **9**(6), 873–895 (1999)
3. Benferhat, S., Saffiotti, A., Smets, P.: Belief functions and default reasoning. Artif. Intell. **122**, 1–69 (2000)
4. Bezzazi, H., Makinson, D., Pérez, R.P.: Beyond rational monotony: some strong non-Horn rules for nonmonotonic inference relations. J. Log. Comput. **7**(5), 605–631 (1997)
5. Doder, D.: A logic with big-stepped probabilities that can model nonmonotonic reasoning of system P. Publications de l'Institut Mathématique, ns. **90**(104), 13–22 (2011)
6. Doder, D., Woltran, S.: Probabilistic Argumentation Frameworks – A Logical Approach. In: Straccia, U., Cali, A. (eds.) Scalable Uncertainty Management SUM 2014. Lecture Notes in Artificial Intelligence, vol. 8720, pp. 134–147. Springer, New York (2014)
7. Doder, D., Ognjanović, Z., Marković, Z., Perović, A., Rašković, M.: A Probabilistic Temporal Logic that Can Model Reasoning About Evidence. FoIKS 2010. LNCS, vol. 5956, pp. 9–24 (2010)
8. Doder, D., Rašković, M., Marković, Z., Ognjanović, Z.: Measures of inconsistency and defaults. Int. J. Approx. Reason. **51**, 832–845 (2010)
9. Doder, D., Perović, A., Ognjanović, Z.: Probabilistic approach to nonmonotonic consequence relations. In: Liu, W. (ed.) Symbolic and Quantitative Approaches to Reasoning with Uncertainty ECSQARU 2011. Lecture Notes in Artificial Intelligence, vol. 6717, pp. 459–471. Springer, New York (2011)
10. Doder, D., Grant, J., Ognjanović, Z.: Probabilistic logics for objects located in space and time. J. Log. Comput. **23**(3), 487–515 (2013)
11. Dung, P.M.: On the acceptability of arguments and its fundamental role in nonmonotonic reasoning, logic programming and n-person games. Artif. Intell. **77**(2), 321–358 (1995)
12. Friedman, N., Halpern, J.Y.: Plausibility measures and default reasoning. J. ACM **48**, 648–685 (2001)
13. Gabbay, D.: Theoretical foundations for non-monotonic reasoning in expert systems. Logics and Models of Concurrent Systems, pp. 439–457. Springer, New York (1985)
14. Grant, J., Parisi, F., Parker, A., Subrahmanian, V.S.: An AGM-style belief revision mechanism for probabilistic spatio-temporal logics. Artif. Intell. **174**, 72–104 (2010)
15. Halpern, J.Y., Pucella, R.: A logic for reasoning about evidence. J. Artif. Intell. Res. **26**, 1–34 (2006)

16. Hunter, A.: A probabilistic approach to modelling uncertain logical arguments. J. Approx. Reason. **54**(1), 47–81 (2013)
17. Ilić-Stepić, A., Ognjanović, Z.: Logics for reasoning about processes of thinking with information coded by p-adic numbers. Studia Logica **103**, 145–174 (2015)
18. Khrennikov, A.: Human subconscious as a p-adic dynamical system. J. Theor. Biol. **193**, 179–196 (1998)
19. Khrennikov, A.: P-adic discrete dynamical systems and collective behaviour of information states in cognitive models. Discrete Dyn. Nat. Soc. **5**, 59–69 (2000)
20. Kraus, S., Lehmann, D., Magidor, M.: Nonmonotonic reasoning, preferential models and cumulative logics. Artif. Intell. **44**(1–2), 167–207 (1990)
21. Lehmann, D., Magidor, M.: What does a conditional knowledge base entail? Artif. Intell. **55**, 1–60 (1992)
22. Li, H., Oren, N., Norman, T.J.: Probabilistic argumentation frameworks. In: Modgil, S., Oren, N., Toni, F. (eds.) Theory and Applications of Formal Argumentation TAFA 2012. Lecture Notes in Computer Science, vol. 7132, pp. 1–16. Springer, New York (2012)
23. Lukasiewicz, T.: Weak nonmonotonic probabilistic logic. Artif. Intell. **168**(1–2), 119–161 (2005)
24. Makinson, D.: General patterns in nonmonotonic reasoning. Handbook of Logic in Artificial Intelligence and Logic Programming. Non Monotonic Reasoning and Uncertain Reasoning, vol. 3, pp. 35–110. Clarendon Press, Oxford (1994)
25. Ognjanović, Z., Marković, Z., Rašković, M., Doder, D., Perović, A.: A probabilistic temporal logic that can model reasoning about evidence. Ann. Math. Artif. Intell. **65**(2–3), 217–243 (2012)
26. Parker, A., Subrahmanian, V.S., Grant, J.: A logical formulation of probabilistic spatial databases. IEEE Trans. Knowl. Data Eng. **19**, 1541–1556 (2007)
27. Perović, A., Ognjanović, Z., Rašković, M., Marković, Z.: A probabilistic logic with polynomial weight formulas. In: Sven Hartmann, L., Kern-Isberner, G. (eds.) Foundations of Information and Knowledge Systems FoIKS 2008. Lecture Notes in Computer Science, vol. 4932, pp. 239–252. Springer, New York (2008)
28. Rašković, M., Marković, Z., Ognjanović, Z.: A logic with approximate conditional probabilities that can model default reasoning. Int. J. Approx. Reason. **49**(1), 52–66 (2008)

Chapter 7
Related Work

Abstract We relate our results to a number of papers of other authors published from the mid-1980s onwards. Although we have tried to provide a comprehensive bibliography, we are aware that it is inevitably incomplete and that some important references might be missing. A lot of efforts have been made to reduce the number of those absent works. We try to arrange our comments in temporal order of appearance of transcripts, with some exceptions if some papers are closely related by their subjects but not by publishing times. Whenever possible we comment journal versions of the papers, although there were preliminary versions in proceedings.

7.1 Papers on Completeness of Probability Logics

As we mentioned in Chap. 2, a lot of recent interest in probability logic was initiated by the paper [49], published in 1986, in which Nilsson presented a procedure for probabilistic entailment which, given probabilities of premises, could calculate bounds on probabilities of the derived sentences. The Nilsson's approach was semantical and stimulated other authors to provide axiomatizations and decision procedures for the logic. In the same year Gaifman studied structures appropriate for higher order probabilities and their connections with modal logics [20]. Geifman defined probability models (similar to our measurable models $LPP_{1,\mathrm{Meas}}$ for the propositional logic with iterations of probability operators) as tuples (called HOP, higher order probability space) of the form

- $\langle W, F, P, PR \rangle$,

where W is a nonempty set, F is a field of subsets of W (events), P is the subjective probability of an agent, while PR associates to every $A \in F$, and every closed interval $\Delta \subset [0, 1]$ an event $PR(A, \Delta)$ that the probability of A is in Δ. Geifman gave a semantical analysis of those models. He presented a list of postulates for HOPs, e.g., $PR(A, [0, 1]) = W$, and (a version of Miller's principle $P(A|P(A) \in [s, t]) \in [s, t]$)

Z. Ognjanović et al., *Probability Logics*,
DOI 10.1007/978-3-319-47012-2_7

- if $C = \bigcap_{i=1}^{n} PR(B_i, \Delta_i)$, and $P(C \cap PR(A, [s, t])) \neq 0$, then

$$s \leq P(A|C \cap PR(A, [s, t])) \leq t$$

and showed that every *HOP* uniquely determines a mapping p which assigns to every $w \in W$ a probability $p(w) : F \to [0, 1]$, $p(w)(A) = \sup\{s : w \in PR(A, [s, 1])\}$. Gaifman then considered valuations that map primitive propositions to F, so that he could work with a formal language extending propositional logic with formulas of the form $PR(\psi, \Delta)$. He proved that nesting of *PR*'s does not add expressivity and that by translating modal $\Box\psi$ to $PR(\psi, [1, 1])$ he could embed the modal system *S*5 into his logic. Finally, he introduced General HOP's by adding an additional argument to *PR* ranging over a set of ordered (or partially ordered) time-points. so that he was able to describe changes of probabilities of events.

Fattorosi-Barnaba and Amati considered a class of calculi P_F that gave probability interpretations to modal operators [15]. They introduced a list of operators of the form M_r and L_r, that are, in our notation, $P_{>r}$, and $P_{\geq 1-r}$, respectively. Note that M_r and L_r are mutually definable. M_0 and L_0 (i.e., $P_{>0}$, and $P_{\geq 1}$) were identified with the possibility operator M, and the necessity operator L, which probably motivated the notation. Their probability Kripke-like models are our $LPP_{1,\mathrm{Meas}}^{\mathrm{Fr}(n)}$ measurable models with probability functions with a finite range $F \subset [0, 1]$, and with iterations of probability operators. Since the range of probabilities is finite, compactness holds, and Fattorosi-Barnaba and Amati provided a finitary axiomatization which is strongly complete. To prove completeness, although the range of probability functions is finite, they first showed that the infinitary inference rule 3 from Chap. 3 is deducible in their system, and then defined measures in the canonical model using sup. In [70, 71] van der Hoeck used the notation similar to ours: his probability operators are of the form $P_r^{>}$, and the other operators are defined using $P_r^{>}$'s. He simplified completeness proof from [15] using only finitary means and proved decidability of the logic by showing that P_F has the finite model property[1] and, since the range of probability functions is finite, only finite number of models should be inspected. Independently, the same logic as P_F was axiomatized in [54]. Van der Hoeck showed that, if $P_1^{\geq}\varphi$ is denoted by $\Box\varphi$, the modal system *KD* is embedded in P_F. He also addressed a serious logical issue caused by finitary axiomatizations of non-compact logics with real-valued probability functions,[2] and mentioned [4] as one of possible ways to overcome that problem.

Alechina gave a strongly complete axiomatization for the class of measurable first-order probability models ($LPP_{1,\mathrm{Meas}}$, in our notation) with $[0, 1]_{\mathbb{Q}}$-valued probabilities [4]. The main novelty is the infinitary rule:

- From $\Sigma \vdash \neg P_{=r}\varphi$, for every $r \in [0, 1]_{\mathbb{Q}}$, infer $\Sigma \vdash \bot$

[1]Every consistent formula is satisfiable in a finite model.

[2]There are consistent but unsatisfiable sets of formulas.

which guarantees that the probabilities of sentences belong to $[0, 1]_{\mathbb{Q}}$. We used similar rules in [55] and [65, 66] to ensure that probabilities are in decidable subsets of $[0, 1]$ and in $[0, 1]_{\mathbb{Q}(\varepsilon)}$, the unit interval of Hardy field, respectively.

Computational aspects of probability logics were discussed in [21]. The paper [40] showed that it is possible to apply a very efficient numerical method of column generation to solve the PSAT satisfiability problem for $LPP_{2,\mathrm{Meas}}$.

In [7] Bacchus introduced a first-order logic, called *Lp*, suitable for representing and reasoning with statistical information, for example statistical statements about the state of the world, e.g., "More than 75 % of all birds fly." This approach is similar to Keisler's logics [41] with probability quantifiers (see Chap. 2). Keisler's $(P\overrightarrow{x} \geq r)\psi(x)$ is Bacchus' $[\alpha]_{\overrightarrow{x}} \geq r$, where $[\alpha]_{\overrightarrow{x}}$ is considered as a probability term. Bacchus added flexibility to the logic by allowing, beside the usual first-order language, conditional probability terms of the form $[\alpha|\beta]_{\overrightarrow{x}}$ and, field terms built up from field constants 0 and ± 1, and variables, using function symbols $+$, $-$ and $\times$, and terms built up from probability and field terms, e.g., a formula is

$$\forall r(r \times [\alpha|\beta]_{\overrightarrow{x}} > [\gamma]_{\overrightarrow{y}}).$$

Quantification is allowed over variables that are evaluated in domains, and over field variables, as in the previous example. The considered *Lp*-models are, similarly to Keisler's, first-order structures with probabilities on their domains. The presented finitary axiom system contains the standard axioms and rules of the first-order calculus with equality, axioms of a totally ordered field, and axiom about probability, e.g.,

- $\forall x_1 \ldots \forall x_n \alpha \to [\alpha]_{\overrightarrow{x}} = 1$,
- $\forall z_1 \forall z_2 [[\alpha]_{\overrightarrow{x}} = z_1]_{\overrightarrow{y}} = z_2 \to [\alpha]_{\langle \overrightarrow{x}, \overrightarrow{y} \rangle} \geq z_1 \times z_2$,
- $[\alpha]_{\overrightarrow{x}} = [\alpha]_{\pi(\overrightarrow{x})}$, for every permutation π of $\overrightarrow{x}$, etc.

Since the axiom system is finitary, the strong completeness theorem, $T \vdash \alpha$ iff $T \models \alpha$ is proved by a way of standard Henkin style procedure. In Bacchus' framework, the price paid to obtain that was

- domains are countable,
- probability functions are finitely additive, and
- the range of probability functions is required to be the unit interval of a totally ordered field, instead of a particular field, e.g., $[0, 1]$, or $[0, 1]_{\mathbb{Q}}$.

As a consequence, there are sentences on the real-valued probability functions that are not provable in *Lp*, e.g.,

$$(\forall r_1 \in [0, 1])([(\forall r_2 < r_1)[\alpha]_{\overrightarrow{x}} > r_2] \to [\alpha]_{\overrightarrow{x}} \geq r_1).$$

Fagin, Halpern, and Megiddo provided a comprehensive study of several formal languages for reasoning about probability [13]. They started with the class $LPP_{2,\mathrm{Meas}}$ of measurable propositional models with real-valued probability functions and without iteration of probability operators. Since they were not able to axiomatize the set of valid For_{LPP_2}-formulas (in the language containing the probability operators $P_{\geq s}$), they extended the language to allow basic probability formulas of the form

$$a_1 w(\alpha_1) + \ldots + a_n w(\alpha_n) \geq s,$$

where a_i's and s are rational numbers, α_i's classical propositional formulas, and $w(\alpha_i)$'s are primitive weight terms denoting probabilities.[3] Probability formulas are Boolean combinations of basic probabilistic formulas. Fagin, Halpern, and Megiddo gave the finitary axiomatic system AX_{MEAS} which is divided into three parts

- the standard axioms and rules for propositional logic,
- axioms for reasoning about probability
  - $w(\alpha) \geq 0$,
  - $w(true) = 1$,
  - $w(\varphi \wedge \psi) + w(\varphi \wedge \neg\psi) = w(\varphi)$, and
  - if $\models \varphi \leftrightarrow \psi$, then $w(\varphi) = w(\psi)$, and
- axioms for reasoning about linear inequalities, e.g.,
  - $x \geq x$,
  - $(a_1 x_1 + \cdots + a_n x_n \geq c) \wedge (a'_1 x_1 + \cdots + a'_n x_n \geq c') \rightarrow ((a_1 + a'_1)x_1 + \cdots + (a_n + a'_n)x_n) \geq (c + c')$,
  - $(t \geq c) \rightarrow (t > d)$, for $c > d$, etc.

To prove the weak completeness, $\vdash A$ iff $\models A$, they showed (using axioms for reasoning about linear inequalities) that for every $w(\varphi) \in \text{Subf}(A)$, $\vdash w(\varphi) \leftrightarrow \sum_{a \in \text{Atoms}(\varphi)} w(a)$, and that, without loss of generality, A can be assumed to be a finite conjunction of (possibly negated) basic probabilistic formulas. That conjunction is satisfiable iff the corresponding linear system of the form given in Theorem 3.11 is satisfiable. So, unsatisfiability of the system implies that $\vdash \neg A$, i.e., that A is inconsistent. In [55, 56, 64] an axiomatization of the logic LPP_2 without linear combinations of primitive weight terms was given, while [63] presented an adaptation of the logic LPP_2 so that the strong completeness was proved for the logic with Boolean combinations of linear inequalities of primitive weight terms. Fagin, Halpern, and Megiddo proved decidability and NP-completeness of PSAT, as it is mentioned in Sect. 3.5. They also discussed conditional probabilities by extending the language to allow polynomial weight formulas, e.g., $2w(p_1 \wedge p_2)w(p_2) + 3w(p2) \geq w(p_1)w(p_2)$, and proved decidability of PSAT. To obtain the weak completeness, they extend the language to a first-order language such that variables can appear in formulas

$$(\forall x)(\exists y)[(3 + x)w(\varphi)w(\psi) + 2w(\varphi \vee \psi) \geq z].$$

The corresponding axiom system $AX_{FO-MEAS}$ contains the standard first-order axiomatization and, additionally, axioms for real closed fields. In [61–63] strong

[3] So, $w(\alpha) \geq s$ is $P_{\geq s}\alpha$.

completeness was proved for the logics with polynomial weight formulas and [0, 1], and $[0,1]_{{}^*\mathbb{R}}$-valued probability functions. For the latter, logic compactness was also showed. Models that are not measurable were also considered in [13], i.e., the probability measure μ is partial and, for some set A, only the inner measure $\mu_*(A) = \sup\{\mu(B) : B \subset A\}$ and the outer measure $\mu^*(A) = \inf\{\mu(B) : B \supset A\}$ induced by μ can be defined. Fagin, Halpern, and Megiddo provided a weakly complete axiom system for this logic, and prove decidability of PSAT.

Boričić and Rašković considered nonmeasurable models in [10]. They extended Heyting propositional logic by probabilistic operators. Since predicates "at lest r" and "at most r" are not mutually expressible in that context, both types of operators $P_{\geq r}$ and $P^{\leq r}$ are included in the formal language. Kripke type models of the form $\langle W, R, (p^x)_{x\in W}, (p_x)_{x\in W}\rangle$, appropriate for intuitionistic logic, were used as the corresponding semantics. Each world is equipped with two partial functions p^x and p_x with a finite range. The functions represent upper and lower probabilities in x and satisfy

- monotonicity: if xRy, and $X \subset W$ is measurable, then $p^x(X) \leq p^y(X)$, and $p_x(X) \geq p_y(X)$, i.e., the upper probabilities cannot increase, and lower probability cannot decrease, which corresponds to the paradigm of increasing knowledge over time in intuitionistic logic,
- subadditivity for upper probabilities: if $X, Y \subset W$ are measurable, then $p^x(X \cup Y) \leq \min\{p^x(X) + p^{x(Y)}, 1\}$, and
- superadditivity for lower probabilities:: if $X, Y \subset W$ are measurable, then $p_x(X \cup Y) \geq \min\{p_x(X) + p_{x(Y)}, 1\}$.

They presented an axiom system containing

- all instances of formulas provable in the Heyting propositional calculus,
- monotonicity axioms: $P^{\leq r}\alpha \to P^{\leq s}\alpha$, for $r \leq s$; $P_{\geq r}\alpha \to P_{\geq s}\alpha$, for $r \geq s$,
- $P^{\leq 1}\alpha$, $P_{\geq 0}\alpha$,
- finite semiadditivity: $(P^{\leq r}\alpha \wedge P^{\leq s}\beta) \to P^{\leq t}(\alpha \vee \beta)$, $t = \min\{1, r+s\}$, and $(P_{\geq r}\alpha \wedge P_{\geq s}\beta \wedge P_{\leq 0}(\alpha \wedge \beta)) \to P_{\geq t}(\alpha \vee \beta)$, $t = \min\{1, r+s\}$,
- Modus Ponens, and
- Necessitation: from α infer $P_{\geq 1}\alpha$

and proved its simple completeness.

Halpern considered two first-order probability logics in [31]. In the first logic probabilities are defined on the domain, similarly as in [7, 41], while in the second logic probabilities are defined on possible worlds, similarly as in $LFOP_{1,\mathrm{Meas}}$, but with the restriction that the measures $\mu(w)$ in all worlds of a model are equal. Thus, formulas expressing probabilities either hold in every world from a model, or they are not satisfiable in that model, i.e., $\models \varphi \to w(\varphi) = 1$ where no function and relation symbols appear in φ except in the argument ψ of a probability term $w(\psi)$, and the models of that second logic posses also properties of $LFOP_{2,\mathrm{Meas}}$-models. Halpern provided axiomatization for both logics, but completeness can be proved only if the domains are bounded in size by some finite n. In Bacchus' opinion it is difficult to justify that assumption even for artificial intelligence applications [7]. Bacchus explained that first, while domain may be finite, it is questionable that there is a fixed upper bound on its size, and second, that there are many domains, interesting in AI applications, that are not finite. The paper [56] introduced a strongly complete axiom system for $LFOP_1$, the first-order probability logic with probabilities defined on possible worlds with unbounded domains.

Abadi and Halpern in [1] provided a deep analysis of decidability issues for the first-order probability logics introduced in [31]. They proved (assuming that probability functions are discrete, i.e., defined for every singleton) that both logics are highly undecidable

- when the probability is on the domain, if the language contains one binary predicate, the validity problem is Π_1^2 complete,
- with equality in the language and with no other symbol, the validity problem for the logic with the probability on the domain, is at least Π_∞^1 hard,
- when the probability is on possible worlds, if the language contains at least one unary predicate, the validity problem is Π_1^2 complete, while
- with equality and only one constant symbol in the language, the validity problem is Π_∞^1 hard,
- in the case of $[0, 1]_{\mathbb{Q}}$-valued probabilities, complexity decreases from Π_1^2 to Π_1^1.

In the case of arbitrary (i.e., not only discrete) probabilities, the corresponding upper bounds of complexity remained undetermined. As it is mentioned above, if the sizes of domains in models are bounded by some fixed n, the validity problem becomes decidable. In [48, 56] decidability of some fragments of first-order probability logics without iterations of probability operators were presented.

Frisch and Haddawy presented in [19] an incomplete iteration procedure which computes increasingly narrow probability intervals for propositional formulas in $LPP_{2,\mathrm{Meas}}$-models (in our notation). The procedure can be stopped at any time yielding partial information about the probability of sentences, and allowing one to make a tradeoff between precision and computational time. The considered formulas are of the form $P(\phi|\xi) \in I$, where ϕ and ξ are classical propositional formulas and I is a closed subinterval of $[0, 1]$. To avoid indefiniteness of the conditional probability $P(\phi|\xi)$ when the probability of the conditioning formula ξ is 0, they stated that in that case $P(\phi|\xi)$ holds in any model. Their proof system consists of a set of sound inference rules, e.g.,

$$\frac{\begin{array}{c} P(\alpha|\delta) \in [x, y] \\ P(\alpha \vee \beta|\delta) \in [u, v] \\ P(\alpha \wedge \beta|\delta) \in [w, z] \end{array}}{P(\beta|\delta) \in [\max(w, u - y + w), \min(v, v - x + z)]}$$
$$\text{provided that } w \leq y, x \leq v, w \leq v$$

Frisch and Haddawy also mentioned the lack of finitary strongly complete axiomatization caused by non-compactness of the logic.

Fagin and Halpern gave in [14] a finitary axiomatization of higher order probabilities (their models extend our $LPP_{1,\mathrm{Meas}}$ with the modal notion of knowledge) in the formal language that mixes the modal operator of knowledge and linear combinations of primitive weight terms. In this more complex system, they use a proof procedure similar to the one from [13], and prove decidability of PSAT. They also considered assumptions about relationships between knowledge and probability. A strongly complete axiomatization of the probabilistic part of this logic was given in [51, 56, 64].

Hajek, Godo, and Esteva [37] proposed a fuzzy logic of probability for which weak completeness was proved. The fuzzy formulas contains propositional variables of the form f_φ, which we read "probable φ." Godo and Marchioni presented in [22] a modal fuzzy logic approach to model probabilistic reasoning in the sense of de Finetti, in which Łukasiewicz implication can be used to express comparative statements.

Providing a strongly complete axiom system for probability logics was an open question that attracted attention of researchers. For example, in [6] Aumann noted that in the context of probability (he considered multi-agent propositional real-valued probabilities, and the language with the probability operators P_i^r, or in our notation $P^i_{\geq r}$m where i denotes the agent i) still there was no satisfactory syntactic definition of consistency which enables a proof of strong completeness. As we already noted, it was solved for propositional and first-order probability logics using the infinitary axiomatizations from [51, 55, 56]. Heifetz and Mongin in [39] noted that there is no hope to have a finitary strongly complete system for real-valued probabilities and gave a finitary weakly complete axiom system for the multi-agent $LPP_{1,\mathrm{Meas}}$. The key rule in their system guarantees compatible degrees of belief for any two sets of formulas that are equivalent in a particularly defined sense. The space limit does not allow to present their inference rule in details, and we refer readers to the original paper. Another weakly complete axiom system for $LPP_{1,\mathrm{Meas}}$ was given in [72]. The initial axiom system Σ_+ contains the infinitary Archimedean rule 3, and is similar to the propositional part of Ax_{LFOP_1} described in Chap. 4. Thus, the weak completeness theorem obtained in [72] is a simple consequence of Theorem 4.4. The second axiom system Σ_L, a subset of Σ_+, was used to prove completeness for more general not measurable models. In the last part of the paper it was shown that it is possible to provide a weakly complete axiomatization for $LPP_{1,\mathrm{Meas}}$ by replacing the infinitary rule with a finitary version

- $ARCH^f$: From $\beta \rightarrow P_{\geq r-\varepsilon(s,\beta,\alpha)}\alpha$, infer $\beta \rightarrow P_{\geq s}\alpha$,

where $\varepsilon(s, \beta, \alpha)$ is an elementary function of r, β and α which can be computed using the corresponding decision procedure (note that, since the logic is decidable, there is a simple finitary axiomatization with only one schema: all instances of valid formulas). A handmade verification of our approach to strongly complete axiomatization of probability logics which is based on the notion of infinitary deductions was given in [73], while in [46] key properties of the completeness proof technique presented in this book were formally verified using the proof assistant Coq. Another approach similar to ours was presented in [24] for coalgebras over measurable spaces, where Goldblatt (1949) used a methodology for infinitary logics he developed in [23].

Halpern and Pucella [35] provided a weakly complete axiomatization for reasoning about linear combinations of upper probabilities. In semantics, uncertainty is represented by a set P of probability measures with the same set of measurable sets of possible worlds. For a measurable set of worlds X, upper probability is defined as is $P^\star(X) = \sup\{\mu(X) \mid \mu \in P\}$. They also showed that the satisfiability problem for the formulas of the logic is NP-complete.

John Burgess in [11] considered the operator of comparative probability, $p \leq q$, read as "p is less probable than q," and gave the corresponding complete axiomatization in a class of models (in our notation) $LPP_{2,\mathrm{Meas}}$. Our papers related to that subject are [57, 61].

Meier presented in [47] an infinitary propositional probability logic (with infinitary formulas) and proved strong completeness for σ-additive probabilities. Due to the cardinality argument, that logic is undecidable. In [51, 56] strong completeness for σ-additive probabilities was given using (in the propositional case decidable) logics with finitary formulas and infinitary proofs.

7.2 Papers on (Infinitary) Modal Logics

Goldblatt [23] and Segerberg [68] addressed the problem of characterization of semantical consequence relations in non-compact modal logics (e.g., in the linear discrete temporal logic with the next and always operators) using infinitary rules. For example, in a Gentzen type framework, Segerberg included the rule

- (Scott's Rule) If $\Gamma \vdash A$, then $\Box\Gamma \vdash \Box A$, where Γ is allowed to be infinite.

They provided completeness proofs with the property that for every instance of an infinitary rule, if the negation of the consequence belongs to a maximal extension Σ, then there is a premiss C of the rule such that $\neg C \in \Sigma$. That approaches could be applied to temporal, dynamic, multi-agent epistemic logics, etc., and generalized the paper [69]. Infinitary modal logics were also discussed in [16].

Finally, we note that some of the papers that discussed relationships between probability logics and modal logics are: [4, 5, 27, 30, 32, 70, 71]. Using the translation $P_{\geq 1}$ to $\Box$, and $P_{>0}$ to $\Diamond$, their results are that, since $P_{\geq 1}(\alpha \to \beta) \to (P_{\geq 1}\alpha \to P_{\geq 1}\beta)$ and $P_{>0}(p \vee \neg p)$, probability logics generalize modal logic KD. Under additional assumptions about probability models, some other stronger modal logics (e.g., reflexive, transitive, symmetric) can be embedded into probability systems. The paper [14] considered probabilistic operators and the modal operator of knowledge.

7.3 Papers on Temporal Probability Logics

Combinations of temporal and probabilistic formalizations are presented in several papers. Some papers that either implicitly or explicitly combine probabilistic and temporal reasoning were motivated by the need to analyze probabilistic programs and stochastic systems [38, 44] provided semantics and axiom systems for logics with temporal formulas that do not involve probabilities explicitly and are interpreted over Markov systems which can simulate the execution of probabilistic programs. A double-exponential space decidability procedure of a probabilistic propositional dynamic logics with explicit probabilities was given in [17], but the completeness problem is not solved. A fragment of that logic was considered in [42] and a PSPACE-decision procedure was provided.

A sound first-order axiomatization (which is not complete) is provided in [26] for a logic for representing branching time, chance, and action.

In [25] a strongly complete dynamic propositional logic of qualitative probabilities was presented. The logic can be used for reasoning about probabilistic processes. In the logic, finitary formulas are used, while the axiom system contains an infinitary rule

- if the set of rationals that are greater or equal to the probability of one proposition is contained in the set of rationals that are greater or equal to the probability of another proposition, then the probability of the first proposition is greater or equal to the probability of the second one.

The completeness proof is similar as in our approach. This logic extends the logic of qualitative probabilities introduced by Segerberg in [67].

7.4 Papers on Applications of Probability Logics

In [43], Kraus, Lehmann and Magidor proposed System P determining the core of nonmonotonic reasoning. There are many semantics for default reasoning that are characterized by those rules. Lehmann and Magidor [45] showed that preferential entailment is characterized by System P. In the same paper, a family of nonstandard ($^*\mathbb{R}$) probabilistic models characterizing the default consequence relation defined by the system P, was proposed.

The idea of using probabilities and infinitesimals in default reasoning can be found in the ε-semantics [2]. This inspired the approach to deal with default information based on belief functions in [9], where mass values are either close to 0 or close to 1.

In [34], Halpern compared nonstandard probability spaces with lexicographic probability systems, which uses sequences of probability measures to represent uncertainty. He showed that those two approaches are equivalent if the state space is finite, otherwise nonstandard probability spaces are more general.

The paper [8] used a special subclass of probability measures, so called big-stepped probabilities, to provide a standard semantics for System P.

In [18], Friedman and Halpern introduced plausibility measures, applied them to default reasoning, and unified many previously proposed results regarding System P into one framework.

In [36], Halpern and Pucella provided a first-order linear time logic (without the until operator) for reasoning about evidence, where evidence is seen as a function from prior beliefs to beliefs after making an observation. They proved the weak completeness theorem and posed as an open question construction of propositional logic for reasoning about evidence. That problem is solved in [59].

7.5 Books About Probability Logics

Finally, we would like to mention several books related to our text.

Ernest Adams (1926–2009) investigated methods to calculate probabilities of conclusions in valid deductions from probable premises in [2], with the main intention to provide a sound philosophical and mathematical background for introduction of operative alternatives to material implication. He argued that the natural way to weaken the classical material implication is to introduce probability semantics. Moreover, in his opinion, the conditional probability $Pr(\alpha|\beta)$ is more suitable probability interpretation of the conditional $\beta \rightarrow \alpha$ than the alternative interpretation $Pr(\beta \rightarrow \alpha)$. He used so-called extreme probabilities (i.e., probabilities that are infinitesimally close to either 0 or 1) to define the ε-interpretation of conditionals. In particular, he proposed the following probability interpretation of $\beta \rightarrow \alpha$:

- a conditional $\beta \rightarrow \alpha$ is satisfied iff $Pr(\alpha|\beta) = 1 - \varepsilon$ for some nonnegative infinitesimal ε.

Adams also analyzed semantical properties of the introduced interpretation of conditionals. His book [3] is an introductory course in probability reasoning emphasizing both technical and philosophical issues.

Theodore Hailperin gave in [28] an excellent and comprehensive overview of origins of the relationship between mathematical logic and probability, and analyzed the notion of consequence in a propositional logic with a probability valued extension of the classical semantics

- ψ is a probability logical consequence of $\varphi_1, \ldots, \varphi_m$, wrt. the nonempty subsets $\alpha_1, \ldots, \alpha_m, \beta \subset [0, 1]$,

iff in all probability models **M** the following is satisfied

- if $P_{\mathbf{M}}(\varphi_1) \in \alpha_1$ and …and $P_{\mathbf{M}}(\varphi_m) \in \alpha_m$, then $P_{\mathbf{M}}(\psi) \in \beta$,

where $P_{\mathbf{M}} : For_C \rightarrow [0, 1]$ satisfies

- if $\models \varphi$, then $P_{\mathbf{M}}(\varphi) = 1$,
- if $\models \varphi \rightarrow \psi$, then $P_{\mathbf{M}}(\varphi) \leq P_{\mathbf{M}}(\psi)$, and
- if $\models \varphi \rightarrow \neg\psi$, then $P_{\mathbf{M}}(\varphi \vee \psi) = P_{\mathbf{M}}(\varphi) + P_{\mathbf{M}}(\psi)$.

He gave a decision procedure which enables determining of optimal intervals of probability values for conclusions and proved

- There is, for any explicitly given probability logical consequence, an effective procedure for determining whether

$$P(\varphi_1) \in [a_1, b_1], \ldots, P(\varphi_m) \in [a_m, b_m] \models P_{\mathbf{M}}(\psi) \in [l, u]$$

is, or is not, a valid consequence relation.

Hailperin extended his approach to conditional probabilities. In another book [29] he offered first-order counterpart of his propositional probability logic, with, again, a large part devoted to historical topics.

The book [33] written by Joseph Halpern contains a comprehensive overview of models for representing uncertainty, and provides philosophical background for the field, while references are given to papers where proofs of statements could be found. A part of the book is devoted to a probability logics-based formalization of reasoning about uncertainty. A number of motivating puzzles are given and analyzed in different frameworks (probability, lower and upper probabilities, Dempster–Shafer belief functions, possibility measure, ranking function, likelihood, plausibility, multi-agent epistemic logic, Bayesian networks, etc.). Halpern considers the notions of independence, entropy, expectation, expected utility, defaults, belief revision, etc.

In a strict mathematical way Jeff Paris (1944) also offers a careful study of different models (probability, plausibility, fuzzy logic, possibility measures, Dempster–Shafer theory, belief networks) to represent degrees of (real-valued) beliefs [60]. His goal is to give a firm mathematical background to the process of constructing computationally feasible expert systems capable to predict the values of beliefs $Bel(A)$, $Bel(B|C)$ from a finite set of identities of the form $Bel(D) = d$, and $Bel(E|F) = e$. Paris analyzed a number of issues, e.g., maximum-entropy principle, techniques for belief revision, etc., but did not consider axiomatization issues.

References

1. Abadi, M., Halpern, J.Y.: Decidability and expressiveness for first-order logics of probability. Inf. Comput. **112**, 1–36 (1994)
2. Adams, E.W.: The Logic of Conditionals: An Application of Probability to Deductive Logic. D. Reidel Publishing Company, Dordrecht (1975)
3. Adams, E.W.: A Primer of Probability Logic. Center for the Study of Language and Information, Stanford (1996)
4. Alechina, N.: Logic with probabilistic operators. In: Proceedings ACCOLADE 1994, pp. 121–138 (1995). http://www.cs.nott.ac.uk/~nza/papers/Alechina:95c.ps.gz
5. Alechina, N.: Modal Quantifiers. Ph.D. thesis, ILLC Dissertation Series 1995-20. Institute for Logic, Language and Computation, Amsterdam (1995). http://www.cs.nott.ac.uk/~nza/papers/Alechina:95d.pdf
6. Aumann, R.J.: Interactive epistemology II: probability*. Int. J. Game Theory **28**, 301–314 (1999)
7. Bacchus, F.: *Lp*, a logic for representing and reasoning with statistical knowledge. Comput. Intell. **6**(4), 209–231 (1990)
8. Benferhat, S., Dubois, D., Prade, H.: Possibilistic and standard probabilistic semantics of conditional knowledge bases. J. Logic Comput. **9**(6), 873–895 (1999)

9. Benferhat, S., Saffiotti, A., Smets, P.: Belief functions and default reasoning. Artif. Intell. **122**, 1–69 (2000)
10. Boričić, B., Rašković, M.: A probabilistic validity measure in intuitionistic propositional logic. Mathematica Balkanica **10**(4), 365–372 (1996). (the paper presenyed at the Conference on Mathematical Logic, Belgrade, 25. 3. 1993.)
11. Burgess, J.: Axiomatizing the logic of comparative probability. Notre Dame J. Form. Logic **51**(1), 119–126 (2010)
12. Doder, D.: Primene infinitarnih logika u verovatnosno-temporalnom rezonovanju i teoriji modela Ph.D. thesis (in Serbian). University of Belgrade (2011). http://elibrary.matf.bg.ac.rs/bitstream/handle/123456789/2467/docddFR.pdf?sequence=1
13. Fagin, R., Halpern, J., Megiddo, N.: A logic for reasoning about probabilities. Inf. Comput. **87**(1–2), 78–128 (1990). (A preliminary version appeard in the Proceedings of the 3rd IEEE Symposium on logic in computer science, 277–291, 1988)
14. Fagin, R., Halpern, J.: Reasoning about knowledge and probability. J. ACM **41**(2), 340–367 (1994). (A preliminary version appeard in the Proceedings of the Second Conference on Theoretical Aspects of Reasoning About Knowledge, 277–294, 1988)
15. Fattorosi-Barnaba, M., Amati, G.: Modal operators with probabilistic interpretations I. Studia Logica **46**(4), 383–393 (1989)
16. Fattorosi-Barnaba, M., Grassotti, S.: An infinitary graded modal logic. Math. Logic Q. **41**(4), 547–563 (1995)
17. Feldman, Y.: A decidable propositional dynamic logic with explicit probabilities. Inf. Control **63**, 11–38 (1984)
18. Friedman, N., Halpern, J.: Plausibility measures and default reasoning. J. ACM **48**(6), 648–685 (2001)
19. Frish, A., Haddawy, P.: Anytime deduction for probabilistic logic. Artif. Intell. **69**, 93–122 (1994)
20. Gaifman, H.: A theory of higher order probabilities. In: Halpern, J.Y. (ed.) Proceedings of the Theoretical Aspects of Reasoning about Knowledge, pp. 275–292. Morgan-Kaufmann, San Mateo (1986)
21. Georgakopoulos, G., Kavvadias, D., Papadimitriou, C.: Probabilistic satisfiability. J. Complex. **4**(1), 1–11 (1988)
22. Godo, L., Marchioni, E.: Coherent conditional probability in a fuzzy logic setting. Logic J. IGPL **14**(3), 457–481 (2006)
23. Goldblatt, R.: A framework for infinitary modal logic. In: Mathematics of Modality. Lecture Notes in Center for the Study of Language and Information (CSLI), vol. 43, pp. 213–229. CSLI Publications (1993)
24. Goldblatt, R.: Deduction systems for coalgebras over measurable spaces. J. Logic Comput. **20**(5), 1069–1100 (2010)
25. Guelev, D.: A propositional dynamic logic with qualitative probabilities. J. Philos. Logic **28**(6), 575–605 (1999)
26. Haddawy, P.: A logic of time, chance and action for representing plans. Artif. Intell. **80**, 243–308 (1996)
27. Haddawy, P., Frisch, A.M.: Modal logics of higher-order probability. In: Shachter, R.D., Levitt, T.S., Kanal, L.N., Lemmer, J.F. (eds.) Proceedings of the Fourth Annual Conference on Uncertainty in Artificial Intelligence UAI 1988, Uncertainty in Artificial Intelligence, vol. 4, pp. 133–148. North-Holland (1990)
28. Hailperin, T.: Sentential Probability Logic: Origins, Development, Current Status and Technical Applications. Lenigh Univeristy Press, Bethlehem (1996)
29. Hailperin, T.: Logic with a Probability Semantics. Lenigh Univeristy Press, Bethlehem (2011)
30. Halpern, J.Y., McAllester, D.A.: Likelihood, probability, and knowledge. Comput. Intell. **5**, 151–160 (1989)
31. Halpern, J.Y.: An analysis of first-order logics of probability. Artif. Intell. **46**, 311–350 (1990). (A preliminary version appeared in the Proceedings of the 11th International Joint Conference on Artificial Intelligence, 1375–1381, 1989)

32. Halpern, J.Y.: The relationship between knowledge, belief, and certainty. Ann. Math. Artif. Intell. **4**, 301–322 (1991)
33. Halpern, J.Y.: Reasoning about Uncertainty. The MIT Press, Cambridge (2003)
34. Halpern, J.Y.: Lexicographic probability, conditional probability, and nonstandard probability. Games Econ. Behav. **68**(1), 155–179 (2010)
35. Halpern, J.Y., Pucella, R.: A logic for reasoning about upper probabilities. J. Artif. Intell. Res. **17**, 57–81 (2002)
36. Halpern, J.Y., Pucella, R.: A logic for reasoning about evidence. J. Artif. Intell. Res. **26**, 1–34 (2006)
37. Hajek, A., Godo, L., Esteva, F.: Fuzzy logic and probability. In: Proceedings of the Eleventh Conference on Uncertainty in Artificial Intelligence, pp. 237–244 (1995)
38. Hart, S., Sharir, M.: Probabilistic temporal logics for finite and bounded models. In: 16th ACM Symposium on Theory of Computing, pp. 1–13. ACM (1984)
39. Heifetz, A., Mongin, P.: Probability logic for type spaces. Games Econ. Behav. **35**(1–2), 31–53 (2001)
40. Jaumard, B., Hansen, P., de Aragao, M.P.: Column generation methods for probabilistic logic. ORSA J. Comput. **3**, 135–147 (1991)
41. Keisler, J.: Probability quantifiers. In: Barwise, J., Feferman, S. (eds.) Model Theoretic Logics, pp. 509–556. Springer, Berlin (1985)
42. Kozen, D.: A probabilistic PDL. J. Comput. Syst. Sci. **30**, 62–178 (1985)
43. Kraus, S., Lehmann, D., Magidor, M.: Nonmonotonic reasoning, preferential models and cumulative logics. Artif. Intell. **44**(1–2), 167–207 (1990)
44. Lehmann, D., Shelah, S.: Reasoning with time and chance. Inf. Control **53**, 165–198 (1982)
45. Lehmann, D., Magidor, M.: What does a conditional knowledge base entail? Artif. Intell. **55**, 1–60 (1992)
46. Maksimović, P.: Development and Verification of Probability Logics and Logical Frameworks. Ph.D. thesis. University of Nice - Sophia Antipolis and Faculty of Technical Sciences, University of Novi Sad. (2013)
47. Meier, M.: An infinitary probability logic for type spaces. Israel J. Math. **192**, 1–58 (2012)
48. Milošević, M., Ognjanović, Z.: A first-order conditional probability logic. Logic J. Interes. Group Pure Appl. Logics **20**(1), 235–253 (2012)
49. Nilsson, N.: Probabilistic logic. Artif. Intell. **28**, 71–87 (1986)
50. Ognjanović, Z.: A logic for temporal and probabilistic reasoning. In: Workshop on Probabilistic Logic and Randomised Computation, ESSLLI 1998, Saarbruecken, Germany (1998)
51. Ognjanović, Z.: Some probability logics and their applications in computer sciences. Ph.D. thesis (in Serbian). University of Kragujevac (1999). http://elibrary.matf.bg.ac.rs/bitstream/handle/123456789/197/phdZoranOgnjanovic.pdf?sequence=1
52. Ognjanović, Z.: A completeness theorem for a first order linear-time logic. Publications de L'Institute Matematique (Beograd), ns. **69**(83), 1–7 (2001). http://elib.mi.sanu.ac.rs/files/journals/publ/89/n083p001.pdf
53. Ognjanović, Z.: Discrete linear-time probabilistic logics: completeness, decidability and complexity. J. Logic Comput. **16**(2), 257–285 (2006)
54. Ognjanović, Z., Rašković, M.: A logic with higher order probabilities. Publications de L'Institute Matematique, n.s. **60**(74), 1–4 (1996). http://elib.mi.sanu.ac.rs/files/journals/publ/80/n074p001.pdf
55. Ognjanović, Z., Rašković, M.: Some probability logics with new types of probability operators. J. Logic Comput. **9**(2), 181–195 (1999)
56. Ognjanović, Z., Rašković, M.: Some first-order probability logics. Theor. Comput. Sci. **247**(1–2), 191–212 (2000)
57. Ognjanović, Z., Perović, A., Rašković, M.: Logics with the qualitative probability operator. Logic J. IGPL **16**(2), 105–120 (2008)
58. Ognjanović, Z., Doder, D., Marković, Z.: A branching time logic with two types of probability operators. In: Benferhat, S., Grant, J. (eds.) Fifth International Conference on Scalable Uncertainty Management SUM-2011, October 10 - 12, 2011, Dayton, Ohio, USA. Lectuer Notes in Computer Science, vol. 6929, pp. 219–232. Springer (2011)

59. Ognjanović, Z., Marković, Z., Rašković, M., Doder, D., Perović, A.: A propositional probabilistic logic with discrete linear time for reasoning about evidence. Ann. Math. Artif. Intell. **65**(2–3), 217–243 (2012)
60. Paris, J.: The Uncertain Reasoner's Companion: A Mathematical Perspective. Cambridge University Press, Cambridge (2006)
61. Perović, A.: Some applications of the formal method in set theory, model theory, probabilistic logics and fuzzy logics. Ph.D. thesis (in Serbian). University of Belgrade (2008). http://elibrary.matf.bg.ac.rs/bitstream/handle/123456789/100/phdAleksandarPerovic.pdf?sequence=1
62. Perović, A., Ognjanovć, Z., Rašković, M., Marković, Z.: How to restore compactness into probabilistic logics? In: Proceedings of the 11th European Conference. JELIA 2008, Dresden, Germany, September 28-October 1, 2008. Lecture Notes in Computer Science, vol. 5293, pp. 338–348. Springer, Berlin (2008)
63. Perović, A., Ognjanovć, Z., Rašković, M., Marković, Z.: A probabilistic logic with polynomial weight formulas. In: Hartmann, S., Kern-Isberner, G. (eds.) Proceedings of the Fifth International Symposium Foundations of Information and Knowledge Systems, FoIKS 2008, Pisa, Italy, 11–15 February 2008. Lecture Notes in Computer Science, vol. 4932, pp. 239 – 252. Springer (2008)
64. Rašković, M., Ognjanović, Z.: Some propositional probabilistic logics. In: Kovačević-Vujičić, V., Mijajlović, Ž. (eds.) Proceedings of the Kurepa's Symposium 1996, Belgrade, pp. 83–90. Scientific review 19-20 (1996)
65. Rašković, M., Ognjanović, Z., Marković, Z.: A logic with conditional probabilities. In: Proceedings of the JELIA 2004. Lecture Notes in Computer Science, vol. 3229, pp. 226–238. Springer (2004)
66. Rašković, M., Marković, Z., Ognjanović, Z.: A logic with approximate conditional probabilities that can model default reasoning. Int. J. Approx. Reason. **49**(1), 52–66 (2008)
67. Segerberg, K.: Qualitative probability in a modal setting. In: Fenstad, E. (ed.) Proceedings 2nd Scandinavian Logic Symposium, pp. 341–352. North-Holland, Amsterdam (1971)
68. Segerberg, K.: A model existence theorem in infinitary propositional modal logic. J. Philos. logic **23**, 337–367 (1994)
69. Sundholm, G.: A completeness proof for an infinitary tense-logic. Theoria **43**, 47–51 (1977)
70. van der Hoeck, W.: Some consideration on the logics $P_F D$. In: Voronkov, A. (ed.) Proceeding of the 2nd Russian Conference on Logic ProgrammingLogic Programming, 1991. Lecture Notes in Computer Science, vol. 592, pp. 474–485. Springer, Berlin (1992)
71. van der Hoeck, W.: Some consideration on the logics $P_F D$. J. Appl. Non-Class. Logics **7**(3), 287–307 (1997)
72. Zhou, C.: A complete deductive system for probability logic. J. Logic Comput. **19**(6), 1427–1454 (2009)
73. Zhou, C.: Probability logic of finitely additive beliefs. J. Logic Lang. Inf. **19**(3), 247–282 (2010)

Erratum to: Extensions of the Probability Logics LPP_2 and $LFOP_1$

Aleksandar Perović, Dragan Doder, Nebojša Ikodinović and Angelina Ilić Stepić

Erratum to:
Chapter 5 in: Z. Ognjanović et al., *Probability Logics*, DOI 10.1007/978-3-319-47012-2_5

The book was inadvertently published without including the coauthor's names Aleksandar Perović, Dragan Doder, Nebojša Ikodinović, and Angelina Ilić Stepić below the chapter title in the first page of Chap. 5. The erratum book has been updated with the changes.

The updated original online version for this book can be found at
10.1007/978-3-319-47012-2_5

A. Perović (✉) · D. Doder · N. Ikodinović · A. Ilić Stepić
Mathematical Institute of the Serbian Academy of Sciences and Arts, Belgrade, Serbia
e-mail: pera@sf.bg.ac.rs

Z. Ognjanovć et al., *Probability Logics*,
DOI 10.1007/978-3-319-47012-2_8

Appendix A
General Notions

A.1 Formal Axiom Systems

In general, a formal system consists of syntax and semantics (interpretation, meaning). Strictly speaking, it may be argued that only the syntax constitutes the formal system. However, no one is constructing a formal system without having in mind some intended interpretation, so we consider it reasonable to treat semantics as a part of a formal system.

A.1.1 Syntax

Syntax consists of language and derivation apparatus.

A.1.1.1 Language

Language consists of: symbols (alphabet) and formation rules.

Usually, we have several types of symbols, including at least one type of symbols for variables. We may have symbols for operations, relations, constants, logical connectives, brackets, comma, etc. For symbols, we may use any kind of objects. The only restriction is that a symbol should not be a sequence of some other symbols, as this would prevent the unique readability of expressions.

While in natural languages the acceptable words are given as a list in a dictionary and grammar prescribes what acceptable sentences are, here we have formation rules which define how different kinds of expressions are to be built as sequences of symbols. There is always at least one type of expressions—formulas. There are two approaches to defining a language. When the main concern is proving theorems about the syntax (so-called meta-theorems), we try to keep the number of symbols and expressions to a minimum. As meta-theorems are usually proved by induction on

Z. Ognjanović et al., *Probability Logics*,
DOI 10.1007/978-3-319-47012-2

the length or complexity of expressions, this reduces the number of clauses, i.e., simplifies the proofs. The price to pay is poor readability. For example, in propositional logic we may have many logical connectives:

- $\neg$ (not),
- $\wedge$ (and),
- $\vee$ (or),
- $\rightarrow$ (implies),
- $\leftrightarrow$ (is equivalent).

They may all be replaced by a single one, e.g.:

- $\uparrow$ (nand, not and)

from which all others may be defined, but the formulas become unreadable. As in this book our main concern is readability, and most meta-theorems have already been proved in journal papers, we shall avoid this practice and try to use a more standard notation.

A.1.1.2 Derivation Apparatus

Derivation apparatus consists of Axioms and Inference (derivation) rules. Axioms are just some chosen formulas. They can be given as an explicit list or as some schemas for constructing formulas of a certain shape: axiom schemata. Inference rules are relations on the set of formulas, i.e., an inference rule is a set of $n+1$-tuples of formulas ($n \in \mathbb{N}$), where first n formulas are called premises and the last formula is called the conclusion (consequence).

From axioms, using inference rules, we get theorems (provable formulas). Each theorem has a proof (derivation) which is a sequence of formulas such that each formula in the sequence is either an axiom or is obtained as a consequence of some inference rule from some previous formulas in the sequence, and the last formula in the sequence is our theorem. The proofs are usually finite, but in this book we allow also infinitary inference rules, where the number of premises is infinite, and, consequently, the proofs will be infinite sequences, in a form of a denumerable ordinal (see Chap. 3).

We say that an axiom system is finitary, if:

- the set of axiom schemata is recursive[1] (i.e., for every formula it is decidable whether it is an axiom instance), and
- relations representing inference rules are recursive (i.e., for a given $n+1$-tuple of formulas it is decidable whether it belongs to the relation).

Syntactic consequence relation, denoted by $\vdash$, is a generalization of provability. We say that a formula α is a syntactic consequence of a set of formulas T, denoted by $T \vdash \alpha$, if there is a proof of α which, in addition to axioms, uses also formulas from T. In other words:

---

[1] The set of axiom schemata can be finite or infinite (in which case it must be effectively specifiable).

- $T \vdash \alpha$ iff there is a sequence of formulas such that for each of the formulas in the sequence either:
  - It is an axiom.
  - It belongs to T.
  - It is obtained from some previous formulas in the sequence using one of the inference rules, and
- the last formula in the sequence is α.

A.1.2 Semantics

Semantics provides an interpretation (meaning) for the syntax. We may have informal semantics, but here we are interested in the formal ones, which are usually some mathematical theories or constructions. For a given syntax, we may have different semantics, but also quite different syntaxes may have essentially the same semantics.

First, we must interpret the language. The interpretation of symbols will be in some set which we call the universe of interpretation. Symbols for variables are interpreted as ranging over this set and, e.g., an operation symbol is interpreted as an operation on the elements of the universe, etc. Then, the expressions of the language are interpreted in the universe. In particular, formulas are interpreted as sentences about the elements of the universe.

This interpretation has to be such that axioms are interpreted as true sentences and inference rules preserve truth, i.e., from true premises they derive true conclusion. In this case, we say that the interpretation is sound and that the given universe with its structure is a model.

Another important property of this pair syntax–semantics, is called (weak) completeness:

- if every sentence true in all models can actually be derived from axioms.

In this sense, we have interpreted a syntactic notion of provability by a semantic notion of truth.

Similarly, corresponding to the notion of syntactic consequence, $T \vdash \alpha$, there is a notion of "semantic consequence", $T \models \alpha$, defined by:

- $T \models \alpha$ iff every model for all formulas from T, is a model for α.

The best agreement between a given syntax and a given semantics is when these two consequence relations coincide and then we say that the syntax is strongly complete for the semantics.

This relation can, dually, be expressed in terms of a syntactic notion of consistency: we say that a set of formulas is consistent, if we cannot derive a contradiction from it.

Strong completeness is equivalent to the following statement:

- A set of formulas is consistent iff it has a model.

A.2 Propositional Calculus

Propositional (or Sentential) calculus is one of the simplest formal theories.

A.2.1 Language

The symbols of the language are:

- denumerable set of primitive propositions (propositional variables) $\phi = \{p, q, r, \ldots\}$,
- classical propositional connectives $\neg$, and $\wedge$, and
- auxiliary symbols: parentheses (and).

There is only one kind of expressions—formulas—which are defined by:

1. Primitive propositions are formulas.
2. If α and β are formulas, so are $\neg\alpha$ and $(\alpha \wedge \beta)$.
3. Formulas are obtained only by a finite number of applications of rules (1) and (2).

We immediately introduce the following abbreviations:

- $(\alpha \vee \beta)$ stands for $\neg(\neg\alpha \wedge \neg\beta)$,
- $(\alpha \rightarrow \beta)$ stands for $(\neg\alpha \vee \beta)$, and
- $(\alpha \leftrightarrow \beta)$ stands for $(\alpha \rightarrow \beta) \wedge (\beta \rightarrow \alpha)$.

If α is a formula then the set $\mathrm{Subf}(\alpha)$ of subformulas of α is defined recursively:

- $\alpha \in \mathrm{Subf}(\alpha)$,
- if $\neg\beta \in \mathrm{Subf}(\alpha)$, then $\beta \in \mathrm{Subf}(\alpha)$, and
- if $\beta \wedge \gamma \in \mathrm{Subf}(\alpha)$, then $\beta, \gamma \in \mathrm{Subf}(\alpha)$.

We write $\mathrm{len}(\alpha)$ to denote the length of (or size) of α, assuming a reasonably succinct encoding. If we denote the cardinality (the number of elements) of a set T by $|T|$, then:

$$|\mathrm{Subf}(\alpha)| \leq \mathrm{len}(\alpha).$$

A.2.2 Derivation Apparatus

An axiom system Ax_{LP} for propositional calculus LP is given by the following axiom schemas:

1. $\alpha \rightarrow (\beta \rightarrow \alpha)$
2. $(\alpha \rightarrow (\beta \rightarrow \gamma)) \rightarrow ((\alpha \rightarrow \beta) \rightarrow (\alpha \rightarrow \gamma))$
3. $(\neg\alpha \rightarrow \neg\beta) \rightarrow (\beta \rightarrow \alpha)$

and the inference rule Modus Ponens (MP):

1. From α and $\alpha \rightarrow \beta$ infer β.

The rule MP can be also represented as follows:

$$\frac{\begin{array}{c}\alpha \\ \alpha \rightarrow \beta\end{array}}{\alpha}$$

Note that the formulas that appear above are schema formulas. They represent an infinite number of "real" formulas that can be obtained by systematic replacing schema formulas by concrete formulas. For example, instances of the first axiom are:

- $p \rightarrow (p \rightarrow p)$ (both α and β are replaced by the primitive proposition p), and
- $p \rightarrow (q \rightarrow p)$ (α is replaced by p and β by the q), etc.

As an example, we give a formal proof of the theorem $\alpha \rightarrow \alpha$ for an arbitrary formula α, which is a sequence of five formulas. On the right-hand side we provide comments.

1. $(\alpha \rightarrow ((\alpha \rightarrow \alpha) \rightarrow \alpha)) \rightarrow ((\alpha \rightarrow (\alpha \rightarrow \alpha)) \rightarrow (\alpha \rightarrow \alpha))$	(Axiom 2)
2. $\alpha \rightarrow ((\alpha \rightarrow \alpha) \rightarrow \alpha)$	(Aksiom 1)
3. $(\alpha \rightarrow (\alpha \rightarrow \alpha)) \rightarrow (\alpha \rightarrow \alpha)$	(MP from (1), (2))
4. $\alpha \rightarrow (\alpha \rightarrow \alpha)$	(Aksiom 1)
5. $\alpha \rightarrow \alpha$	((MP from (3), (4))

A.2.3 Semantics

The intended informal interpretation of this formal theory is that variables denote some propositions or sentences for which we assume only that they are either true or false. The connectives are interpreted as intuitive logical operations: "not" and "and". The main formal interpretation is the two-element Boolean algebra of True and False or 1 and 0. The universe in which we interpret variables is $\{0, 1\}$ and connectives are interpreted as operations on that set given by the usual truth tables.

α	β	$\neg\alpha$	$\alpha \wedge \beta$
1	1	0	1
1	0	0	0
0	1	1	0
0	0	1	0

For any assignment of values 0 or 1 to variables, we can calculate, in a unique way, the value of any formula. We say that the classical propositional connectives of LP are truth-functional. On the other hand, in this book we consider probability and modal operators that are not truth-functional, i.e., in the general case it is not possible to calculate truth values of formulas from truth values of subformulas.

A formula that gets the value 1 for each assignment is called a tautology.

Another interpretation, offered by Boole himself, is a Boolean algebra of subsets of some set. If p denotes some proposition, its interpretation will be the set of all cases (possible worlds) in which this proposition is true. The interpretation of connectives is naturally the intersection for $\wedge$ (as the set of cases when both propositions are true), and for $\neg$ the complement with respect to the set of all cases (possible worlds)—which is the universe here. This interpretation is well suited for probabilistic logic: the probability of a proposition will be the measure of its interpretation. When the number of possible cases is finite, this reduces to the familiar fraction: the number of positive cases divided by the number of all possible cases.

A.2.4 Completeness Proof

The weak completeness theorem proves that the set of tautologies coincides with the set of theorems of the formal system. The strong completeness theorem proves that an arbitrary (possibly infinite) set of formulas T is consistent if and only if it has a model, i.e., there is an assignment of values 0 and 1 to variables so that each formula from the set T gets value 1.

The main steps in proving strong completeness of the axiom system Ax_{LP} with respect to the above-mentioned formal interpretations are:

- Soundness: every instance of axioms schemata is a tautology, while applications of the inference rule MP on tautologies derive tautologies (i.e., MP preserves the validity).
- Deduction theorem: if T is a set of formulas and α and β are formulas, then

$$T \cup \{\alpha\} \vdash \beta \text{ iff } T \vdash \alpha \rightarrow \beta.$$

- Lindenbaum's theorem – every consistent set T of formulas can be extended to a maximal consistent set $\mathcal{T}$ (for every formula α, $\mathcal{T}$ contains either α or $\neg\alpha$) in the following way: let $\alpha_0, \alpha_1, \ldots$be a list of all propositional formulas; a sequence $(T_i)_{i\in\mathbb{N}}$ of consistent extensions of T is constructed such that $T_0 = T$, $T_{i+1} = T_i \cup \{\alpha_i\}$, if $T_i \cup \{\alpha_i\}$ is consistent, otherwise $T_{i+1} = T_i \cup \{\neg\alpha_i\}$; $\mathcal{T} = \bigcup_i T_i$,
- Construction of a model for a consistent set T of formulas:
  - an assignment I to primitive propositions is defined such that $I(p) = 1$ iff $p \in \mathcal{T}$, and
  - for every formula α it can be proved that $I(\alpha) = 1$ iff $\alpha \in \mathcal{T}$.
- Since all formulas from T belongs to $\mathcal{T}$, I is a model of T.

Obviously, weak completeness is a consequence of strong completeness:

- We want to prove: if $\models \alpha$, then $\vdash \alpha$.
- It is equivalent to: if $\nvdash \alpha$, then $\not\models \alpha$, i.e.,
- if $\neg\alpha$ is consistent, then $\neg\alpha$ has a model,

which follows from the strong completeness theorem. Furthermore, since the axiom system Ax_{LP} is finitary, the strong completeness theorem implies another important property—compactness:

- For every set T of its formulas, T has a model iff every finite subset of T has a model.

The point is that compactness follows from strong completeness just in case when logic is finitary. As we explain in this book, probabilistic logics are inherently non-compact, and we need some kind of infiniteness to obtain strongly complete axiomatizations (see Sect. 3.3).

A.3 Object Language and Meta-Language

When we consider formal systems, two different languages are involved :

- the formal language mentioned in Sect. A.1.1 is called the object language, and
- the language used to talk about a formal system is called the meta-language.

While the object language is very precise, as we describe above, the meta-language (although mathematized) is not always so formal. This division is also reflected when we consider statements:

- theorems in a formal system are formulas in the object language, and their proofs are given the precise meaning of special sequences of formulas in the object language, while
- theorems about properties of a formal system (e.g., the completeness theorem) are expressed and proved in the meta-language.

In this book, different object languages are used for different probability logics. Usually, the classical propositional (or first order) language is extended with some probability operators, while different conditions restrict the formation rules, e.g., we consider formulas with(out) iterations of probability operators.

A.4 Probability

If $W \neq \emptyset$, then H is an algebra of subsets of W, if $H \subset \mathbb{P}(W)$ such that:

- $W \in H$, and
- if $A, B \in H$, then $W \setminus A \in H$ and $A \cup B \in H$.

A function $P : H \to [0, 1]$ is a finitely additive probability measure, if the following conditions hold:

- $P(W) = 1$, and
- $P(A \cup B) = P(A) + P(B)$, whenever $A \cap B = \emptyset$.

For W, H and P described as above, the triple $\langle W, H, P \rangle$ is called a (finitely additive) probability space.

We also say that an algebra H is a σ-algebra, if:

- $\bigcup_{i \in \mathbb{N}} A_i \in H$ whenever $A_i \in H$ for every $i \in \mathbb{N}$,

while a probability measure P is σ-additive, if:

- $P(\bigcup_{i \in \mathbb{N}} A_i) = \sum_{i \in \mathbb{N}} P(A_i)$, whenever $A_i \in H$ and $A_i \cap A_j = \emptyset$ for all $i \neq j$.

In Kolmogorov's approach the conditional probability of B given A is defined using the notion of probability:

$$P(B|A) = \frac{P(B \cap A)}{P(A)}$$

for every A such that $P(A) > 0$.

On the other hand, in the approach proposed by de Finetti, coherent conditional probability is the primitive notion. Let W be a non empty set, H an algebra of subsets of W, and $H^0 = H \setminus \{\emptyset\}$. Then, $P : H \times H^0 \to [0, 1]$, is a conditional probability if the following holds:

- $P(A, A) = 1$, for every $A \in H^0$,
- $P(\cdot, A)$ is a finitely additive probability on H for any given $A \in H^0$, and
- $P(C \cap B, A) = P(B, A) \cdot P(C, B \cap A)$, for all $C \in H$ and $A, B, A \cap B \in H^0$.

Index

A
Abadi, 110, 192
Abduction, 45
Adams, 62, 196
Additive
  group, 5
Additivity, 30, 41, 43, 51, 65, 82
  finite, 12, 53, 55, 57, 66, 79, 86, 95, 112, 126, 146, 154, 166, 208
  finitely additive probability measure, 116
  non-additivity, 27, 28, 35, 62
  rule, 67
  σ, 4, 13, 14, 53, 56, 58, 63, 65, 66, 81, 95, 96, 194, 208
Alechina, 188
Algebra, 112, 207
  Boolean, 63, 206
  Heyting algebra, 153
  of subsets, 79, 91, 92, 95, 116, 126
  σ, 63, 66, 67, 208
Alphabet, 201
Amati, 188
Approximate
  conditional probability, 156
  default, 173
Archimedean
  axiom, 83
  property, 65, 85
  rule, 82, 114
Argumentation, 184
Aristotle, 20
Ars Conjectandi, 24
Assignment, 205, 206
Atom, 67, 81, 97, 100–102, 144, 169, 183
  measure, 97
Atomic formula, 1, 51, 65, 111, 161, 184
Aumann, 193
Axiom, 1, 8, 26, 67, 202
  Archimedean, 83
  continuity, 59
  continuity at 0, 65
  Δ_0-collection, 8
  Δ_0-separation, 8
  empty set, 8
  equivalence, 50
  extensionality, 8
  finitary system, 2
  infinity, 8
  instance, 82, 83
  K, 10, 38, 88
  monotonicity, 65, 191
  necessitation, 65
  of choice, 8
  ordered pair, 8
  powerset, 8
  probability, 13, 55, 68, 83
  regularity, 6, 8
  schema, 3, 9, 82
  Segerberg, 68
  union, 8
Axiom system, 1, 4, 14, 50, 55, 68, 86, 201
  $Ax_{LPCP_2^{[0,1]_{\mathbb{Q}(\varepsilon)},\approx}}$, 157
  Ax_{LFOP_1}, 113, 114
  $Ax_{LPP_1^{\mathrm{LTL}}}$, 127
  Ax_{LPP_2}, 82, 86, 93, 96
  $Ax_{LPP_{2,\leq}}$, 151
  $AX_{L_{\mathbb{Q}_p}}$, 161
  strongly complete, 7
  finitary system, 3, 84
  infinitary system, 13, 83
  strongly complete, 86
  weakly complete, 67, 68, 86
Axiomatization, 9, 12, 54

Z. Ognjanović et al., *Probability Logics*,
DOI 10.1007/978-3-319-47012-2

classical, 13
complete, 9
finitary, 2, 3, 84–86
Hamblin, 67
infinitary, 13
Keisler's logic, 65
Kolmogorov, 11, 48, 58, 60
modal logic, 9
not complete, 26
probability, 58, 61, 65, 67
Reichenbach, 54–56
strongly complete, 2, 85, 86
weakly complete, 85

B
Bacchus, 189
Banach space, 159
Barcan formula, 118
Barwise, 7, 62
Basic probability formula, 78, 156, 179, 190
Basic weight formula, 99, 104
Bayes, 32
Bernoulli, 24
Big-stepped probability, 183, 195
Bolzano, 37
Boole, 40, 46, 68
Borel, 51, 57
Branching time logic, 12, 125
Burgess, 67, 194

C
Canonical model, 87, 91, 93, 95, 129, 139, 157
weak, 95
Cantelli, 58
Carathéodory, 56
theorem, 95, 96
Cardano, 20
Cardinal
infinite, 8
Cardinality, 6, 57
Carnap, 4, 50, 63
Compactness, 2, 11, 84, 86, 127
non-compactness, 3, 77, 84, 85, 194
theorem, 2, 5, 84, 95, 136
theorem of Barwise, 8, 66, 140
Completeness, 11, 12, 14, 86, 92, 128
incompleteness theorem, 3
$LFOP_{1,\mathrm{Meas}}$, 116
$LFOP_{1,\mathrm{Meas,Neat}}$, 117
$LFOP_{1,\mathrm{Meas},\sigma}$, 117
$LPFOP_{1,\mathrm{Meas,All}}$, 117
$LPP^{\mathrm{LTL}}_{1,\mathrm{Meas}}$, 130
$LPP^{I}_{2,\mathrm{Meas}}$, 154
$LPP_{2,\mathrm{Meas}}$, 93
$LPP_{2,\mathrm{Meas,All}}$, 95
$LPP_{2,\mathrm{Meas,Neat}}$, 96
$LPP_{2,\mathrm{Meas},\sigma}$, 95
proof, 13, 48, 66, 68, 95
strong, 2, 3, 5, 13, 84, 85, 93, 95, 96, 116, 117, 128, 130, 158
theorem, 1, 66, 84, 87, 93, 206
weak, 1, 5, 67, 84
Complexity, 3, 11, 13, 130
PSAT, 3, 97
Conditional probability, 12, 19, 29, 31, 33, 36–38, 40, 41, 43–45, 50, 53, 59, 107, 196, 208
approximate, 7, 13, 14, 156, 166
coherent, 13
de Finetti, 13, 156, 208
Kolmogorov, 12, 156, 208
operator, 14
Conditionals, 43, 196
Consequence, 11, 93
necessary, 40
probability, 64
probability of, 23
relation, 1, 11, 56, 64, 166, 202
semantical, 1, 2, 81, 116, 203
syntactical, 1, 2, 83, 116, 202
Consistency, 3, 8, 50, 52
Consistent
belief, 52
formal system, 3
formula, 1
maximal set, 84, 87, 90, 94, 206
set of formulas, 2, 64, 66, 84, 85, 87, 89, 93, 95, 96
Cournot, 40, 54
Cox
David, 62
Richard, 61

D
De Finetti, 52, 156, 193, 208
De Moivre, 29
De Morgan, 38
Decidability, 3, 7, 9, 11–13, 64, 67, 68, 86, 121, 122, 128, 142, 168
$LPP^{I}_{2,\mathrm{Meas}}$, 154
LPP^{LTL}_{1}, 130
propositional probability logic, 2
PSAT, 3, 13, 77, 97, 124, 154

Decision procedure, 168
    probability logic, 2, 41
Deduction, 26, 44, 45, 83
    theorem, 87, 93, 94, 96, 206
Deductive logic, 48
Deductive system, 55
Deductively closed set, 84, 90, 91
Default, 14, 195
    reasoning, 9, 14, 20, 107, 165
    rule, 166, 183
Dempster, 62, 197
Dirac, 62
Disjunctive normal form, 97, 100, 101, 103, 104, 168
    complete, 81, 100
Domain, 12, 63
    constant, 13, 113
    first-order, 64, 111

E

ε-semantics, 195, 196
Esteva, 193
Evidence, 27, 44, 48, 50, 51, 61, 68, 175
    reasoning about, 14
    total, 61
Expectation, 20, 21, 24, 31, 32, 50, 197
    conditional, 66
    mathematical, 52, 58

F

Fagin, 189, 192
Fattorosi-Barnaba, 188
Fermat, 20
Fine, 62
Finite additivity, 77, 79, 86, 95
First-order
    calculus, 63
    language, 51, 63, 64
    logic, 2, 3, 5, 48, 65
    modal models, 13
    model, 65, 83
    probability logic, 2, 13
    probability quantifier, 83
    structure, 5, 64
Forcing relation, 153
Formation rule, 201
Formula, 204
    classical, 13, 78–81
    Δ_0, 4
    finite, 2
    infinitary, 5, 8
    infinite, 4
    LPP_2, 78–81
    probability, 64, 78
    Σ_1, 4
Free variable, 37, 63–65
Frege, 47
Friedman, 195
Frisch, 192
Function symbol, 5
Fuzzy logic, 193, 197

G

Gabbay, 165
Gaifman, 63, 187
Gamble, 20, 25
    problem, 20
Gentzen, 3, 194
Gödel, 3, 4, 8, 48, 62, 162
Godo, 193
Goldblatt, 193, 194

H

Haddawy, 192
Hailperin, 26, 28, 31, 34, 35, 38–41, 43, 50, 56, 62, 68, 196
Hajek, 193
Halpern, 110, 189, 191–193, 195–197
Hamblin, 66
Heifetz, 193
Henkin, 3, 62, 87, 94, 115, 189
Heuristics, 3, 13, 98
    bee colony optimization, 107
    genetic algorithms, 98
    variable neighborhood search, 107
Hilbert, 4
    space, 159
Hoover, 65
Huygens, 20
Hyperfinite
    model, 66
    numbers, 6, 7
Hypertime interval, 6

I

Implication, 46, 48, 51, 54
    intuitionistic, 153
    Łukasiewicz, 193
    material, 9, 46, 196
    probability, 54, 56
    strict, 9, 46
Incompleteness theorem, 3
Inconsistency, 50, 84, 85

Inconsistent
  infinitary set of formulas, 94
  set of formulas, 84, 85
Independence, 19, 31, 34, 38, 40, 41, 44, 162, 197
  stochastical, 43
  trials, 57
Induction, 43–45, 56
  transfinite, 3, 87
Inference, 1, 34, 39, 46, 48, 51, 54
  Archimedean rule, 114
  generalization, 114
  infinitary rule, 4, 5, 13, 23, 83, 86, 94, 151, 194, 202
  necessary, 27, 39
  necessitation rule, 10, 114
  plausible, 61
  probabilistic, 39, 61
  Q_F-rule, 141
  rule, 1, 3, 10, 14, 38, 82, 83, 202
  rule of induction, 55
  uncertain, 34
Infinitesimal, 4, 6, 7, 14, 23, 47, 64, 167, 195
Interpretation, 203
  argument, 26
  probability, 20, 44, 48, 52, 54, 61
    for modal operators, 188
    of conditionals, 196
  propositional, 80, 91
  sylogism, 46
Intuitionistic
  implication, 153
Intuitionistic logic, 2, 12, 14, 152
  $LPP^I_{2,\mathrm{Meas}}$, 152

J
Jaynes, 61

K
Keisler, 11, 62, 64, 83, 110, 189
Keynes, 48, 49, 51, 52, 54, 159
Knowledge, 22, 24, 39, 49, 162, 192, 194
  common, 162
  mathematical, 47
  partial, 41
  uncertain, 68, 107
  updating, 60
Kolmogorov, 11, 12, 40, 48, 58, 63, 156, 208
Kraus, 166, 195
Krauss, 63
Kripke, 10, 62, 121
  first-order models, 13
  models, 2, 10, 13, 67, 83, 118, 126, 152, 153, 188, 191

L
λ-calculus, 162
Lambert, 33
Language, 201
  analysis, 47
  arithmetic, 4
  countable, 2, 80
  finite, 2, 66
  first-order, 5, 51, 64
  formal, 1, 9, 14, 23, 80, 85, 87
  infinitary, 5, 63, 64
  meta, 207
  meta-level, 11, 56
  modal, 9, 11
  object, 207
  object-level, 2, 11, 56, 86
  probability, 11, 12, 78
  propositional, 9
  symbolic, 40, 41
Laplace, 35
Lehmann, 166, 195
Leibnitz, 21
Likelihood, 45
Lindenbaum's theorem, 87, 90, 94, 115, 206
Linear programming, 43, 68, 69, 99
Linear system, 13, 97, 98, 100, 101
Linear time logic, 12, 125
Linear weight formula, 13
Loeb, 7
  measure, 7
  process, 66
Logic
  branching time, 3, 125
  classical, 2, 3, 12, 49
  deductive, 48
  finitary, 12, 66
  first order probability logic $LFOP_1$, 110
  first-order, 2, 3, 5, 12, 48, 67
  infinitary, 5, 12, 64, 65, 194
  intuitionistic, 152
  $L_{\mathbb{A}P}$, 65
  $LFOP_1$, 110
  linear time, 3, 125
  $L_{\omega_1\omega}$, 5, 8, 9, 63
  $L_{\omega P}$, 64
  $L_{\omega_1 P}$, 64
  $LPCP_2^{[0,1]_{\mathbb{Q}(\varepsilon)},\approx}$, 156, 168
  LPP_1, 122
  LPP_2, 78

$LPP_2^{A,\omega_1,\mathrm{Fin}}$, 137
$LPP_2^{\mathrm{Fr}(n)}$, 135
$LPP_{2,\mathrm{Meas}}^I$, 152
$LPP_{2,P,Q,O}$, 141
$LPP_{2,\preceq}$, 151
LPP_1^{LTL}, 125
$L_{\mathbb{Q}_p}$, 160
many-valued, 46
mathematical, 1, 11, 14, 48
modal, 3, 9, 46, 66, 67, 82, 87, 118
non-compact, 3, 194
PK, 68
Port-Royal, 21, 24
probability, 2, 3, 11, 12, 21, 22, 24, 33, 37–39, 41, 43, 44, 46, 48–51, 54, 62, 64, 77, 118
propositional, 2, 12, 46, 82, 83, 86, 204
PWF, 158
spatiotemporal, 183
$S5U$, 67
temporal, 2, 3, 12, 125, 178, 194
Łukasiewicz, 193

M
MacColl, 45
Magidor, 166, 195
Marchioni, 193
Measurable
model, 13, 79, 117
set, 56
Megiddo, 189
Meier, 194
Meta-level, 11
language, 207
theorem, 201, 203, 206
Modal
model, 13, 69, 153
operator, 2, 66, 188
Modal logic, 3, 9, 13, 46, 62, 66, 67, 118, 187
axiomatization, 9
infinitary, 194
K, 9, 38, 68
KD, 194
necessitation, 10
PK, 68
$S5$, 67
$S5U$, 67
T, 67
Model, 203
intuitionistic Kripke, 153
strong, 139
weak, 138
Modus Ponens, 10, 82, 114, 205
Mongin, 193
Monotonicity, 64
cautious, 166
rational, 166

N
Necessitation rule, 10, 65, 82, 114
Nilsson, 68, 187
Non-isolated type, 85
non-Archimedean, 85
Nonmonotonic reasoning, 165
Nonstandard
analysis, 5–7
probability measure, 166, 183
universe, 6
NP-completeness, 97, 98, 190, 194

O
Object-level, 11
language, 207
theorem, 201–203, 205
Operator
$\Box$, 9, 10, 118, 194
$\bigcirc$, 126, 179
$\preceq$, 151
$\gtrsim$, 67
U, 126, 179
$CP_{\approx r}$, 156, 166
$CP_{\geq s}$, 118, 156
$CP_{\leq s}$, 118, 156
necessity, 9
Q_F, 141
temporal, 126, 179
Ordinal, 3
countable, 83
infinite, 3

P
Pacioli, 20
p-adic
norm, 160
numbers, 9, 12–14, 159, 180
Paris, 197
Partially ordered set, 3, 9, 11, 14
Pascal, 20
Peirce, 44
Plato, 20
Plausibility, 45, 61, 67, 195, 197
Plausible reasoning, 20, 61

Poincaré, 52
Pólya, 61
Polynomial weight formula, 3, 13, 158, 190
Popper, 61
Poretskiy, 46
Possibility measure, 197
Possible world, 1, 2, 10, 12, 13, 66, 67, 79, 110, 111, 117, 152, 153, 169, 191, 192, 194, 206
Primitive proposition, 10, 78, 204
Probability, 19, 21, 22, 101, 207
  $\preceq$, 151
  σ additive, 4, 13, 53, 56, 58, 63, 65, 66, 81, 95, 96, 207
  additive, 30, 53, 59
  axiom, 67
  axiomatization, 11, 48, 50, 55, 60, 61
  best possible bounds, 68
  boundaries, 28, 38, 43
  coherent, 208
  complex valued, 62
  conditional, 7, 13, 14, 50, 196, 208
  $CP_{\approx r}$, 156, 166
  $CP_{\geq s}$, 118, 156
  $CP_{\leq s}$, 118, 156
  finitely additive, 55, 57, 66, 77, 79, 86, 95, 207
  first-order logic, 2, 51, 65
  iterations, 50
  justification logics, 162
  logic, 2, 3, 6, 11, 12, 22, 54, 62, 64, 78, 110, 118, 122, 125, 135, 137, 156, 158, 160, 168
  lower bound, 34, 62, 82, 162
  measurable model, 79, 154
  measure, 2, 39, 79, 207
  model, 10, 63, 65, 79, 112
  negative, 62
  non-additive, 27, 28, 35, 62
  non-standard valued, 62, 85, 107, 166, 183
  operator, 2, 10–12, 14, 66, 67, 78
  optimal bounds, 11
  plausability, 61
  propositional logic, 2, 77
  Q_F operator, 14
  qualitative operator, 14, 67
  quantifier, 64
  range of, 4, 5, 9, 12
  real-valued, 3, 65
  space, 208
  temporal, 33, 125
  upper bound, 82, 162
Proof, 202
  finitary, 85
  infinitary, 3
Propositional formula, 78, 204
PSAT, 3, 13, 98, 124, 189–191, 193
PSPACE, 130, 159, 195
Pucella, 193, 196

Q
Quantifier
  $Px > r$, 64

R
Ramsey, 52
Rational monotonicity, 166
Recursive, 3, 4, 9, 14, 86
  enumerable, 2
Reichenbach, 54
Rényi, 62
Retroduction, 45
Robinson, 5, 64
Rule
  default, 183
  formation, 201
  Inference, 82
  inference, 202
  infinitary, 83
  of induction, 55
  ω, 4
Russell's paradox, 48

S
Satisfiability, 1–3, 37, 80
  finite, 5, 77, 84
  relation, 65, 79, 112
Savage, 61
Scott, 63, 194
Segerberg, 62, 67, 194, 195
Semantics, 1, 2, 10, 203
  ε-, 195, 196
  $LFOP_1$, 111
  LPP_1^{LTL}, 126
  LPP_2, 79
  $LPP_{2,\text{Meas}}^I$, 153
Set
  admissible, 4, 7, 138
  countable, 57
  hyperfinite, 6
  unordered, 3, 9, 11, 14
Shafer, 27, 31, 33, 35, 62, 197
σ

σ-incomplete, 148
additive, 208
algebra, 208
Soundness, 1, 86, 206
Suppes, 61
Syntax, 201
$LFOP_1$, 110
LPP_2, 78

T
Tarski, 4, 62, 64
Tartaglia, 20
Tautology, 10, 82, 206
Temporal logic, 125, 178, 194
Term, 111
probabilistic, 178
rigid, 113, 120
Theorem
meta-, 201
object-, 202
Truth assignment, 63
Truth value of an $LFOP_1$-formula, 112
Turing, 3

U
Undecidability, 4, 13
Unsatisfiable, 2, 84
set of formulas, 85

V
Valid formula, 1, 2, 80
Valuation, 10, 79
first order, 112
value of an $LFOP_1$-term, 112
Van der Hoeck, 188
Venn, 43
Von Mises, 54

W
Wigner, 62
World
possible, 1, 2, 10, 12, 13, 66, 67, 79, 110, 111, 117, 152, 153, 169, 191, 192, 194

The manufacturer's authorised representative in the EU is Springer Nature Customer Service Centre GmbH, Europaplatz 3, 69115 Heidelberg, Germany. If you have any concerns regarding our products, please contact ProductSafety@springernature.com

Printed and bound by CPI Group (UK) Ltd, Croydon, CR0 4YY
01/10/2026
02225861-0011